ONE WEEK LOAN

Designing Capable and Reliable Products

Designing Capable and Reliable Products

J.D. Booker
University of Bristol, UK

M. Raines
K.G. Swift
School of Engineering
University of Hull, UK

OXFORD AUCKLAND BOSTON JOHANNESBURG MELBOURNE NEW DELHI

Butterworth-Heinemann
Linacre House, Jordan Hill, Oxford OX2 8DP
225 Wildwood Avenue, Woburn, MA 01801-2041
A division of Reed Educational and Professional Publishing Ltd

℞ A member of the Reed Elsevier plc group

First published 2001

British Library Cataloguing in Publication Data
A catalogue record for this book is available from the British Library

ISBN 0 7506 5076 1

Library of Congress Cataloging in Publication Data
A catalog record for this book is available from the Library of Congress

Typeset by Academic & Technical Typesetting, Bristol
Printed and bound by MPG Ltd, Bodmin, Cornwall

Contents

Preface

In manufacturing companies the cost of quality can be around 20% of the total turnover. The largest proportion of this is associated with costs due to failure of the product during production or when the product is in service with the customer. Typically, such failure costs are due to rework, scrap, warranty claims, product recall and product liability claims, representing lost profit to the company. A lack of understanding of variability in manufacture, assembly and service conditions at the design stage is a major contributor to poor product quality and reliability. Variability is often detected too late in the design and development process, if at all. This can lead to design changes prior to product release, which extend the time to bring the product to market or mean the incursion of high costs due to failure with the customer.

To improve customer satisfaction and business competitiveness, companies need to reduce the levels of non-conformance and attendant failure costs associated with poor product design and development. Attention needs to be focused on the quality and reliability of the design as early as possible in the product development process. This can be achieved by understanding the potential for variability in design parameters and the likely failure consequences in order to reduce the overall risk. The effective use of tools and techniques for designing for quality and reliability can provide this necessary understanding to reduce failure costs.

Various well-known tools and techniques for analysing and communicating potential quality and reliability problems exist, for example Quality Function Deployment (QFD), Failure Mode and Effects Analysis (FMEA) and Design of Experiments (DOE). Product manufacturing costs can be estimated using techniques in Design for Assembly (DFA) and Design for Manufacture (DFM). For effective use, these techniques can be arranged in a pattern of concurrent product development, but do not specifically question whether component parts and assemblies of a design can be processed capably, or connect design decisions with the likely costs of failure. Quality assurance registration with BS EN ISO 9000 does not necessarily ensure product quality, but gives guidance on the implementation of the systems needed to trace and control quality problems, both within a business and with its suppliers.

Chapter 1 of this book starts with a detailed statement of the problem, as outlined above, focusing on the opportunities that exist in product design in order to reduce failure costs. This is followed by a review of the costs of quality in manufacturing companies, and in particular how failure costs can be related to design decisions and the way products later fail in service. An introduction to risk and risk assessment provides the reader with the underlying concepts of the approaches for designing capable and reliable products. The chapter ends with a review of the key principles in designing for quality and reliability, from both engineering design research and industrial viewpoints.

Capable design is part of the Design for Quality (DFQ) concept relating to quality of conformance. Chapter 2 presents a knowledge-based DFQ technique, called Conformability Analysis (CA), for the prediction of process capability measures in component manufacture and assembly. It introduces the concepts of component manufacturing capability and the relationships between tolerance, variability and cost. It then presents the Component Manufacturing Variability Risks Analysis, the first stage of the CA methodology from which process capability estimates can be determined at the design stage. The development of the knowledge and indices used in an analysis is discussed within the concept of an 'ideal design'. The need for assembly variability determination and the inadequacy of the DFA techniques in this respect is argued, followed by an introduction to assembly sequence diagrams and their use in facilitating an assembly analysis. The Component Assembly Variability Risks Analysis is then presented, which is the second stage of the CA methodology. Finally explored in this chapter is a method for linking the variability measures in manufacturing and assembly with design acceptability and the likely costs of failure in service through linkage with FMEA.

The use of CA has proved to be beneficial for companies introducing a new product, when an opportunity exists to use new processes/technologies or when design rules are not widely known. Design conformance problems can be systematically addressed, with potential benefits, including reduced failure costs, shorter product development times and improved supplier dialogue. A number of detailed case studies are used to demonstrate its application at many different levels.

Chapter 3 reports on a methodology for the allocation of capable component tolerances within assembly stack problems. There is probably no other design effort that can yield greater benefits for less cost than the careful analysis and assignment of tolerances. However, the proper assignment of tolerances is one of the least understood activities in product engineering. The complex nature of the problem is addressed, with background information on the various tolerance models commonly used, optimization routines and capability implications, at both component manufacturing and assembly level. Here we introduce a knowledge-based statistical approach to tolerance allocation, where a systematic analysis for estimating process capability levels at the design stage is used in conjunction with methods for the optimization of tolerances in assembly stacks. The method takes into account failure severity through linkage with FMEA for the setting of realistic capability targets. The application of the method is fully illustrated using a case study from the automotive industry.

Product life-time prediction, cost and weight optimization have enormous implications on the business of engineering manufacture. Using large *Factors of Safety* in a deterministic design approach fails to provide the necessary understanding of the nature of manufacture, material properties, in-service loading and their variability. Probabilistic approaches offer much potential in this connection, but have yet to be taken up widely by manufacturing industry. In Chapter 4, a probabilistic design methodology is presented providing reliability estimates for product designs with knowledge of the important product variables. Emphasis will be placed on an analysis for static loading conditions. Methods for the prediction of process capability indices for given design geometry, material and processing route, and for estimating material property and loading stress variation are presented to augment probabilistic design formulations. The techniques are used in conjunction with FMEA to facilitate the setting of reliability targets and sensitivity analysis for redesign purposes. Finally, a number of fully worked case studies are included to demonstrate the application of the methods and the benefits that can accrue from their usage.

Chapter 5 discusses the important role of the product development process in driving the creation of capable and reliable products. Guidance on the implementation problems and integrated use of the main tools and techniques seen as beneficial is a key consideration. The connection of the techniques presented in the book with those mentioned earlier will be explored, together with their effective positioning within the product development process. Also touched on are issues such as design reviews, supplier development and Total Quality Management (TQM) within the context of producing capable and reliable products.

The book provides effective methods for analysing mechanical designs with respect to their capability and reliability for the novice or expert practitioner. The methods use physically significant data to quantify the engineering risks at the design stage to obtain more realistic measures of design performance to reduce failure costs. All core topics such as process capability indices and statistical modelling are covered in separate sections for easy reference making it a self-contained work, and detailed case studies and examples are used to augment the approaches. The book is primarily aimed at use by design staff for 'building-in' quality and reliability into products with application of the methods in a wide range of engineering businesses. However, the text covers many aspects of quality, reliability and product development of relevance to those studying, or with an interest in, engineering design, manufacturing or management. Further, it is hoped that the text will be useful to researchers in the field of designing for quality and reliability.

The authors are very grateful to Mr Stan Field (formerly Quality Director at British Aerospace Military Aircraft & Aerostructures Ltd) and to Mr Richard Batchelor of TRW for their invaluable support and collaboration on this work. Thanks are also extended to Mr Bob Swain of the School of Engineering for his help with the preparation of many drawings. The Engineering & Physical Sciences Research Council, UK (Grant Nos GR/J97922 and GR/L62313), has funded the work presented in this book.

J.D. Booker, M. Raines, K.G. Swift
School of Engineering, University of Hull, UK
May 2000

Notation

a_p	Additional assembly process risk
C_p	Process capability index (centred distributions)
C_{pk}	Process capability index (shifted distributions)
C_v	Coefficient of variation
D	FMEA Detectability Rating
f	Frequency
$f(x)$	Function of x, probability density function
$F(x)$	Cumulative distribution function
f_p	Fitting process risk
g_p	Geometry to process risk
h_p	Handling process risk
k	Number of classes
k_p	Surface engineering process risk
K	Stress intensity factor
K_c	Fracture toughness
Kt	Stress concentration factor
L	Loading stress
L_n	Nearest tolerance limit
m	Number of components in the system
m_p	Material to process risk
n	Number of components in an assembly stack, number of load applications
N	Population
N_p	Number of data pairs
O	FMEA Occurrence Rating
p	Probability of failure per application of load
P	Probability, probability of failure
Pc	Product cost
q_a	Component assembly variability risk
q_m	Component manufacturing variability risk
r	Correlation coefficient
R	Reliability
R_n	Reliability at nth application of load
s	Principal stress

s_p	Surface roughness to process risk
S	FMEA Severity Rating, strength
Su	Ultimate tensile strength
Sy	Uniaxial yield strength
t	Bilateral tolerance
T	Unilateral tolerance
t_p	Tolerance to process risk
V	Variance
w	Class width
z	Standard deviation multiplier, Standard Normal variate
$\phi(\ )$	Function of
Φ_{SND}	Function of the Standard Normal Distribution
μ	Mean
σ	Standard deviation
σ'	Standard deviation estimate for a shifted distribution
Σ	Sum of
τ_u	Ultimate shear strength
τ_y	Shear yield strength

Abbreviations

AEM	Assembly Evaluation Method
BS	British Standard
BSI	British Standards Institute
CA	Conformability Analysis
CAPRA	Capability and Probabilistic Design Analysis
CDF	Cumulative Distribution Function
DFA	Design for Assembly
DFM	Design for Manufacture
DFMA	Design for Manufacture and Assembly
DFQ	Design for Quality
DMP	Design-Make-Prove
DOE	Design of Experiments
FMEA	Failure Mode and Effects Analysis
FS	Factor of Safety
FTA	Fault Tree Analysis
GNP	Gross National Product
ISO	International Organization of Standards
LEFM	Linear Elastic Fracture Mechanics
LR	Loading Roughness
MA	Manufacturing Analysis
PDF	Probability Density Function
PDS	Product Design Specification
PIM	Product Introduction Management
ppm	Parts-per-million
PRIMA	Process Information Map
QFD	Quality Function Deployment
QMS	Quality Management System
RPN	Risk Priority Number
RSS	Root Sum Square
SAE	Society of Automotive Engineers
SM	Safety Margin
SND	Standard Normal Distribution
SPC	Statistical Process Control
SSI	Stress–Strength Interference
TQM	Total Quality Management

Introduction to quality and reliability engineering

1.1 Statement of the problem

In order to improve business performance, manufacturing companies need to reduce the levels of non-conformance and attendant failure costs stemming from poor product design and development. Failure costs generally make up the largest cost category in a manufacturing business and include those attributable to rework, scrap, warranty claims, product recall and product liability claims. This represents lost profit to a business and, as a result, it is the area in which the greatest improvement in competitiveness can be made (Russell and Taylor, 1995).

The effect of failure cost or 'quality loss' on the profitability of a product development project is shown in Figure 1.1. High levels of failure cost would produce a loss on sales and would probably mean that the project fails to recover its initial level of investment.

In an attempt to combat high quality costs and improve product quality in general, companies usually opt for some kind of quality assurance registration, such as with BS EN ISO 9000. Quality assurance registration does not necessarily ensure product quality, but gives guidance on the implementation of the systems needed to trace and control quality problems, both within a business and with its suppliers. The adoption of quality standards is only the first step in the realization of quality products and also has an ambiguous contribution to the overall reduction in failure costs. A more proactive response by many businesses has been to implement and support long-term product design and development strategies focusing on the engineering of the product.

It has been realized for many years that waiting until the product is at the end of the production line to measure its quality is not good business practice (Crosby, 1969). This has led to an increased focus on the integration of quality into the early design stages of product development (Evbuomwan *et al.*, 1996; Sanchez, 1993). Subsequently, there has been a gradual shift away from the traditional 'on-line' quality techniques, such as Statistical Process Control (SPC), which has been the main driver for quality improvement over the last 50 years, to an 'off-line' quality approach using design tools and techniques.

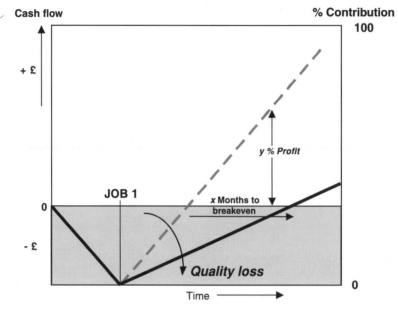

Figure 1.1 Effect of quality loss on the profitability of a product development project

The focus on quality improvement in design is not misplaced. Studies have estimated that the majority of all costs and problems of quality are created in product development. Focusing on the generation of product faults in product development, we find that typically 75% originate in the development and planning stages, but compounding the problem, around 80% of faults remain undetected until final test or when the product is in use (see Figure 1.2). The consequences of a design fault can be crippling: massive recalls, costly modifications, loss of reputation and sales, or even going out of business! Engineers and designers sometimes assume that someone else is causing product costs, but it is the details of how a product is designed that generates its costs in nearly every category (Foley and Bernardson, 1990).

The most significant cost savings can result from changes in product design rather than, say, from changes in production methods (Bralla, 1986). The costs 'fixed' at the planning and design stages in product development are typically between 60 and 85%, but the costs actually incurred may only be 5% of the total committed for the project. Therefore, the more problems prevented early on, through careful design, the fewer problems that have to be corrected later when they are difficult and expensive to change (Dertouzos *et al.*, 1989). It is often the case that quality can be 'built in' to the product without necessarily increasing the overall cost (Soderberg, 1995). However, to achieve this we need to reduce the 'knowledge gap' between design and manufacture as illustrated in Figure 1.3.

Design is recognized as a major determinant of quality and therefore cost. It is also a driving factor in determining the 'time to market' of products (Welch and Dixon, 1992). Historically, designers have concerned themselves with product styling, function and structural integrity (Craig, 1992). Now the designer has the great responsibility of ensuring that the product will conform to customer requirements,

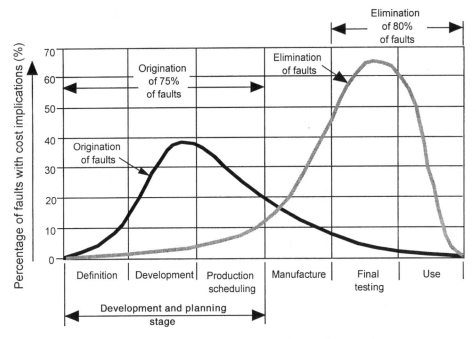

Figure 1.2 Origination and elimination of faults in product development (DTI, 1992)

comply to specification, meet cost targets and ensure quality and reliability in every aspect of the product's use, all within compressed time scales.

From the above, it is clear that the designer needs to be aware of the importance of the production phase of product development. As far as quality is concerned, the designer must aim to achieve the standards demanded by the specification, but at the same time should be within the capabilities of the production department. Many designers have practical experience of production and fully understand the limitations and capabilities that they must work within. Unfortunately, there are also many who do not (Oakley, 1993). From understanding the key design/manufacture interface issues, the designer can significantly reduce failure costs and improve business competitiveness. One of the most critical interface issues in product development is that concerning the allocation of process capable tolerances.

There is probably no other design improvement effort that can yield greater benefits for less cost than the careful analysis and assignment of tolerances (Chase and Parkinson, 1991). The effects of assigning tolerances on the design and manufacturing functions are far reaching, as shown in Figure 1.4. Product tolerances affect customer satisfaction, quality inspection, manufacturing and design, and are, therefore, a critical link between design, manufacture and the customer (Gerth, 1997; Soderberg, 1995). They need to be controlled and understood!

Each product is derived from individual pieces of material, individual components and individual assembly processes. The properties of these individual elements have a probability of deviating from the ideal or target value. In turn, the designer defines allowable tolerances on component characteristics in anticipation of the manu-facturing variations, but more often than not, with limited knowledge of the cost

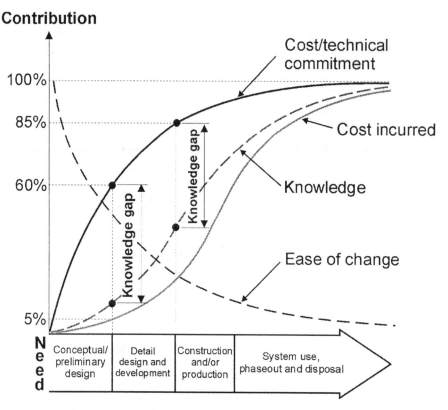

Figure 1.3 Commitment and incursion of costs during product development and the 'knowledge gap' principle (adapted from Fabrycky, 1994)

implication or manufacturing capability in order to meet the specification (Craig, 1992; Korde, 1997). When these variations are too large or off target, the usability of the product for its purpose will be impaired (Henzold, 1995). It therefore becomes important to determine if a characteristic is within specification, and, if so, how far it is from the target value (Vasseur *et al.*, 1992).

Improperly set tolerances and uncontrolled variation are one of the greatest causes of defects, scrap, rework, warranty returns, increased product development cycle time, work flow disruption and the need for inspection (Gerth and Hancock, 1995). If manufacturing processes did not exhibit variation, quality problems would not arise, therefore reducing the effects of variability at the design stage, in a cost-effective way, improves product quality (Bergman, 1992; Kehoe, 1996).

A significant proportion of the problems of product quality can directly result from variability in manufacturing and assembly (Craig, 1992). However, the difficulties associated with identifying variability at the design stage mean that these problems are detected too late in many cases, as indicated by a recent study of engineering change in nine major businesses from the aerospace, industrial and automotive sectors (Swift *et al.*, 1997). On average, almost 70% of product engineering rework was due to quality problems, that is failure to satisfy customer expectations and to

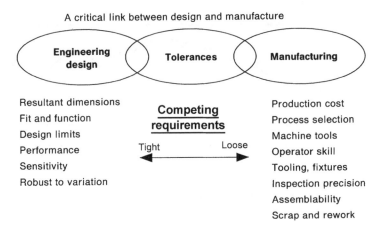

A critical link between design and manufacture

| Engineering design | Tolerances | Manufacturing |

Resultant dimensions
Fit and function
Design limits
Performance
Sensitivity
Robust to variation

Competing requirements

Tight   Loose

Production cost
Process selection
Machine tools
Operator skill
Tooling, fixtures
Inspection precision
Assemblability
Scrap and rework

Figure 1.4 Tolerances – the critical link between design and manufacture (Chase and Parkinson, 1991)

anticipate production variability on the shop floor. The need for more than 40% of the rework was not identified until production commenced.

The reasons for the rework, described in Figure 1.5, can be classified into four groups:

- Customer driven changes (including technical quality)
- Engineering science problems (stress analysis errors, etc.)
- Manufacturing/assembly feasibility and cost problems
- Production variability problems.

This indicates that customer related changes occurred throughout concept design, detailing, prototyping and testing with some amendments still being required after production had began. Engineering science problems, which represented less than 10% of the changes on average, were mostly cleared before production commenced. The most disturbing aspect is the acceptance by the businesses that most of the manufacturing changes, and more so manufacturing variability changes, were taking place during production, product testing and after release to the customer. Because the cost of change increases rapidly as production is approached and passed, the expenditure on manufacturing quality related rework is extremely high. More than 50% of all rework occurred in the costly elements of design for manufacture and production variability.

Further evidence of the problems associated with manufacturing variability and design can be found in published literature (Lewis and Samuel, 1991). Here, an investigation in the automotive industry showed that of the 26 quality problems stated, 12 resulted from process integrity and the integrity of assembly. Process integrity was defined as the correct matching of the component or assembly design to either the current manufacturing process or subsequent processes. Integrity of assembly was defined as the correct matching of dimensions, spatial configuration of adjacent or interconnecting components and subassemblies.

Variability associated with manufacturing and assembly has historically been considered a problem of the manufacturing department of a company (Craig,

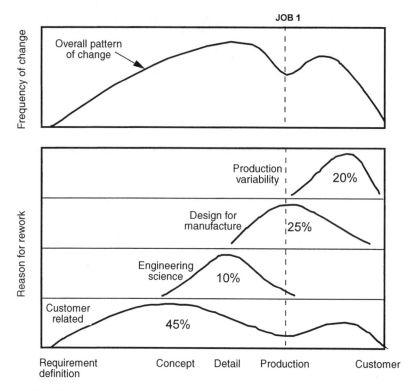

Figure 1.5 Disposition of rework in product development (Swift *et al.*, 1997)

1992). It is now being recognized that there is a need to reduce such variations at the design stage, where its understanding and control may lead to (Leaney, 1996a):

- Easier manufacture
- Improved fit and finish
- Less work in progress
- Reduced cycle time
- Fewer design changes
- Increased consistency and improved reliability
- Better maintainability and repairability.

Variation is an obvious measure for quality of conformance, but it must be associated with the requirements set by the specification to be of value at the design stage. Unfortunately, difficulty exists in finding the exact relationship between product tolerance and variability. Approximate relationships can be found by using process capability indices, quality metrics which are interrelated with manufacturing cost and tolerance (Lin *et al.*, 1997)[*].

The first concern in designing process capable products is to guarantee the proper functioning of the product, and therefore to satisfy technical constraints. Dimensional

[*] It is recommended at this stage of the text that the reader unfamiliar with the basic concepts of variation and process capability refer to Appendix I for an introductory treatise on statistics, and Appendix II for a discussion of process capability studies.

characteristics reflect the spatial configuration of the product and the interaction with other components or assemblies. Tolerances should be allocated to reflect the true requirements of the product in terms of form, fit and function in order to limit the degradation of the performance in service (Kotz and Lovelace, 1998). Ideally, designers like tight tolerances to assure fit and function of their designs. All manufacturers prefer loose tolerances which make parts easier and less expensive to make (Chase and Parkinson, 1991).

Tolerances alone simply do not contain enough information for the efficient manufacture of a design concept and the designer must use process capability data when allocating tolerances to component characteristics (Harry and Stewart, 1988; Vasseur *et al.*, 1992). Process capability analysis has proven to be a valuable tool in this respect, and is most useful when used from the very beginning of the product development process (Kotz and Lovelace, 1998).

If the product is not capable, the only options available are to either: manufacture some bad product, and sort it out by inspection; rework at the end of the production line; narrow the natural variation in the process; or widen the specification to improve the capability. Post-production inspection is expensive and widening the specification is not necessarily desirable in some applications as this may have an impact on the functional characteristics of the product. However, in many cases the tolerance specification may have been set somewhat arbitrarily, implying that it may not be necessary to have such tight tolerances in the first place (Kotz and Lovelace, 1998; Vasseur *et al.*, 1992). Making the product robust to variation is the driving force behind designing capable and reliable products, lessens the need for inspection and can reduce the costs associated with product failure.

Variability must become the responsibility of the designer in order to achieve these goals (Bjørke, 1989). An important aspect of the designer's work is to understand the tolerances set on the design characteristics, and, more importantly, to assess the likely capability of the characteristics due to the design decisions.

Industry is far from understanding the true capability of their designs. Some comments from senior managers and engineers in the industry give an indication of the cultural problems faced and the education needed to improve design processes in this respect.

We will have difficulty meeting those tolerances – it is 'bought-in' so we'll get the supplier to do the inspection.

$C_{pk} = 1.33$! We do much better than that in the factory. We're down to $C_{pk} = 0.8$!

I don't see how we make this design characteristic at $C_{pk} = 1.5$. Let's kill it with 100% inspection.

The components are not going to be process capable, but we can easily set the tolerance stack at ±0.1 mm when we build the assembly machine. Our assembly machine supplier uses robots.

I can see that this design is not likely to be capable, but my new director has said we are to use this design solution because it has the lowest part count.

I can't spend any more time on design. I see the problems, but it will cost the department too much if I have to modify the design.

I have been told that we must not use any secondary machining operations to meet the tolerance requirements. It just costs too much!

Good design practice does not simply mean trying to design the product so that it will not fail, but also identifying how it might fail and with what consequences (Wright, 1989). To effectively understand the quality of conformance associated with design decisions requires undertaking a number of engineering activities in the early stages of product development. In addition to understanding the capability of the design, the designer must consider the severity of potential failures and make sure the design is sufficiently robust to effectively eliminate or accommodate defects. Effective failure analysis is an essential part of quality and reliability work, and a technique useful in this capacity is Failure Mode and Effects Analysis (FMEA). (See Appendix III for a discussion of FMEA, together with several key tools and techniques regarded as being beneficial in new product development.)

FMEA is a systematic element by element assessment to highlight the effects of a component, product, process or system failure to meet all the requirements of a customer specification, including safety. FMEA can be used to provide a quantitative measure of the risk for a design. Because FMEA can be applied hierarchically, through subassembly and component levels down to individual dimensions and characteristics, it follows the progress of the design into detail listing the potential failure modes of the product, as well as the safety aspects in service with regard to the user or environment. Therefore, FMEA provides a possible means for linking potential variability with consequent design acceptability and associated failure costs. The application of a technique that relates design capability to potential failure costs incurred during production and service would be highly beneficial to manufacturing industry.

Conceivably, a number of new issues in product design and development have been discussed in this opening section, but in summary:

- Understanding and controlling the variability associated with design characteristics is a key element of developing a capable and reliable product
- Variability can have severe repercussions in terms of failure costs
- Designers need to be aware of potential problems and shortfalls in the capability of their designs
- There is a need for techniques which estimate process capability, quantify design risks and estimate failure costs.

Next, we review the costs of quality that typically exist in a manufacturing business, and how these are related to the way products fail in service. The remainder of the chapter discusses the important elements of risk assessment as a basis for design. This puts in context the work on designing for quality and reliability, which are the main topics of the book.

1.2 The costs of quality

The costs of quality are often reported to be between 5 and 30% of a company's turnover, with some engineering businesses reporting quality costs as high as 36%

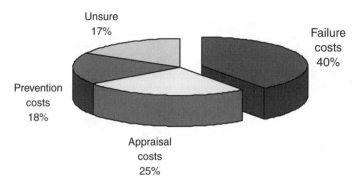

Figure 1.6 The costs of quality in UK industry (Booker, 1994)

(Dale, 1994; Kehoe, 1996; Maylor, 1996). This figure can be as high as 40% in the service industry! (Bendell *et al.*, 1993). In general, the overall cost of quality in a business can be divided into the following four categories:

- **Prevention costs** – These are costs we expect to incur to get things right first time, for example quality planning and assurance, design reviews, tools and techniques, and training.
- **Appraisal costs** – Costs which include inspection and the checking of goods and materials on arrival. Whilst an element of inspection and testing is necessary and justified, it should be kept to a minimum as it does not add any value to the project.
- **Failure costs** – *Internal failure costs* are essentially the cost of failures identified and rectified before the final product gets to the external customer, such as rework, scrap, design changes. *External failure costs* include product recall, warranty and product liability claims.
- **Lost opportunities** – This category of quality cost is impossible to quantify accurately. It refers to the rejection of a company product due to a history of poor quality and service, hence the company is not invited to bid for future contracts because of a damaged reputation.

Up to 90% of the total quality cost is due to failure, both internal and external, with around 50% being the average (Crosby, 1969; Russell and Taylor, 1995; Smith, 1993). A survey of UK manufacturing companies in 1994 found that failure under the various categories was responsible for 40% of the total cost of quality, followed by appraisal at 25%, and then prevention costs at 18%. This is shown in Figure 1.6. Of the companies surveyed, 17% were unsure where their quality costs originated, but indicated that these costs could be attributable to failure, either internally or externally.

Many organizations fail to appreciate the scale of their quality failures and employ financial systems which neglect to quantify and record the true costs. In many cases, the failures are often costs that are logged as 'overheads'. Quality failure costs represent a direct loss of profit! Organizations may have financial systems to recognize scrap, inspection, repair and test, but these only represent the 'tip of the iceberg' as illustrated in Figure 1.7.

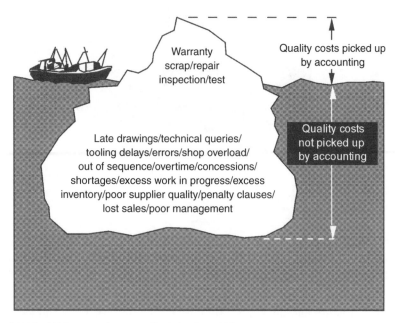

Figure 1.7 The hidden cost of poor quality (Labovitz, 1988)

A company should minimize the failure costs, minimize appraisal costs, but be prepared increase investment on prevention. Some quality gurus promote the philosophy of zero defects. Whilst this is obviously the ultimate goal, the prevention costs can become prohibitive. It is possible to determine the optimum from a cost point of view. This value may not be constant across the different business sectors, for instance a machine shop may be prepared to accept a 1% scrap rate, but it is doubtful that the public would accept that failure rate from a commercial aircraft! Each must set its own objectives, although 4% has been stated in terms of a general target as a percentage of total sales for manufacturing companies (Crosby, 1969). A simple quality–cost model that a business can develop to define an optimum between quality of conformance and cost is illustrated in Figure 1.8.

1.2.1 Cases studies in failure costs

External failure costs and lost opportunities are potentially the most damaging costs to a business. Several examples commonly quoted in the literature are given below.

This first example applies to UK industry in general. The turnover for UK manufacturing industry was in the order of £150 billion in 1990 (Smith, 1990). If the total quality cost for a business was likely to be somewhere in the region of 20%, with failure costs at approximately 50% of the total, it is likely that about £15 billion was wasted in defects and failures. A 10% improvement in failure costs would have released an estimated £1.5 billion into the economy. IBM, the computer manufacturer, estimated that they were losing about $5.6 billion in 1986 owing to costs of non-conformance and its failure to meet quality standards set for its products and

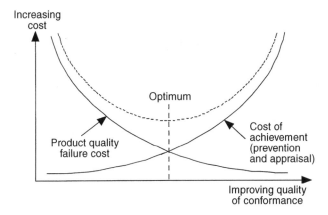

Figure 1.8 The optimization of quality costs

services. A further $2 billion was estimated as being lost as a result of having poor working processes. IBM proceeded on the basis that the company had $7.6 billion of potential savings to be obtained getting things 'right first time' (Kruger, 1996).

In the US over the last 30 years, there has been an increasing trend in product liability claims and associated punitive damages. For example, after a legal battle, General Motors had to pay a plaintiff's family $105 million, when the plaintiff was killed when a poorly positioned petrol tank in a truck exploded during an accident (Olson, 1993). The UK motorcycle industry in the 1970s suffered greatly due to their Japanese counterparts not only producing more cost-effective bikes, but also of higher quality. The successful resurgence of Triumph only recently was based on matching and even bettering the Japanese on the quality of it products.

Most producers believe in the adage 'quality pays' in terms of better reputation and sales, customer loyalty, lower reject rates, service and warranty costs. They should also realize that 'safety pays' in terms of reducing the legal exposure and the tremendous costs that this can incur, both directly and indirectly, for example from compensation, legal fees, time and effort, increased insurance premiums, recalls and publicity (Wright, 1989). Few manufacturers understand all the cost factors involved, and many take a shortsighted view of the actual situation with regard to the costs of safety.

Measures to minimize safety problems must be initiated at the start of the life cycle of any product, but too often determinations of criticality are left to production or quality control personnel who may have an incomplete knowledge of which items are safety critical (Hammer, 1980). Any potential non-conformity that occurs with a severity sufficient to cause a product or service not to satisfy intended normal or reasonably foreseeable usage requirements is termed a 'defect' (Kutz, 1986). The optimum defect level will vary according to the application, where the more severe the consequences of failure the higher the quality of conformance needs to be.

The losses that companies can face are influenced by many factors including market sector, sales turnover and product liability history. It is not easy to make a satisfactory estimate of the product liability costs associated with quality of non-conformance, and

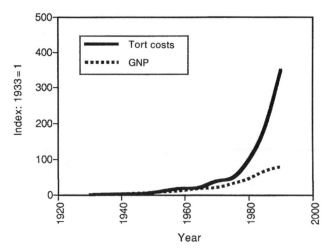

Figure 1.9 US tort cost escalation compared with GNP growth (Sturgis, 1992)

losses due to safety critical failures in particular are subject to wide variation (Abbot, 1993).

It is known that product liability costs in the US have risen rapidly in recent years and this trend is set to continue. It has been predicted that US tort costs would reach $300 billion by the year 2000, which would then represent 3.5% of US economic output (Sturgis, 1992). Tort is a term used in common law for a civil wrong and for the branch of law dealing with the liability for these wrongs. It is an alleged wrongful act for which the victim can bring a civil action to seek redress. Examples of individual torts are negligence, nuisance, and strict liability. Tort cost growth far outstripped Gross National Product (GNP) growth since 1930, increasing 300 times over this period, as shown in Figure 1.9, compared with a 50-fold increase for GNP.

The way in which tort costs are moving provides valuable evidence of the costs of 'getting it wrong' (Sturgis, 1992). Product liability experts believe that while the US system has its differences, as lawyers in the UK become more attuned to the US system, a similar situation may occur here also. Some background to the situation in the UK is shown in Figure 1.10, indicating that product liability costs could reach £5 billion annually. A further escalation in product liability claims could result in higher insurance premiums and the involvement of insurance companies in actually defining quality and reliability standards and procedures (Smith, 1993).

Insurance appears to be the safest solution for companies to defend themselves against costly mistakes, but there are problems, notably the cost of the premiums and the extent of the cover provided (Wright, 1989). The insurance sector must address some of the above issues in assessing their exposure (Abbot, 1993). Case histories provide some insight into the costs that can accrue though.

For a catastrophic failure in the aerospace industry with a high probability of loss of life, which relates to an FMEA Severity Rating $(S) = 10$, a business could quite possibly need insurance cover well in excess of £100 million. This will allow for costs due to failure investigations, legal actions, product recall and possible loss of

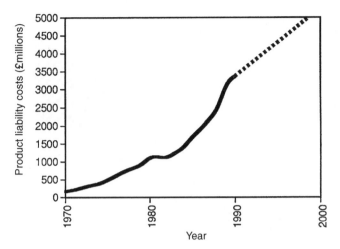

Figure 1.10 Trend of product liability costs in the UK (Sturgis, 1992)

business. High failure costs are not only associated with the aerospace industry. Discussions relating to the automotive sector suggest that for a failure severity of $S = 9$, complete failure with probable severe injury and/or loss of life, a business could well face the need for cover in excess of £10 million. Less safety critical business sectors and lower severity ratings reduce the exposure considerably, but losses beyond £1 million have still been recorded (Abbot, 1993). The relationship between safety critical failures and potential cost is summarized in Figure 1.11. It is evident that as failures become more severe, they cost more, so the only approach available to the designer is to reduce the probability of occurrence.

Little progress towards reducing product liability exposure can be made by individuals within a business unless top management are committed to marketing safe products. Safety is one aspect of the overall quality of a product. While most producers and suppliers realize the importance of quality in terms of sales and reputation, there is sometimes less thought given to the importance of safety in terms of legal liabilities. Where 'quality' is mentioned it should be associated with 'safety'. Management strategy should be based on recognizing the importance of marketing a safe product, and the potential costs of failing to do so, and that failures will occur and plans must be made for mitigating their effect (Wright, 1989).

1.2.2 Quality–cost estimating methods

Quality–cost models can help a business understand the influence of defect levels on cost during product development. More importantly, designers should use models to predict costs at the various stages. These results make the decision-making process more effective, particularly at the design stage (Hundal, 1997). The estimation of quality costs in the literature is commonly quoted at three quite different levels:

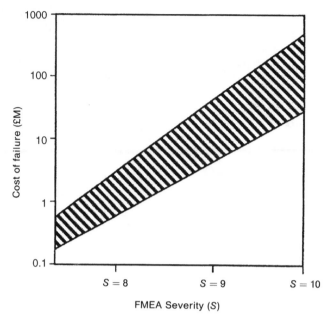

Figure 1.11 The potential cost of safety critical failures

- Economic quality–cost models which are 'global' or 'macro-scaling' top down methods, show general trends in quality costs which are predicted based on varying some notion of time or quality improvement. The model in the latest standard BS 6143, as shown in Figure 1.12, is such a model. It is perhaps closest to the view of some quality experts, but surprisingly infers that prevention costs reduce substantially with increased quality improvement. A model published recently also combines failure and appraisal costs, two distinct categories (Cather and Nandasa, 1995). Quality managers believe that many of the widely publicised quality–cost models are inaccurate and may even be of the wrong form (Plunkett and Dale, 1988). A valid model that could be used to audit business performance and predict the effects of change would be most helpful. However, since the modelling of quality costs range from inaccurate to questionable, they are unlikely to provide a rigorous basis for product engineering to connect design decisions to the costs of poor quality.
- 'Micro-scaling' or bottom up approach to quality costs, where it is possible to calculate the cost of losses involved in manufacture and due to returns and/or claims. This method requires a great deal of experience and relies on the availability of detailed cost data throughout a product's life-cycle. While this is a crucial activity for a business, it is also not a practical approach for estimating the quality cost for product in the early stages of product development.
- The Conformability Analysis (CA) method presented in this book and Taguchi's Quality Loss Function (Taguchi *et al.*, 1989) are what might be called 'meso-scaling' or quality–cost scaling. Here past failure costs are scaled to new requirements allowing for changes in design capability. It gives more precision than the global approach, but would clearly lack the accuracy of the bottom up approach possible

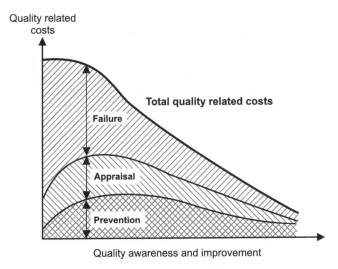

Figure 1.12 Global quality–cost model (BS 6143, 1990)

once a product is in production. However, these methods are more useful at the design stage.

We have already seen elements of the CA approach when considering the costs due to safety critical failures. A further insight into the way that failure costs can be estimated for non-safety critical failures is also used to support the CA methodology. Estimates for the costs of failure in this category are based on the experiences of a sample of industrial businesses and published material as follows.

Consider a product whose product cost is Pc. The costs due to failure at the various stages of the product's life cycle have been investigated, and in terms of Pc, they have been found to be (Braunsperger, 1996; DTI, 1992):

$0.1\,Pc$ – internal failure cost due to rework at the end of the production line
   Pc – external failure cost for return from customer inspection
$10\,Pc$ – external failure cost for warranty return due to failure with customer in use.

The relationship is commonly known as the '10× rule' and is shown in Figure 1.13. The 10× rule demonstrates how a fault, if not discovered, will give rise to ten times the original elimination costs in a later phase of the life-cycle. In other words, products must be designed in such a way that scarcely any defects develop or if they do, they can be identified as early as possible in the product development process and rectified (Braunsperger, 1996). Other surveys have found that these costs could be even higher as shown in Figure 1.14.

Suppose a particular fault in a product is not detected through internal tests and inevitably results in a failure severity $S = 5$. If around 80% of failures are found by customer testing and 20% are warranty returns, then the expected cost on average for one fault will be $2.8Pc$, from Figure 1.13. If the product has been designed such that $C_{pk} = 1.33$, or in other words, approximately 30 parts-per-million (ppm) failures are expected for the characteristic which may be faulty, then for a product costing £100 the probable cost of failure per million products produced would be £8400.

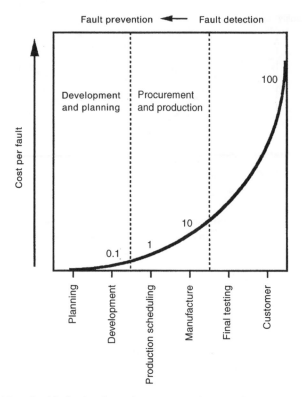

Figure 1.13 The 10× rule of fault related costs by percentage (DTI, 1992)

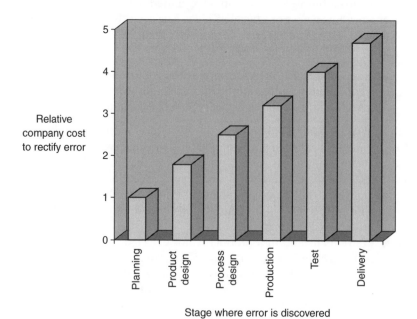

Figure 1.14 Cost escalation of rectifying errors at down stream stages (Ostrowski, 1992)

At $C_{pk} = 1$, or approximately 1300 ppm failures expected, the probable cost of failure per million products would be £364 000. At $C_{pk} = 0.8$ (or 8000 ppm) the probable cost of failure would increase by an order of magnitude to over £2.2 million. These failure costs do not take into account the costs associated with damaged company reputation and lost opportunities which are difficult to assess, but do indicate that failure cost estimates associated with product designs are possible. This aspect of the CA methodology is further developed in Chapter 2.

1.3 How and why products fail

1.3.1 Failure mechanisms

We have already established that variability, or the lack of control and understanding of variability, is a large determinant of the quality of a product in production and service and, therefore, its success in avoiding failure. In addition, understanding the potential failure mechanisms and how these interact with design decisions is necessary to develop capable and reliable products (Dasgupta and Pecht, 1991). It is helpful next to investigate the link between the causes and modes of failure and variability throughout the life-cycle of a mechanical product.

Mechanical failure is any change or any design or manufacturing error that renders a component, assembly or system incapable of performing its intended function (Ullman, 1992). However, it is also possible to suggest several key aspects of failure (Bignell and Fortune, 1992):

- A failure is said to occur when disappointment arises as a result of an assessment of an outcome of an activity.
- Failure can be a shortfall of performance below a standard, the generation of undesirable effects or the neglect of an opportunity.
- Failure can occur in a variety of forms, namely: catastrophic or minor, overwhelming or only partial, sudden or slow.
- Failure can arise in the past, present or future.
- Failure will be found to be multi-causal, and to have multiple effects.

For the product to fail there must be some failure mechanism caused by lack of control of one or more of the engineering variables involved. Most mechanisms of mechanical failure can be categorized by one of the following failure processes (O'Connor, 1995; Rao, 1992; Sadlon, 1993):

- Overload – static failure, distortion, instability, fracture
- Strength degradation – creep failure, fatigue, wear, corrosion.

Figure 1.15(a) shows the results of an investigation to find the frequency of failure mechanisms in typical engineering components and aircraft components. By far the most common failure mechanism is fatigue. It has been suggested that around 80% of mechanical failures can, in fact, be attributable to fatigue (Carter, 1986). Failures caused by corrosion and overload are also common. Although the actual stress rupture mode of failure is cited as being uncommon, overload and brittle fracture

Mechanism	Percentage of failures	
	Engineering components	Aircraft components
Corrosion	29	3
Fatigue	25	61
Brittle fracture	16	–
Overload	11	18
High temperature corrosion	7	2
SCC/ corrosion fatigue/ HE	6	8
Creep	3	–
Stress rupture	–	1
Wear/abrasion/ erosion	3	7

(a) Frequency of failure mechanisms

Cause	Percentage of failures	
	Engineering components	Aircraft components
Improper material selection	38	–
Fabrication imperfections	15	17
Faulty HT	15	–
Design errors	11	16
Unanticipated service conditions	8	10
Uncontrolled environmental conditions	6	–
Inadequate inspection/QC	5	–
Material mix	2	–
Inadequate maintenance	–	44
Defective material	–	7
Unknown	–	6

(b) Frequency of causes of failure

Key: SCC - Stress corrosion cracking, HE - Hydrogen embrittlement, HT - Heat treatment, QC - Quality control

Figure 1.15 The frequency and causes of mechanical failure (Davies, 1985)

failures may also be categorized as being rupture mechanisms, distinct from strength degradation.

Figure 1.15(b) provides some insight into the reasons for the mechanical failures experienced. Some root causes of failure are found to be improper material selection, fabrication imperfections and design errors. Other causes of failure are ultimately

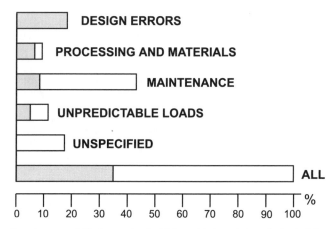

Figure 1.16 Designer's responsibility for mechanical failures (designer's share is shaded) (Larsson *et al.*, 1971)

related to variability in production and service conditions. In general, several primary root causes of failure can be suggested (Ireson *et al.*, 1996; Villemeur, 1992):

- Design errors
- Production errors (manufacturing, assembly)
- Handling/transit damage
- Misuse or operating errors
- Adverse environmental conditions.

From the above, there is a strong indication that the design errors are responsible for many mechanical failures found in the field. This is supported by another study, the results of which are shown in Figure 1.16. Through making poor or uneducated design decisions, the designer alone accounted for around 35% of the failures of those investigated.

Historically, the failure of products over their life-cycles, sometimes termed the 'bath-tub curve', can be classified into three distinct regions as shown in Figure 1.17. A detailed breakdown of the attributable factors in each region are also given below (Kececioglu, 1991).

Infant mortality period – Quality failures dominate and occur early in the life of the product. In detail, these can be described as:

- Poor manufacturing techniques including processes, handling and assembly practices
- Poor quality control
- Poor workmanship
- Substandard materials and parts
- Parts that failed in storage or transit
- Contamination
- Human error
- Improper installation.

Useful life period – Stress related failures dominate and occur at random over the total system lifetime – caused by the application of stresses that exceed the design's

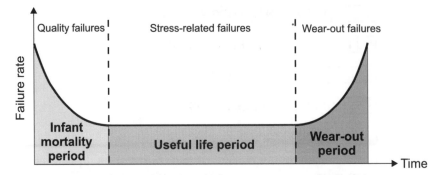

Figure 1.17 'Bath-tub' curve showing typical life characteristic of a product (Priest, 1988)

strength (most significant period as far as reliability prediction activities are concerned). In detail, these can be described as:

- Interference or overlap of designed-in strength and experienced stress during operation
- Occurrence of higher than expected random loads
- Occurrence of lower than expected random strengths
- Defects
- Misapplication or abuse
- 'Acts of God'.

Wearout period – Failure occurs when the product reaches the end of its effective life and begins to degenerate and wear out. In detail, these can be described as:

- Ageing
- Wear
- Fatigue
- Creep
- Corrosion
- Poor servicing or maintenance.

1.3.2 The link between variability and failure

There exists a relationship between the failure characteristics of a product over its life-cycle, as described by the three periods of the bath-tub curve above, and the phenomenon of variability. It has already been established that the potential for variation in design parameters is a real aspect of product engineering. Subsequently, three major sources of undesirable variations in products can be classified, these being (Clausing, 1994):

- Production variations
- Variations in conditions of use
- Deterioration (variation with time and use).

The above fits in with the overall pattern of failure as described by Figure 1.17. The first two and sometimes even all three parts of the bath-tub curve are closely connected to variations.

The manufacturing process has a strong impact on component behaviour with respect to failure, and production variabilities arising from lack of precision or deficiencies in manufacturing processes lead to failures concentrated early in the product's life (Klit *et al.*, 1993; Lewis, 1996). A common occurrence is that correctly designed items may fail as a result of defects introduced during production or simply because the specified dimensions or materials are not complied with (Nicholson *et al.*, 1993). Production defects are second only to those product deficiencies created by inadequate design as causes of accidents and improper production techniques can actually create hazardous characteristics in products (Hammer, 1980).

Modern equipment is frequently composed of thousands of components, all of which interact within various tolerances. Failures often arise from a combination of drift conditions rather than the failure of a specific component (Smith, 1993). For example, typically an assembly tolerance exists only to limit the degradation of the assembly performance. Being 'off target' may involve later warranty costs because the product is more likely to break down than one which has a performance closer to the target value (Vasseur *et al.*, 1992). This again is related to manufacturing variation problems, and is more difficult to predict, and therefore less likely to be foreseen by the designer (Smith, 1993).

Variations in a product's material properties, service loads, environment and use typically lead to random failures over the most protracted period of the product's expected life-cycle. During the conditions of use, environmental and service variations give rise to temporary overloads or transients causing failures, although some failures are also caused by human related events such as installation and operation errors rather than by any intrinsic property of the product's components (Klit *et al.*, 1993). Variability, therefore, is also the source of unreliability in a product (Carter, 1997). However, it is evident that if product reliability is determined during the design process, subsequent manufacturing, assembly and delivery of the system will certainly not improve upon this inherent reliability level (Kapur and Lamberson, 1977).

Wearout attracts little attention among designers because it is considered less relevant to product reliability than the other two regions, although degradation phenomena are clearly important for designs involving substantial operating periods (Bury, 1975; Pitts and Lewis, 1993).

Many of the kinds of failures described above may be reduced by either decreasing variations or by making the product robust against these variations (Bergman, 1992). For example, the smaller the variability associated with the critical design parameters, the greater will be the reliability of the design to deal with unforeseen events later in the product's life-cycle (Suh, 1990). As stated earlier, with reference to an individual manufacturing process, variation is an obvious measure for quality performance and the link between product failure, capability and reliability is to a large degree embedded within the prediction of variation at the design stage. However, two factors influence failure: the robustness of the product to variability, and the severity of the service conditions (Edwards and McKee, 1991). To this end, it has been cited that the quality control of the environment is much more important than quality control of the manufacturing processes in achieving high reliability (Carter, 1986).

In order to quantify the sometimes intangible elements of variability associated with the product design and the safety aspects in service requires an understanding of 'risk'. The assessment of risk in terms of general engineering practice will be discussed next. This will lead to a better understanding of designing for quality and reliability, which is the main focus of the book.

1.4 Risk as a basis for design

The science of risk, and the assessment and management of risk, is a very complex subject and one that covers a wide diversity of disciplines. Society is becoming more aware of the risks related to increased technological innovation and industrialization. Recent reports in the media about environment (global warming), health (BSE) and technological (nuclear waste processing) risks have played their part in focusing attention on the problem of how to assess risk and what makes an acceptable risk level. As a result, risk and risk related matters are becoming as important as economic issues on the political agenda. There is, therefore, an increasing need for a better understanding of the topic and tools and techniques that can be used to help assess product safety and support the development of products and processes that are of essentially low risk (EPSRC, 1999). To meet these needs, a new British Standard, BS 6079 (1999), has been published to give guidance to businesses on the management of risks throughout the life of a project.

The term 'risk' is often used to embrace two assessments:

- The frequency (or probability) of an event occurring
- The severity or consequences of the event on the user/environment.

The product of these two conditions equals the risk:

$$\text{Risk} = \text{Occurrence} \times \text{Severity}$$

We can demonstrate the notions of risk and risk assessment using Figure 1.18. For a given probability of failure occurrence and severity of consequence, it is possible to map the general relationship of risk and what this means in terms of the action required to eliminate the risk.

For example, if both occurrence and severity are low, the risk is low, and little or no action in eliminating or accommodating the risk is recommended. However, for the same level of occurrence but a high severity, a medium level of risk can be associated with concern in some situations. The level of occurrence, for some unknown reason, changes from low to medium and suddenly we are in a situation where the risk requires priority action to be eliminated or accommodated in the product.

The aim of a risk assessment is to develop a product which is 'safe' for the proposed market. A safe product is any product which, under normal or reasonably foreseeable conditions of use, including duration, presents no risk or only the minimum risk compatible with the product's use and which is consistent with a high level of protection for consumers (DTI, 1994). In attempting to protect products against failure in service and, therefore, the user or environment, difficulty exists in ascertaining the

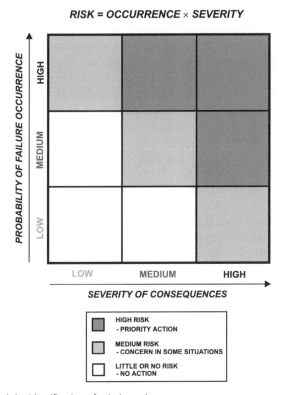

RISK = OCCURRENCE × SEVERITY

Figure 1.18 Risk and the identification of priority action

degree of protection most suitable for a given application. First, the choice should be expressed in terms of risk and probability of failure as shown above, but the determination of an acceptable level for a product depends on many factors, such as (Bracha, 1964; Karmiol, 1965; Welling and Lynch, 1985):

- Safety
- The costs of failure
- Criticality of function that it supports
- Complexity – number of component parts, subsystems
- Operational profile – duty cycle or time it operates
- Environmental conditions – exposure to various environmental conditions
- Number of units to be produced
- Ease and cost of replacement
- 'State of the art' or present state of engineering progress
- Market sector/consumer category.

Suggestions have also been made as to the number of people that are affected by the risk at any one time, and, in principle, it should be possible to link the acceptable level of individual risk to the number of people exposed to that risk (Niehaus, 1987). For example Versteeg (1987) provides risk levels associated with three areas: acceptable risk, reduction desired and unacceptable risk, and the number of people exposed to the risk. This further compounds the problem of assigning acceptable risk targets, but implies that safety is of paramount importance.

Businesses make decisions that affect safety issues, but which are only considered implicitly. In an increasingly complex world, the resulting decisions are not always appropriate because the limits of the human mind do not allow for an implicit consideration of a large number of factors. Formal analyses are needed to aid the decision-making process in these complex situations. However, the application of a formal analysis to safety issues raises new questions. The risks perceived by society and by individuals cannot be captured by a simple technical analysis. There are many reasons for this; however, it is clear that decision making needs to account for both technical and public values (Bohnenblust and Slovic, 1998). The decision to accept risk is not based on the absolute notion of one acceptable risk level, but has some flexibility as the judgement depends on the cost/benefit and the degree of voluntariness (Vrijling *et al.*, 1998). The notion of safety is often used in a subjective way, but it is essential to develop quantitative approaches before it can be used as a functional tool for decision making (Villemeur, 1992). A technique which 'quantifies' safety is FMEA.

1.4.1 The role of FMEA in designing capable and reliable products

In light of the above arguments, it has been found that there are two key techniques for delivering quality and reliability in new products: process capability analysis and FMEA (Cullen, 1994). FMEA is now considered to be a natural tool to be used in quality and reliability improvement and it has been suggested that between 70 and 80% of potential failures could be identified at the design stage by its effective use (Carter, 1986). For example, performing a comprehensive FMEA well will alleviate late design changes (Chrysler Corporation *et al.*, 1995).

FMEA was first mentioned at the start of this chapter. It is recommended that the reader unfamiliar with FMEA refer to Appendix III and several other references provided to gain a firm understanding of its application in product design. In general, an FMEA does the following (Leitch, 1995):

- Provides the designer with an understanding of the structure of the system, and the factors which influence quality and reliability
- Helps to identify items that are of high risk through the calculation of the Risk Priority Number (RPN), and so gives a means of deciding priorities for corrective action
- Identifies where special effort is needed during manufacture, assembly or maintenance
- Establishes if there are any operational constraints resulting from the design
- It gives assurance to management and/or customers that quality and reliability are being or have been properly addressed early in the project.

Of the many characteristics of a product defined by the dimensions and specifications on a drawing, only a few are critical to fulfilling the product's intended function.

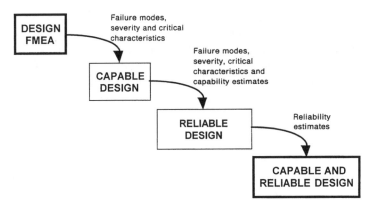

Figure 1.19 The FMEA input into designing capable and reliable products

Hence, a critical characteristic is defined as one in which high variation could significantly affect product safety, function or performance (Liggett, 1993). In order to assess the level of importance of the characteristics in a design, a process of identifying the critical characteristics and then using special symbols on the detailed drawing is commonly used. For example, the symbol '▼' is used by some companies to indicate that a particular characteristic should be controlled during manufacture using SPC. The identification process is facilitated by the use of a design FMEA using multi-disciplined teams.

As seen in Figure 1.19, the important results from an FMEA in terms of designing capable and reliable products are the potential failure modes, severity rating and critical characteristics for the design. By identifying the capability of the critical characteristics, and the potential failure mode, a statistical analysis can then be performed to determine its reliability. The FMEA Severity Rating (S) is crucial for setting capability and reliability targets because it is a useful indication of the level of the safety required for the application. Although subjective in nature, the effective use of an FMEA in the design process is advocated as it brings significant benefits.

In summary, to reduce risk at the product design stage requires that we do not speculate without supporting evidence on the causes, consequences and solutions for an actual or potential design problem. This means making predictions, where appropriate, based on evidence from testing, experience or other hard facts using statistical probabilities, not vague guesses. The use of FMEA to evaluate all the potential risks of failure and their consequences, both from normal use and foreseeable misuse, is a key element in designing capable and reliable products (Wright, 1989).

1.5 Designing for quality

The improvement of the quality of a design is seen as the primary need of industry, but to facilitate this we need appropriate methods for predicting quality and evaluating the long-term quality of an engineer's design (Mørup, 1993; Russell and Taylor, 1995; Shah, 1998; Taguchi *et al.*, 1989). However, there is relatively little work published in the field of Design for Quality (DFQ) compared to Design for Assembly (DFA), for example, and little methodology exists as yet (Bralla, 1996). One possible reason for this is that DFQ methods should have the objective of selecting the 'technically perfect' from a number of alternative solutions (Braunsperger, 1996). The word quality, therefore, implies a relative rather than a precise standard from which the designer has to work (Nixon, 1958). This requires a cultural shift of thought in design activities.

Techniques such as FMEA, DFA and Quality Function Deployment (QFD) can enhance the success of a product, but alone they will not solve all product development issues (Andersson, 1994; Jenkins *et al.*, 1997a; Klit *et al.*, 1993). They provide useful aids in the process of quality improvement, but they do not ensure product quality (Andersson, 1994). There exists an important need for DFQ techniques to aid design and support the product development process (Andreasen and Olesen, 1990). In addition, it has been cited that in order to make further reductions in product development time requires new progress in these techniques (Dertouzos *et al.*, 1989).

A substantial review of DFQ and the framework for its application has been proposed by Mørup (1993). This identifies eight key elements in DFQ which are placed under the headings of preconditions, structured product development and supporting methods/tools and techniques as illustrated in Figure 1.20. In thinking about DFQ, Mørup states that it is convenient to divide product quality into two main categories:

- 'Big **Q**' which is the customer/user perceived quality
- 'Little **q**' which relates to our efforts in creating big **Q**.

Product quality is a vector with several types of quality elements. The **Q** vector relates to issues including reputation, technology, use, distribution and replacement. The term **q** can also be considered as a vector with elements related to variability in component manufacture, assembly, testing, storage, product transport and installation. The notion of little **q** can also be expressed as an efficiency, related to efforts in meeting **Q**. The issues of **q** are met when the product meets those systems that are used to realize quality **Q**. The maintenance of **Q** relies upon the ability of a business to understand and control the variability which might be associated with the process of product realization. The quality in a product is not directly connected to cost. Every single **Q** element Q_i has corresponding **q** elements that contribute to cost. **Q** is fundamentally connected to selling price.

As can be seen from the above, the DFQ and the **Q/q** concepts are extremely broad in perspective. The general model may be used to drive the considerations of the important issues throughout the stages of production development and in the design of individual components and assemblies. The **q** element of quality described by Mørup is adopted in the CA methodology presented in Chapter 3 of this book.

The link between customer wants/perceived quality, **Q**, and quality of conformance, **q**, is a vague area. A large number of problems created at the design/manufacture interface are also caused by technical quality problems, for example wrong material specification, wrong dimensions, etc. These are essentially design communication deficiencies and so are amenable to an appraisal by a methodology of sorts. The flow of information through these quality disciplines is shown in Figure 1.21 where an analogy is made to the design of a simple hole in a plate. Further investigation of technical

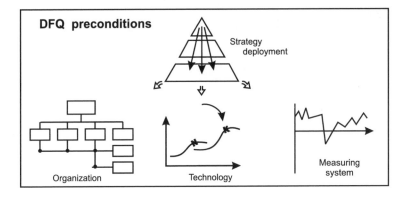

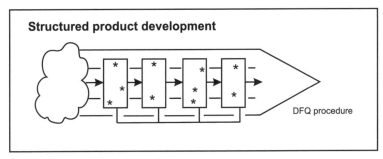

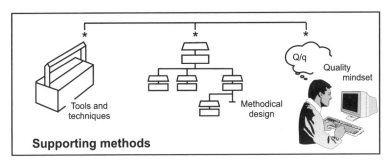

Figure 1.20 Preconditions for and main elements of DFQ (adapted from Mørup, 1993)

quality, **Qq**, should be also performed if we are to gain a further understanding of DFQ. For example, the measuring and monitoring of design drawing errors using SPC attribute techniques has been a major step forward in reducing the design changes for an aerospace company.

It is also possible to categorize the different types of DFQ techniques that are required to analyse the several types of quality highlighted above. These are (Mørup, 1993):

- Specification techniques which aid product developers in formulating quality objectives and specifications → **Q**
- Synthesis techniques which aid the designer in generating ideas and in detailing solutions → **Qq**

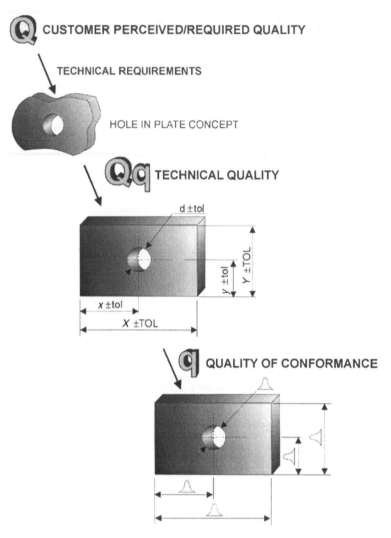

Figure 1.21 Hole in plate analogy to quality and the **Q/q** concepts

- Verification techniques which verify and evaluate the quality of solutions in relation to the specification → **q**.

There is a need for verification techniques in DFQ that can be used in the early and critical product development phases, where the quality is determined, i.e. can be applied on abstract and incomplete product models (Mørup, 1993). The CA methodology is largely a verification technique that aims to achieve this.

The DFQ approach at the verification level has many elements in common with Design for Manufacture (DFM) techniques. DFM helps create a product design that eases the task of manufacturing and lowers manufacturing cost. This is achieved by invoking a series of guidelines, principles and recommendations at the design stage and providing an understanding of the characteristics, capabilities and limitations of the manufacturing processes employed (Bralla, 1998; Kalpakjian, 1995). These design rules, or 'producibility' guidelines, are more generally applied at the component level than the assembly level, although DFA is sometimes, rather confusingly, associated with DFM (Leaney, 1996b; Russell and Taylor, 1995). Producibility guidelines are commonly developed by companies for designing products that are similar in nature to the ones for which the guidelines were written. They are therefore limited in their application because they may not apply to innovative design or where process capabilities are taken to the extreme limits by customer requirements.

Designing for robustness has also been associated with the DFM guidelines (Russell and Taylor, 1995). Robust design has different meanings to different engineering communities. For example, the three descriptions below focus on three different but connected aspects of product design:

- Robust design creates performance characteristics that are very insensitive to variations in the manufacturing process, and other variations related to the environment and time (Lewis, 1996; Sanchez, 1993).
- Robust design is the design of a product or process that results in functionally acceptable products within economic tolerances (Taguchi *et al.*, 1989).
- Robust design improves product quality by reducing the effects of variability (Phadke, 1989).

The first definition focuses on a process orientated design, the second the economic aspects of the design, and the third the impact of variability on the product in use. Although robust design is mentioned as a DFM guideline, no guidance for how to achieve 'robustness' is given. The definition of robust design must be made clear, and more importantly detailed guidance must be given to the designer on what to do to achieve robustness in a practical way. DFM techniques do not specifically answer this question. However, a robust design can be defined as a capable design in the context of the work presented later.

As stated earlier, awareness is growing that cost and quality are essentially designed into products (or not!) in the early stages of product engineering. The designer needs to know, or else needs to be able to, predict the capability of the process used to produce the design and to ensure the necessary tolerance limits are sufficiently wide to avoid manufacturing defects. Furthermore, the designer must consider the severity of potential failures and to make sure the design is sufficiently robust effectively to eliminate or accommodate defects. The major benefit of doing this is to reduce the potential for failure costs. Alternatively, in seeking to control or reduce cost, the safety of the product can be jeopardized, for example allowing production cost to dominate a design decision to the extent that the product does not meet customer expectation. The design phase of a product is crucial because it is here that the product's configuration and, therefore, much of its potential for harm are determined. Those engaged in product design must give a high priority to the elimination or control of hazards associated with the product. This may have to be to the detriment of ease of manufacture, styling, user convenience, price and other marketing factors (Wright, 1989).

The experiences of industrial collaborators and surveys of UK businesses suggest that failure costs are the main obstacle to reducing the costs of quality. There is, therefore, a need for design methods and guidelines to give businesses the foresight to identify product characteristics that depict potential costly failures, particularly in the case of bought out components and assemblies. Designers need models to predict costs at the various design stages. The provision of these techniques is essential for future business competitiveness. Through DFQ we can change over from merely preventing and eliminating quality problems to actively incorporating the level of quality expected by the consumer into the product. At the same time we can see that the target of high product quality is to a great extent compatible with the target of low costs, and thus with the creation of good business (Mørup, 1993). We need a culture, professionalism and techniques for Design for Quality.

1.6 Designing for reliability

The pursuit of product quality over the last few decades has been intense in many companies around the world, but in many sectors of industry, reliability is considered to be the most important quality attribute of the product (Kehoe, 1996). As consumers become more aware than ever of quality, their expectations for reliability are also increasing. Even equipment without obvious safety concerns can have important reliability implications. Most products can, therefore, benefit from the use of sound reliability techniques (Burns, 1994). Reliability prediction, in turn, has the benefit that it gives a quantifiable estimate of the likely reliability that can be assessed to see if this is appropriate for the market (Stephenson and Wallace, 1996).

Reliability prediction is undoubtedly an important aspect of the product design process, not only to quantify the design in terms of reliability, but also to determine the critical design parameters that go to produce a reliable product. To this end, it is necessary to have a mathematical, quantitative measure of reliability defined by probability (Leitch, 1995). It is, however, a controversial aspect of reliability engineering in terms of accuracy and validity because it relies on detailed knowledge about sometimes unknown design parameters (Burns, 1994; Carter, 1986; O'Connor, 1995). In addition, the practical 'engineering' of reliability in the product is still seen as a science to many designers and evaluating concept designs for reliability is especially difficult for inexperienced staff (Broadbent, 1993). A fundamental reason for this is the supporting use of statistics and probability theory. Designers and engineers have consistently turned a blind eye to the advantages of using these methods; however, they can be trained to use them without being rigorously schooled in their mathematical foundation (Morrison, 1997). Reliability prediction will remain a controversial technique until the statistical methods for quantifying design parameters becomes embedded in everyday engineering practice. Using statistically based design techniques will not solve the whole problem, although it will be much closer to the desired end result (Carter, 1997).

The reliability of a product is the measure of its ability to perform its intended function without failure for a specified time in a particular environment. Reliability engineering has developed into two principal areas: part and system. Part reliability is concerned with the failure characteristics of the individual part to make inferences about the part population. This area is the focus of Chapter 4 of the book and dominates reliability analysis. System reliability is concerned with the failure characteristics of a group of typically different parts assembled as a system (Sadlon, 1993).

There are currently three main approaches used in the pursuit of a reliable product (Stephenson and Wallace, 1996):

- Reliability prediction (includes probabilistic design)
- Design techniques (for example, FMEA or Fault Tree Analysis (FTA))
- Pre-production reliability testing (prototyping).

Reliability prediction techniques are a controversial but effective approach to designing for reliability, as discussed later. FTA (Bignell and Fortune, 1984; Straker, 1995) and FMEA are established techniques that aim to determine the potential causes and effects of failures in components and systems. Although subjective in nature, they are useful at the system level where many interactions of components take place. Finally, prototype testing is a pre-production exercise performed on a few model products tested to determine if they meet the specification requirements. All three approaches have their value in the product development process, although the aim of the first two must surely be to reduce the third, prototype testing, which can be an expensive undertaking. However, you could never remove the necessity for some kind of prototype testing, especially where the aim is to verify the integration of parts and subassemblies. From the above, it would seem that designing for reliability depends on both quantitative results and qualitative processes (Ben-Haim, 1994). However, the collection of reliability tasks might look intimidating causing one to wonder if the benefit gained by performing all is worth the effort (Burns, 1994).

Virtually all design parameters such as dimensional characteristics, material properties and service loads exhibit some statistical variability and uncertainty that influence the adequacy of the design (Rice, 1997). Variability may arise from material quality, adverse part geometry and environmental factors (Weber and Penny, 1991). Historically, the designer has catered for these variabilities by using large factors of safety in a deterministic design approach. In probabilistic design, statistical methods are used to investigate the combination and interaction of these parameters, having characterized distributions, to estimate the probability of failure. A key requirement is detailed knowledge about the distributions involved to enable plausible results to be produced. The amount of information available at these early stages is limited, and the designer makes experienced judgements where information is lacking. This is why the deterministic approach is still popular, because many of the variables are taken under the 'umbrella' of one simple factor. If knowledge of the critical variables in the design can be estimated within a certain confidence level, then the probabilistic approach becomes more suitable. It is essential to quantify the reliability and safety of engineering components and probabilistic analyses must be performed (Weber and Penny, 1991).

There are numerous applications for probabilistic design techniques in mechanical engineering. A number of important applications exist in design optimization and reliability engineering, specifically where it would be useful to explore the level of random failure, resulting from the interaction of the distributions of loading stress and material strength, discussed in detail in Chapter 4 of the book. The approach is shown in Figure 1.22 again for a simple hole in a plate concept. All the design parameters are shown as statistical distributions rather than unique or nominal values. The final failure prediction through an appropriate probabilistic and failure analysis model reflects the distributional nature of the design parameters.

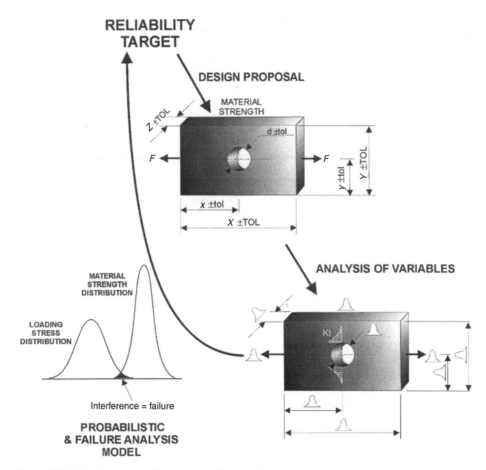

Figure 1.22 Hole in a plate analogy of probabilistic design

More plausible representations of stress and strength distributions for a given situation will enable meaningful failure predictions to be produced, and will be particularly useful where test to failure is not a practical proposition, where weight minimization and/or material cost reduction is important, or where development time is crucial. Engineering experience indicates that many devices are overdesigned, that is, they feature excess weight or excess occupied space. When weight and/or space is at a premium, a more realistic design process is required that permits relatively accurate predictions of device performance. A probabilistic design process, taking into account of the uncertainties with typical design inputs, provides the required realism (Bury, 1975).

In a probabilistic approach, design decisions must reduce the probability of unwanted performance to acceptable levels. In a deterministic approach, the designer can only assure that the performance remains within an acceptable domain (Ben-Haim, 1994). As the deterministic approach provides no firm basis for dealing with variability, it is not pertinent to a reliability approach (Morrison, 1997; Shigley and Mischke, 1989). As an example of this, Haugen (1968) shows that the failure probability for a particular problem can vary from zero to unity because of the variability of the design parameters, but with the factor of safety selected remaining constant. Factors of safety can lead to either an unconservative design with unacceptable high failure rates, or a very conservative design that provides the required performance with unnecessarily high product costs (Rice, 1997).

Deterministic design is still appealing because of its simplicity in form and application, but since factors of safety are not performance related measures, there is no way of indicating if the design is near optimum (Haugen, 1980). With increasing concern over minimizing the cost of failure, the probabilistic design approach will become more important (Dieter, 1986). Probabilistic design gives the designer a better feel of just how conservative or unconservative the design is (Ullman, 1992). In order to determine this, however, it is important to make decisions about the target reliability level (Ditlevsen, 1997).

To be able to evaluate design reliability estimates using probabilistic methods, the designer needs much more information than for a deterministic evaluation (Fajdiga *et al.*, 1996). It can be argued that probabilistic design can be used only when all the needed statistical data is available and it would be dangerous to design to a reliability target when the data is suspect (Shigley and Mischke, 1989). Because of the lack of statistical data for the strength of materials used and the applied loads in particular, design concepts based on the factor of safety will still dominate the design of some products (NASA, 1995; Zhu, 1993). However, the probabilistic approach allows us to perform a sensitivity analysis of the design with respect to the various design parameters to give an idea of the impact of the variability of dimensions, material strength and loads on performance, and this makes design optimization possible (Kapur and Lamberson, 1977). Probabilistic design is another way of thinking about the design problem which must surely be an improvement over using large factors of safety (Loll, 1987).

Probabilistic methods have gained increased interest in engineering as judged from the growing community of reliability engineers and from the increasing number of conferences on the subject (Ditlevsen, 1997). Some practitioners in the UK, however, either seem to lose confidence with statistical and probabilistic methods or are just not aware of them. At present, only larger companies seem to be aware of their importance (Howell, 1999). Some advocates of a statistical approach to engineering design even claim that this is why large chunks of manufacturing have moved to countries like Japan who embrace the use of such techniques. A comment in 1995 by Margetson gives an indication of the situation related to the UK:

> It is essential to introduce probabilistic design methods into engineering design procedures. I feel that the UK will be faced with a severe skill gap. I also feel there is a lack of appreciation of the need and time scale to introduce the required procedures to engineers ... the Japanese have identified probabilistic design as a key technical area.

A problem may lie in the knowledge that is required as an essential input to such approaches, being both statistical in nature and from the authors' own experiences, often difficult to obtain and interpret generally. Unfortunately, experienced designers will not use statistical methodology, although statistical methods should play an important role in the design and manufacture of reliable products (Amster and

Table 1.1 Competing issues in deterministic and probabilistic design approaches

Deterministic design	Probabilistic design
Dominated design for 150 years	Application for approximately 40 years
Design parameters treated as unique values	Design parameters treated as random variables
Underlying empirical and subjective nature	Small samples used to obtain statistical
Factor of safety dominates determination of reliability	distributions
Ignorance about the problem being multi-variable	An understanding of statistics is required
Calculations simple and data widely available	Knowledge scattered throughout engineering texts
Inherent overdesign – wasteful, costly and ineffective	Output as a probability of failure or reliability
Culture of deterministic design still exists in industry	Better understanding of the effects of variability

Hooper, 1986; Carter, 1997). A principal drawback of the probabilistic design approach is that it requires a good knowledge of probability and statistics, and not every design engineer has this knowledge (Kapur and Lamberson, 1977). Summarizing the above, the main characteristics of the deterministic and probabilistic design approaches are shown in Table 1.1.

The provision for reliability must be made during the earliest design concept stage (Dieter, 1986). The more problems prevented early on through careful design, the fewer problems that have to be corrected later through a time-consuming and often confusing process of prototype (Dertouzos *et al.*, 1989). A principal necessity then is to design to a reliability goal without an inordinate amount of component and prototype testing (Mischke, 1989). This can only be achieved by a rigorous appraisal of the design as honestly and early as possible in the product development process.

1.7 Summary

The designer's job is to try to capture customer expectations and translate as many of these expectations as possible to the final product. The functional requirements of the design become detailed into dimensional tolerances or into attributes of the component or assembly. The ability of the manufacturing process, by which these products are made or assembled, to consistently provide dimensions within tolerance may be called its conformance to design. Understanding and controlling the variability associated with these design attributes then becomes a key element of developing a quality product.

Designers rarely fully understand the manufacturing systems that they are designing for, and subsequently they do not understand the variability associated with the design characteristics. Variability can have severe repercussions in terms of failure costs, appearing in production due to rework and scrap, and warranty costs (or worse!) when the product fails in service. There is need to try to anticipate the variability associated with the manufacturing processes used to produce the final product early in the design process. The designer needs to know, or else be able to predict the capability of the process and to ensure the necessary tolerance limits are sufficiently wide to avoid manufacturing defects. However, this has previously been difficult to achieve on concept design or where little detail exists.

Quality assurance systems demand not just tolerances, but process capable tolerances and characteristics that limit the potential failure costs incurred. The level of process capability is not necessarily mandated, but clearly the more severe the consequences of a defect, the lower the probability of occurrence of that defect should be. This 'risk' may be seen to bear a failure cost penalty and the expected cost must be limited. It should be clear that techniques such as SPC can only be applied to enhance the intrinsic capability of a process. Further, any inspection process to remove defective items can be only a compensation for a process with inadequate capability, or perhaps a design which has been specified with tolerances which are just too tight, or assemblies which are not quite practical.

From the above discussion, it follows that the quality and conformance to tolerance of the product characteristics should be 'designed in' and not left to the process engineer and quality engineer to increase to the required level. In order to do this, designers need to be aware of potential problems and shortfalls in the capability of their designs. They therefore need a technique which estimates process capability and quantifies design risks.

Reliability prediction techniques are a controversial but fundamental approach for designing for reliability. A key objective of these methods is to provide the designer with a deeper understanding of the critical design parameters and how they influence the adequacy of the design in its operating environment. These variables include dimensions, material properties and in-service loading. A key requirement is detailed knowledge about the distributions involved to enable plausible results to be produced in an analysis. It is largely the appropriateness and validity of this input information (and failure theory) that determines the degree of realism of the design process and the ability to accurately predict the behaviour and therefore the success of the design. The determination of an absolute value of reliability is impossible. In developing a product, a number of design schemes or alternatives should be generated to explore each for their ability to meet the target requirements. Evaluating and comparing designs and choosing the one with the greatest predicted reliability, or quality for that matter, will provide the most effective design solution.

The costs of quality experienced by manufacturing businesses are dominated by costs associated with failure, for example rework, scrap, design changes, warranty and product liability claims. These costs represent lost profit and potentially impact on the future opportunities of the business. Decisions made during the design stage of the product development process account for a large proportion of the problems that incur failure costs in production and service. It is possible to relate these failure costs back to the original design intent where variability, and the lack of understanding of variability, is a key failure costs driver. Current quality costing models are useful for identifying general trends in a long-term improvement programme, but are of limited use in the identification of the failure costs associated with actual design decisions. An important aspect of the discussion in the next section is to demonstrate a method whereby the estimation of failure costs at the design stage is useful in selecting the most cost effective, from a failure point of view, from a number of alternatives.

In the next chapter, we introduce the concepts of component manufacturing capability and the relationships between tolerance, variability and cost. The Component Manufacturing Variability Risks Analysis is then introduced, the first stage of the CA methodology, from which process capability estimates can be determined at the design stage. The development of the knowledge and indices used in the analysis are discussed within the concept of an 'ideal design'. The need for assembly variability determination and the inadequacy of the DFA techniques in this respect is argued, followed by an introduction to assembly sequence diagrams and their use in facilitating an assembly analysis. The Component Assembly Variability Risks Analysis is discussed next, which is the second stage of the CA methodology. Finally, a cost–quality model is presented, used to determine the failure costs associated with non-safety critical and safety critical applications in production and service through linkage with FMEA. Comprehensive guidance on the application of the technique is given and a number of case studies and example analyses are used to illustrate the benefits of the approach.

2

Designing capable components and assemblies

2.1 Manufacturing capability

One of the basic expectations of the customer is conformance to specification, that is, the customer expects output characteristics to be on target with minimum variation (Abraham and Whitney, 1993; Garvin, 1988). Assessing the capability of designs early in product development therefore becomes crucial. The designer must aim to achieve the standards demanded by the specification, but at the same time not exceed the capabilities of the production department. This may not be an easy task because the determinants of quality are frequently difficult to identify at the design stage. The designer must decide how the specifications are affected by features of the production process, as well as by the choice of materials, design scheme, etc. If the designer knows that a certain manufacturing process can only achieve very poor dimensional tolerances, for example, a decision must be made to decide whether such an process can be used for the new product. If not, an alternative solution must be devised (Oakley, 1993).

The available information about manufacturing processes is both extensive and diverse, but it is not easily accessible for designers, as the structure is not suited for design applications (Sigurjonsson, 1992). The designer's job is made easier if standard rules are available for the candidate manufacturing method. A good set of design rules can indicate process capable tolerances for a design. Life is much more problematic when rules are not available or an unfamiliar manufacturing process is to be used. Many designers have practical experience of production methods and fully understand the limitations and capabilities that they must work within. Unfortunately, there are also many that do not have this experience and, quite simply, do not appreciate the systems that they are supposed to be designing for (Oakley, 1993). For example, three aspects of engineering drawing that are often overlooked, but are vital in design practice, are dimensions, tolerances and specification of surface finish (Dieter, 1986). The designer needs to understand when required tolerances are pushing the process to the limit and to specify where capability should be measured and validated.

Tolerances alone simply do not contain enough information for the efficient manufacture of a design concept (Vasseur *et al.*, 1992). At the design stage, both qualitative manufacturing knowledge about candidate manufacturing processes for design features, and quantitative manufacturing knowledge on the influences of design parameters on production costs, are needed (Dong, 1993). Only with strong efforts to integrate the product design with the selection of the manufacturing process can the desirable performance characteristics be produced with a minimum of variability and minimum cost (Lewis, 1996).

The goal of this first stage of Conformability Analysis (CA), the Component Manufacturing Variability Risks Analysis, is the provision of support in the early stages of the detailed design process for assessing the tolerance and surface finish capability associated with component manufacture. 'Risk' in this sense is a measure of the chances of not meeting the specification. Recognizing that the relationship between a design and its production capability is complex and not easily amenable to precise scientific formulation, the methodology described has resulted largely from knowledge engineering exercises in manufacturing businesses, including those with expertise in particular manufacturing processes. For example, Poeton in Gloucester, UK, specialize in supplying surface coating technology in the form of design guidance and processing techniques. The Poeton engineers had a major consultation role in the determination of the main issues related to surface engineering process variability. To support other aspects of the analysis, such as the tolerance capability analysis, manufacturing tolerance data was readily available, albeit in a form not useful to designers. During the evolution of CA, many alterations and improvements were made through exhaustive consultation and validation in industry. The development of only the current aspects of the analysis will be discussed in detail here. However, some background topics will be reviewed before proceeding.

2.1.1 Variability factors in manufacturing

Designers have always had to deal with the fact that parts cannot be made perfectly; or if they could, they would not remain perfect for long during use. So defining the 'ideal' component is only one aspect of the designer's job (Hopp, 1993). The designer must also decide by how much a still acceptable component can be from the ideal. A component can vary from the ideal in many ways: in its geometry, its material properties, surface finish – a virtually unlimited list (Alexander, 1964).

In analysing a design at the concept development or early detailing stage, it is only necessary to focus on the main variabilities associated with manufacturing processes. In this way the analysis performs at a level of abstraction which facilitates a rapid assessment without being complex or difficult to comprehend. Engineering judgement and experience are needed to identify potential variability risks or 'noise' associated with manufacturing and assembly routes (Phadke, 1989). Before introducing the main aspects of the component manufacturing variability risks analysis, it is worth exploring variability issues in general. There are two main kinds of variability:

- Common-cause or inherent variability is due to the set of factors that are inherent in a machine/process by virtue of its design, construction and the nature of its operation, for example positional repeatability, machine rigidity, which cannot be removed without undue expense and/or process redesign. When only common-cause variability is present, the process is performing at its best possible level under the current process design.

- Assignable-cause or special-cause variability is due to identifiable sources which can be systematically identified and eliminated (see below).

The last sources of variation highlighted are of paramount concern for allocating and analysing mechanical tolerances (Harry and Stewart, 1988).

At the most detailed level, a variation can belong to the basic design properties: form, dimension, material and surface quality for the components, and structure for the relations between components (Mørup, 1993). For example, the levels of dimensional tolerance accuracy and surface roughness associated with industrial manufacturing processes vary widely. In general, tolerances reflect the accuracy specifications of the design requirements. These will inherently reflect the variability in the manufacturing process. That is the design of the tolerance for a certain component characteristic will influence the variability in the measurements of that characteristic. Hence, a tightened tolerance requires higher precision manufacturing devices, higher technical skills, higher operation attention and increased manufacturing steps (Jeang, 1995).

The general factors influencing variation include the following (Dorf and Kusiak, 1994; Mørup, 1993):

- Tool and functional accuracy
- Operator
- Set-up errors
- Deformation – due to mechanical and thermal effects
- Measurement errors
- Material impurities
- Specifications
- Equipment
- Method or job instructions
- Environment.

Further variations may arise from the working of the material during the manufacturing process or from deliberate or unavoidable heat treatment (Bolz, 1981). In general, the undesirable and sometimes uncontrollable factors that cause a functional characteristic to deviate from its target value are often called noise factors and are defined below (Kapur, 1993):

- *Outer noise* – environmental conditions such as temperature, humidity, different customer usage
- *Inner noise* – changes in the inherent properties of the system such as deterioration, wear, corrosion
- *Product noise* – piece to piece variation due to manufacturing variation and imperfections.

An example of assignable-cause variability is given in Figure 2.1. Two cases of milling the same component to finished size are shown. In the first case, the component is

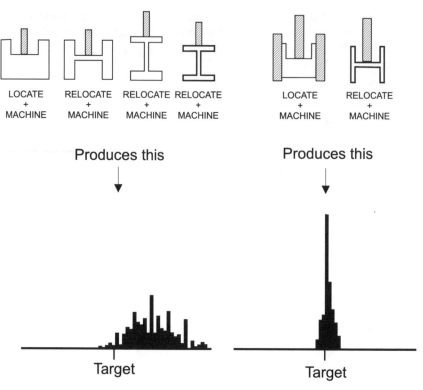

Figure 2.1 Example of process capability improvement (adapted from Leaney, 1996a)

relocated a number of times in the tooling to produce the finished product. The tolerance capability of a key dimension is analysed and the results given in the form of a histogram. In the second case, the number of relocations is reduced and simultaneous operations carried out on the component in the form of gang milling. A significant capability improvement is experienced in the second case as shown by the comparatively low process spread and accumulation around the target value. This is mainly due to the set-up and operation procedures for the manufacture of the component. It is evident from this that a good process set-up ensures better product quality (Hallihan, 1997).

2.1.2 Cost–tolerance relationships

Ideally, engineers like tight tolerances to assure fit and function of their design. Designers often specify unnecessarily tight tolerances due to the lack of appreciation of cost and due to the lack of confidence in manufacturing to produce component parts which conform to specification (Phadke, 1989). However, a tightened tolerance requires higher precision and more expensive manufacturing processes, higher technical skills, higher operator attention, increased manufacturing steps and is more time consuming to achieve (Jeang, 1995; Soderberg, 1995). The configuration and material of a part, as well dimensions, tolerances and surface finishes, can also change the amount of work required in part manufacture (Dong, 1993).

All manufacturers, on the other hand, prefer loose tolerances which make parts easier and less expensive to produce (Chase and Parkinson, 1991). The choice of tolerance is therefore not only related to functional requirements, but also to the manufacturing cost. It has been argued that among the effects of design specifications on costs, those of tolerances are perhaps the most significant (Bolz, 1981). Figure 2.2(a) shows the relationship between product tolerance and approximate relative cost for several manufacturing processes. Each tolerance assignment results in distinct manufacturing costs (Korde, 1997).

Cost–tolerance functions are used to describe the manufacturing cost associated with a process in achieving a desired level of tolerance. The shape of the cost–tolerance curve has been suggested by several researchers. Cagan and Kurfess (1992) propose a hyperbolic cost function. Speckhart (1972) and Dong (1997) suggest an exponential cost function. The inverse quadratic form was advocated by Spotts (1973). A comparison of various researchers' cost models is provided in Figure 2.2(b). Determining the parameters of the cost–tolerance models described, however, is by no means a trivial task (Dong, 1993).

2.1.3 Process capability and tolerances

Product tolerance can significantly influence product variability. Unfortunately, we have difficulty in finding the exact relationship between them. An approximate relationship can be found from the process capability index, a quality metric interrelated to manufacturing cost and tolerance (Lin *et al.*, 1997). The random manner by which the inherent inaccuracies within a manufacturing process are generated produces a pattern of variation for the dimension which resembles the Normal distribution (Chase and Parkinson, 1991; Mansoor, 1963) and therefore process capability indices, which are based on the Normal distribution, are suitable for use. See Appendix IV for a detailed discussion of process capability indices.

There are many variations on the two basic process capability indices, C_p and C_{pk}, these being the most commonly used. A process capability index for a shifted distribution at $C_{pk} = 1.33$ is still regarded to be the absolute minimum (which relates to around 30 ppm failing the specification). This increases to $C_{pk} = 1.66$ (which relates to approximately 0.3 ppm failing) for more safety critical characteristics (Kotz and Lovelace, 1998). Motorola stipulate $C_{pk} = 1.5$ (or approximately 3 ppm) to their suppliers as an absolute minimum (Harry and Stewart, 1988). However, one weakness of the process capability index is that there is no apparent basis for specifying the optimal value (Taguchi *et al.*, 1989). In practice, most designers do not worry about the true behaviour of a manufacturing process and compensate for the lack of knowledge with larger than necessary process capability indices (Gerth, 1997).

The precise capability of a process cannot be determined before statistical control of the actual process has been established. Designers seldom have sufficient data by which to specify the variability of the manufacturing processes and therefore

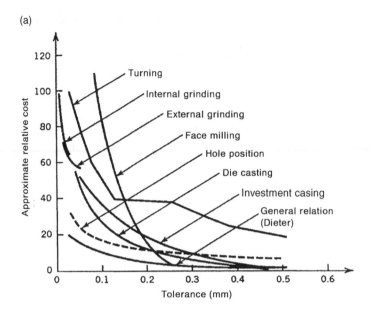

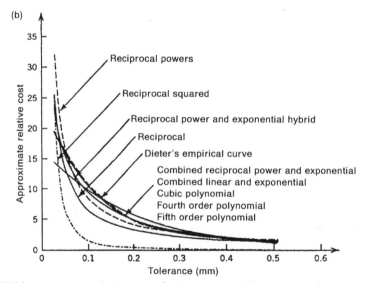

Figure 2.2 (a) Tolerance versus production costs of various processes; (b) comparison of cost-tolerance models (Dong, 1993)

during the design phase, the designer must use the best available process capability data for similar processes (Battin, 1988; Chase and Parkinson, 1991). It is far easier, not to mention less costly, to create robust designs based on known process capabilities than it is to track down and subsequently reduce sources of variation during the manufacturing phase (Harry and Stewart, 1988).

2.2 Component Manufacturing Variability Risks Analysis

In the development of the Component Manufacturing Variability Risk Index, q_m, it was found to be helpful to consider a number of design/manufacture interface issues, including:

- Material to process compatibility
- Component geometry to process limitations
- Process precision and tolerance capability
- Surface roughness and detail capability.

In the formulation of q_m, it has been assumed that there is a basic level of variability associated with an 'ideal' design for a specific manufacturing process, and that the factors listed above can be estimated independently and manipulated according to some function:

$$q_m = f(m_p, g_p, t_p, s_p) \qquad (2.1)$$

where:

$$q_m = \text{component manufacturing variability risk}$$

$$m_p = \text{material to process risk}$$

$$g_p = \text{geometry to process risk}$$

$$t_p = \text{tolerance to process risk}$$

$$s_p = \text{surface roughness to process risk.}$$

It then becomes possible to define the specific variability risk indices relating to equation 2.1 through the use of a number of charts. Reference to the guidance notes associated with each chart is recommended.

The material to process risk, m_p, uses manufacturing knowledge in the assessment of the component material to process compatibility, for example machinability rating or formability rating (see Figure 2.3). These values are greater than unity only when the material to be used in conjunction with the manufacturing process is not defined on the process capability maps used. A thorough discussion of process capability maps is given in Section 2.2.1.

The geometry to process risk, g_p, uses knowledge of known geometries that exhibit variability during component manufacture to support the risks analysis, for example parting lines on castings and long unsupported sections (see Figure 2.4).

The tolerance to process risk chart for t_p (see Figure 2.5) uses knowledge of the dimensional tolerance capability for a number manufacturing processes and material combinations in the form of process capability maps (Figure 2.6).

Related to tolerance, but distinct from the tolerance to process risk, the surface roughness to process risk, s_p, is a function of the surface roughness specification. Figure 2.7 shows the chart used initially. Again, knowledge of the surface roughness values for a number of manufacturing processes is used to facilitate an analysis as provided in Figure 2.8. A discussion of the generation procedure for the surface roughness chart is given in Section 2.2.2.

Initially, knowledge of the chosen manufacturing process is required. If the type of material is stated on the process capability map, assume $m_p=1$, otherwise proceed to 'A'.

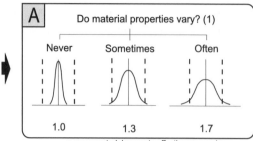

A   Do material properties vary? (1)

Never   Sometimes   Often

1.0   1.3   1.7

= mean material property affecting processing

B	Values of 'B' relating to material compatibility for those processes where a material is not defined on the process capability maps given (2)				
PROCESS **MATERIAL**	IMPACT EXTRUSION COLD FORMING COLD EXTRUSION	SHEET METALWORK		MACHINING	POWDER METAL SINTERING
		CUTTING BLANKING FINE BLANKING	BENDING/DRAWING ROLL FORMING SPINNING		
FREE CUTTING				1.2	
MILD	1.2	1.4	1.2	1.3	1.1
MED. CARBON	1.4	1.5	1.4	1.4	1.1
HIGH CARBON	1.7	1.7	1.5	1.5	1.1
ALLOY	1.7	1.5	1.5	1.5	1.2
TOOL				1.7	1.3
CAST				1.4	
STAINLESS	1.7	1.5	1.4	1.6	1.1
MALLEABLE				1.1	1.0
NODULAR				1.2	
PEARLITIC				1.5	
GREY				1.4	
ALUMINIUM	1.0	1.0	1.0	1.0	1.5
COPPER	1.0	1.2	1.2	1.2	1.7
MAGNESIUM	1.1	1.3	1.7	1.1	
NICKEL	1.5	1.4	1.4	1.6	1.4
ZINC	1.0	1.1	1.5	1.1	
TIN	1.0	1.2	1.7	1.5	
LEAD	1.0	1.2	1.7	1.4	
TITANIUM		1.5	1.5	1.7	1.4
PLASTICS				1.5	

(STEELS covers FREE CUTTING through STAINLESS; IRONS covers MALLEABLE through GREY; NONFERROUS ALLOYS covers ALUMINIUM through TITANIUM.)

Notes

① Variations in material properties create out of tolerance variations during processing. Variations can relate to uncertainty in material composition, flaws, cracks, discontinuations, as well as non-uniform processing characteristics.

② There is a material to process compatibility risk for impact extrusion, cold forging, cold extrusion, sheet metalworking, machining and powder metal sintering processes because their respective process capability maps relate to the ideal material case.

$$m_p = \boxed{A} \times \boxed{B}$$

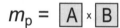

Figure 2.3 Material to process risk chart, m_p

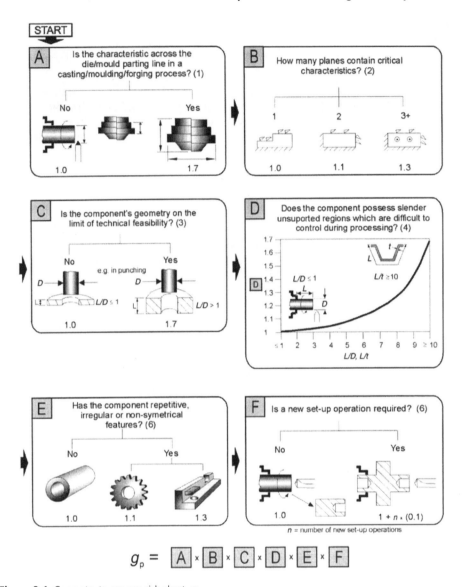

Figure 2.4 Geometry to process risk chart, g_p

<u>Notes</u>

① Allowances should be made for tolerances across the parting line. Allowance may also be required for mismatch of the dies/moulds in some casting and forging processes. Flash thickness allowances may also be required in closed die forging. Note, this only applies to casting, moulding and forging processes.

② Refers to the number of orthogonal axes on which the critical characteristics lie, and which cannot be achieved by processing from a single direction.

③ The processing of components that are on the limits of technical feasibility is likely to result in out of tolerance variation. High forces and flow restriction in metalworking and metal cutting processes can lead to instability. Also, material flow in casting processes, where abnormal sections and complex geometries are present, can lead to variability problems and defects.

④ Slender unsupported regions with large length to thickness ratios are highly liable to distortion during processing.

⑤ Repetitive, irregular or non-symmetrical features require greater process control and complex set-up or tooling requirements. This can be an added source of variability.

⑥ There is all increased risk of variation each time a new set-up operation is required due to changes in the orientation of the part or tooling.

A

Divide the characteristic's design tolerance, which must be in the correct format either +, –, ± or 'TOTAL' as shown in the process capability map for the manufacturing process to be used, by the product of m_p and g_p. This then gives the adjusted tolerance:

$$\text{Adjusted tolerance} = \frac{\text{Design tolerance}}{m_p \times g_p}$$

Using the process capability map for the manufacturing process in question, select the appropriate value of 'A' from the map corresponding to the adjusted tolerance and associated characteristic dimension.

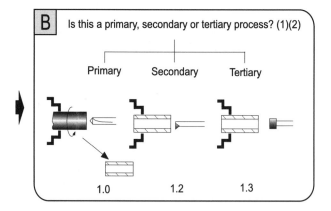

<u>Notes</u>

① A primary process is the basic shape generation method for the component.

② Secondary and tertiary processing normally infers material removal processes, e.g. turning, grinding, honing. Adequate machining allowances must be provided for at the primary/secondary processing stage when a secondary/tertiary process is used respectively.

$$t_p = \boxed{A} \times \boxed{B}$$

Figure 2.5 Tolerance to process risk chart, t_p

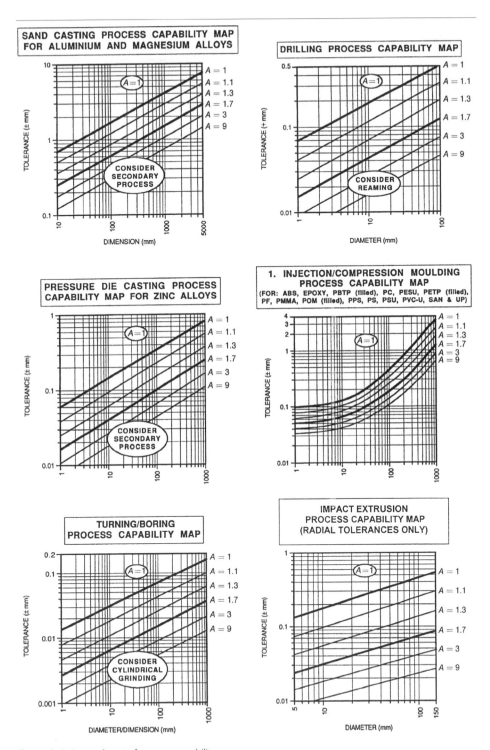

Figure 2.6 A sample set of process capability maps

A

Referring to the surface roughness chart for the manufacturing process to be used, select the appropriate value of 'A' corresponding to the characteristic's design surface roughness.

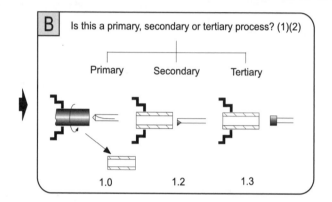

Notes

① A primary process is the basic shape generation method for the component.

② Secondary and tertiary processing normally infers material removal processes, e.g. turning, grinding, honing. Adequate machining allowances must be provided for at the primary/secondary processing stage when a secondary/tertiary process is used respectively.

$$S_p = \boxed{A} \times \boxed{B}$$

Figure 2.7 Surface roughness to process risk chart, s_p

Validation studies in manufacturing businesses and discussion with experts led to the view that knowledge used to define m_p, g_p, t_p and s_p, could be structured such that q_m may be formulated as:

$$q_m = t_p \times s_p \tag{2.2}$$

where:

$$t_p = f(\text{design tolerance}, m_p, g_p) \tag{2.3}$$

(See Figure 2.5 for the complete formulation of t_p.)

A link between the material used and the geometry of the component is compounded in the formulation for the tolerance to process risk as shown in equation 2.3. It is recognized that increasing material incompatibility and geometry complexity

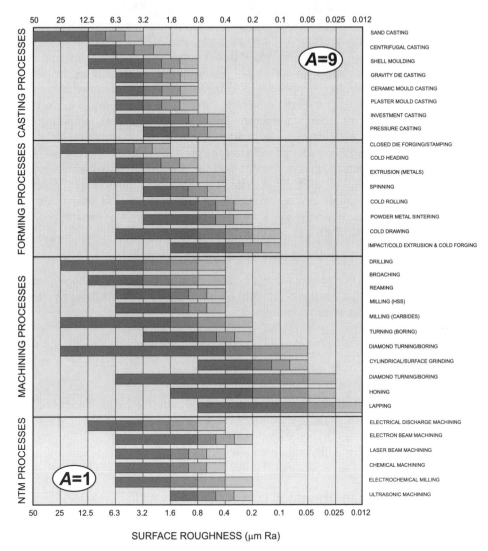

Figure 2.8 Surface roughness risks for a number of manufacturing processes

has the effect of increasing the variability associated with achieving the dimensional tolerance requirement. The above equation relates this notion to the tolerance process risk by dividing the design tolerance for the characteristic, as stated by the designer, by the product of m_p and g_p. The risk index for surface roughness, s_p, usually defaults to unity unless a surface roughness requirement is considered critical, for example a valve seat or lubricated surface.

The underlying notion of the Component Manufacturing Variability Risk Index, q_m, is that an ideal design exists for a component where the risk index is unity, indicating that variability is in control. Risk indices greater than unity exhibit a greater potential for variability during manufacture. The resulting value of q_m indicates the risk of out of tolerance/surface roughness capability when compared to an ideal situation. For the ideal design of a component and processing route, each of the quantities is unity and therefore in all cases $q_m \geq 1$. Tolerance, surface roughness, material and geometry designed into a component, which is not matched with the ideal, have an effect on variability. For example, in die casting there is a higher risk when processing copper alloys compared with the tolerance capability resulting from processing zinc-based materials, largely due to temperature effects.

Additionally, there are other manufacturing processes (for example, tempering and nitriding) that must be considered in the analysis if used in the product's manufacturing route. These processes are carried out after the primary/secondary processes have been used to manufacture the component and are treated as post-manufacturing processes. The potential for variability in the final component due to these processes is great, due to the possible combination of high temperatures and unsymmetrical sections, which are particularly likely to cause out of tolerance variations. This introduces an additional factor to consider, based on:

- Surface engineering processes (bulk and surface heat treatment/coating processes).

Surface engineering processes are usually performed after the primary shape generation of the component, or post-manufacturing, therefore q_m defaults to k_p when it is considered in the manufacturing route:

$$q_m = k_p \tag{2.4}$$

where:

$$k_p = \text{surface engineering process risk.}$$

Figure 2.9 shows the surface engineering process risk chart, k_p. It includes the key variability issues related to these types of process.

Validation of the predictions for process capability through the use of the component manufacturing variability risks analysis, q_m, is given later.

2.2.1 Process capability maps

As can be seen from the above, central to the determination of q_m is the use of the process capability maps which show the relationship between the achievable tolerance and the characteristic dimension for a number of manufacturing processes and material combinations. Figure 2.6 shows a selection of process capability maps used in the component manufacturing variability risks analysis and developed as part of the research. There are currently over 60 maps incorporated within the analysis covering processes from casting to honing. The full set of process capability maps is given in Appendix IV.

Data on the tolerance capability of the manufacturing processes covered were compiled from international standards, knowledge engineering in specialist businesses

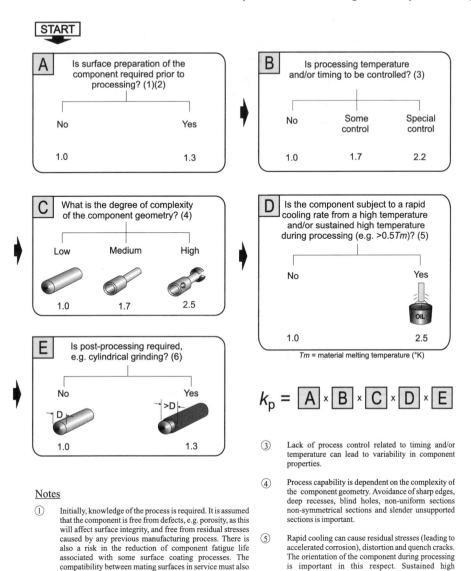

Figure 2.9 Surface engineering process risk chart, k_p

and engineering texts. A selection of references used to generate the maps is given in the Bibliography. The data used in the creation of the maps usually comes in the form of tables, such as that given in Figure 2.10 for machining using turning and boring. International Organization of Standards (ISO) tolerance grades are commonly used as a straightforward way of representing the tolerance capability of a

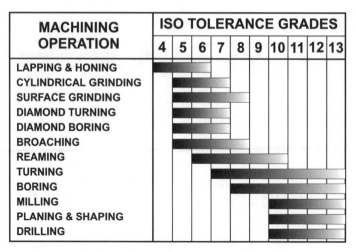

Figure 2.10 ISO tolerance grades for machining processes (adapted from Green, 1992)

manufacturing process. The lower the tolerance grade, the more difficult the attainment using the particular manufacturing process.

The tolerance grades are interpreted using standard tables (BS EN 20286, 1993) for conversion into dimensional tolerances. However, the tolerance grades do not take into consideration different materials machined or the complexity of the component being processed.

Both unilateral and bilateral tolerances are encountered in practice. A unilateral tolerance permits variation in only one direction from a nominal or target value; a bilateral tolerance permits variations in both directions. Most tolerances used, unless stated otherwise, in the generation of the maps are bilateral or $\pm t$, where 't' is half of the unilateral tolerance, T. Bilateral tolerances are a common way of representing manufacturing process accuracy, although some processes are more suited to other tolerance representations. For example, forging requires that the total tolerance or unilateral tolerance is divided $+\frac{2}{3}T$, $-\frac{1}{3}T$, and drilling has a positive tolerance only, $+T$, the negative tolerance from target being negligible. This is catered for in the representation of the tolerance data in the process capability maps.

After plotting the tolerance data, it is useful, in the first instance, to set the boundary conditions as $A = 1$ corresponding to a dimension/tolerance combination that is of no risk, and $A = 1.7$ on the interface of acceptable/special control region. The data used in the creation of the maps spans these two conditions, that is, the region where the process consistently produces the required tolerance. This is shown in Figure 2.11 for the turning/boring data taken from the ISO tolerance grades and many other references. The risk index $A = 1.7$ was taken from initial work in this area, where the empirical values for the component manufacturing variability risks determined were compared to historical c_{pk} data (Swift and Allen, 1994).

The intermediate values for 'A' are derived from the 'squared' relationship that is analogous to that of the relative cost/difficulty trend exhibited by manufacturing processes and their tolerance capability (see Figure 2.12).

A target process capability value, $C_{pk} = 1.33$, is aligned to the risk value at $A = 1.7$. Values for 'A' greater than 1.7 indicated on the maps continue with the squared

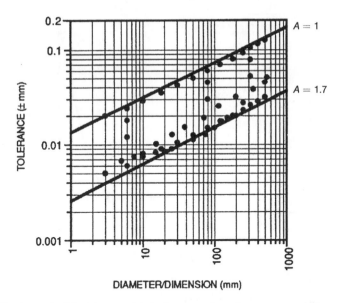

Figure 2.11 Employment of the tolerance data in the generation of a process capability maps

relationship, therefore $(1.7)^2 = 3$ and $(3)^2 = 9$. It follows then that risk indices of 'A' greater than 1.7 would not be process capable. In essence, the spacing of the lines $A = 1$ to $A = 9$ represent decreasing percentages of the tolerance band at any given dimension as the value of 'A' increases. However, the log–log axis as used on the maps show the difference as a linear step. A further development of the use of the maps is that the 'A' values can be interpolated between $A = 1$ and $A = 9$ values bounded on the map. This ultimately improves the accuracy in determining the risk value. Therefore, to determine the tolerance risk value 'A', look along the horizontal axis until the characteristic dimension is found, and locate the adjusted tolerance on the vertical axis. Read off the 'A' value in the zone at which these lines intersect on the map by interpolating as required between the zone bands, $A = 1$ to $A = 9$.

The knowledge contained in the maps is also useful in determining the tolerance requirement at an early stage in the detailed design process. In this capacity, the region of process capable tolerance is bounded by two bold lines at $A = 1$ and $A = 1.7$ on the maps. Of course, this does not take into consideration the material and geometry effects initially, for example parting line allowances. Reference to Swift and Booker (1997) can be made for approximate parting line allowances. In most cases, guidance is also given on the maps for the need of a secondary process if the dimension/tolerance combination defined gives a risk index greater than 3 (which is considered to be out of manufacturing control).

Constructing process capability maps from manufacturing data

The procedure shown in Figure 2.13 can be used by a company to construct their own process capability maps. It is necessary only when a new or specialized manufacturing process is to be used, which is not contained in Appendix IV, and when data from the

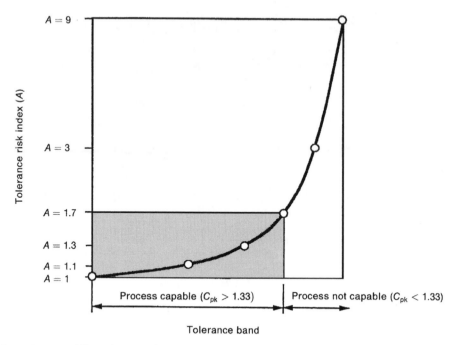

Figure 2.12 Modelling tolerance risk using a 'squared' relationship

machine tool supplier is not available. Of course, this activity may become prohibitive when the costs involved and/or time in performing such studies are high.

The data used to generate the maps is taken from a simple statistical analysis of the manufacturing process and is based on an assumption that the result will follow a Normal distribution. A number of component characteristics (for example, a length or diameter) are measured and the achievable tolerance at different conformance levels is calculated. This is repeated at different characteristic sizes to build up a relationship between the characteristic dimension and achievable tolerance for the manufacture process. Both the material and geometry of the component to be manufactured are considered to be ideal, that is, the material properties are in specification, and there are no geometric features that create excessive variability or which are on the limit of processing feasibility. Standard practices should be used when manufacturing the test components and it is recommended that a number of different operators contribute to the results.

2.2.2 Surface roughness chart

Figure 2.8 shows the range of surface roughness values likely for various manufacturing processes. The ranges determined are bounded within the risk index, A, in the same way as the process capability maps, because a similar cost–surface finish relationship exists, as suggested for tolerance and cost. This is shown in Figure 2.14 for several machining processes. The finer the surface finish required, the longer the manufacturing time, thereby increasing the cost (Kalpakjian, 1995).

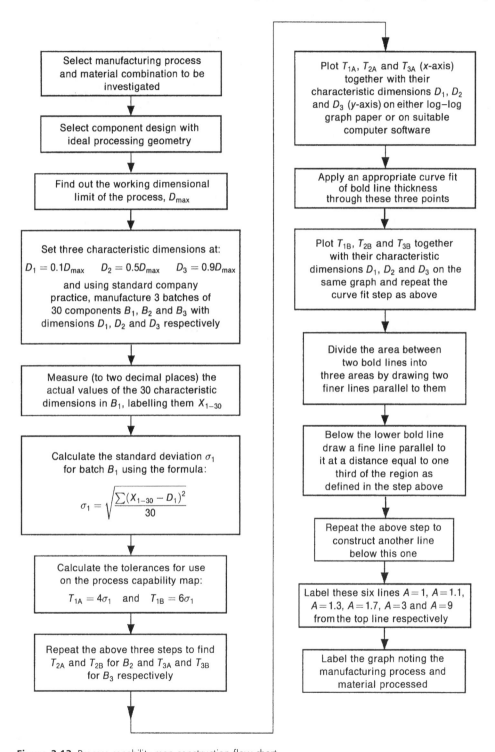

Figure 2.13 Process capability map construction flow chart

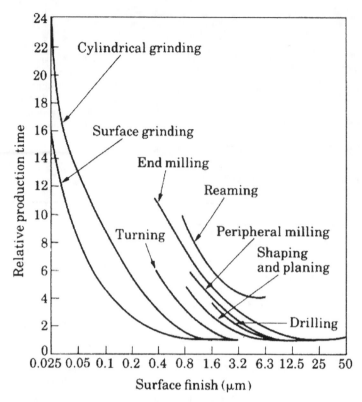

Figure 2.14 Relative manufacturing time as a function of surface roughness for several machining processes (BS 1134, 1990)

For a given manufacturing process, shown on the vertical axis in Figure 2.8, and design surface roughness, shown on the horizontal axis, the risk index 'A' on the shaded band to the right is taken as representative of the risk of obtaining the desired surface roughness. At $A = 1$, this corresponds to a surface roughness that is easily achieved by the manufacturing process. The intermediate values of 'A' ($A = 1.3$, $A = 1.7$, $A = 3$) correspond to increasing technical difficulty/cost to achieve the design surface roughness. $A = 9$ corresponds to a surface roughness that is technically unattainable using the manufacturing process. A secondary (and possibly a tertiary) process may be required to achieve the design surface roughness. All surface roughness values are in μm Ra, which is the most common format.

2.2.3 Validation of the Component Manufacturing Variability Risks Analysis

Validation of the Component Manufacturing Variability Risk, q_m, is essential to CA in determining C_{pk} estimates for component characteristics at the design stage. Collecting component parts from various industrial sources with known statistical histories was central to this. The components were taken from a number of collaborating companies which had produced the components and had measured a critical characteristic using SPC, therefore the process capability indices C_{pk} and C_p could be determined.

The critical characteristic on each component was analysed, q_m calculated from the analysis and the value obtained was plotted against the process capability indices, C_{pk} and C_p, for the characteristic in question. See Appendix V for descriptions of the 21 components analysed, including the values of C_{pk} and C_p from the SPC data supplied. Note that some components studied have a zero process capability index. This is a default value given if the process capability index calculated from the SPC data had a mean outside either one of the tolerance limits, which was the case for some of the components submitted. Although it is recognized that negative process capability indices are used for the aim of process improvement, they have little use in the analyses here. A correlation between positive values (or values which are at least within the tolerance limits) will yield a more deterministic relationship between design capability and estimated process capability.

Figure 2.15(a) shows the relationship between q_m and C_{pk} for the component characteristics analysed. Note, there are six points at $q_m = 9$, $C_{pk} = 0$. The correlation coefficient, r, between two sets of variables is a measure of the degree of (linear) association. A correlation coefficient of ± 1 indicates that the association is deterministic. A negative value indicates an inverse relationship. The data points have a correlation coefficient, $r = -0.984$. It is evident that the component manufacturing variability risks analysis is satisfactorily modelling the occurrence of manufacturing variability for the components tested.

Confidence limits are also drawn on Figure 2.15(a) to give boundaries of C_{pk} for a given q_m determined from the analysis, which are within 95%. The relationship between q_m and C_{pk} is described by a power law after linear regression giving:

$$C_{pk} = \frac{4.093}{q_m^{2.071}} \tag{2.5}$$

This can be approximated to:

$$C_{pk} \approx \frac{4}{q_m^2} \tag{2.6}$$

Note that the 'squared' relationship which was initially used to model the degree of difficulty in obtaining more capable tolerances for a given manufacturing route and product design is being returned by the power law. Similarly, a relationship between the process capability index C_p and q_m for the components analysed is shown in Figure 2.15(b). The data points have a correlation coefficient, $r = -0.956$, and the corresponding power law is given by:

$$C_p = \frac{3.981}{q_m^{1.332}} \tag{2.7}$$

This can be approximated by the following equation:

$$C_p \approx \frac{4}{q_m^{4/3}} \tag{2.8}$$

It is possible, therefore, to determine an estimate for C_{pk} from the formulation given above in equation 2.6, within some confidence. It is assumed that the C_{pk} values given

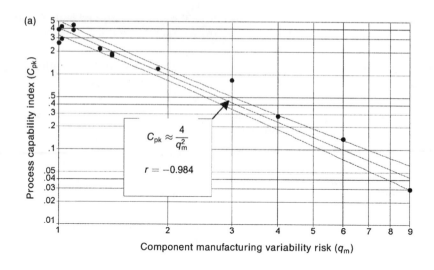

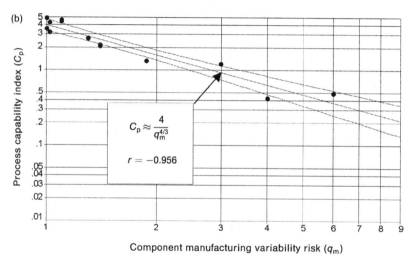

Figure 2.15 Empirical relationships between (a) q_m and C_{pk} and (b) q_m and C_p (with 95% confidence limits)

for the component characteristics were based on the mean values for a number of batches produced and that a confidence limit of 95% is adequate for the outcome for the prediction of C_{pk}. These values will be later transferred to the quality–cost model for the prediction of the failure costs for given levels of process capability and failure severity (see Section 2.5).

2.3 Assembly capability

When considering assembly processes, many problems are due to relatively minor design details. It must be remembered that a product is often an assembly of sheet metal, pipes, wires and a variety of other components held together by nuts, bolts and rivets. The attention to detail cannot be overemphasized at this stage of product design (Jones, 1978). The basic question we need to ask is, 'Can we assemble it every time, not just the once, and if not, why?'

Assembly variation (along with manufacturing variation) is a major contributor to poor quality and increased costs. For example, when the assembly of a poorly designed and poorly made product is attempted, faults such as accumulated tolerance error, incompatible dimensions and difficult part installation become apparent. At the same time, the cost of recovering from these problems during the late phase of production is high (Kroll, 1993). Industry has recognized the need to reduce assembly variations (Craig, 1992).

The three main sources of variation in mechanical assemblies are (Chase *et al.*, 1997):

- Dimensional variations (lengths, angles)
- Form and feature (flatness, roundness, angularity)
- Kinematic variations (small adjustments between mating parts).

The above are all closely linked to manufacturing variability depending on the characteristic associated with the product. Manufacturing is an important aspect of assembly too. The design of products for assembly requires careful consideration of many factors that influence the functionality and manufacturability. For example, while stability and relative precision of part positions are often essential for the functional performance of assemblies, these same requirements may make the product difficult to manufacture. At the same time, dimensional clearance among parts is essential to create paths for assembly operations (Sanderson, 1997). In general, close (or numerically small) tolerances must be maintained on parts that are to be assembled with other parts, and the closer the tolerances, the greater the ease of assembly (Kutz, 1986).

'Assemblability' is a measure of how easy or difficult it is to assemble a product. Tolerances affect the assemblability of a product, which in turn affects the cost of the product because of the scrap cost, and wasted time and energy. It is important to predict the probability of successful assembly of the parts so that the tolerance specifications can be re-evaluated and modified if necessary to increase the probability of success and lower the production cost associated with assembly (Lee *et al.*, 1997). While close tolerances reduce assembly costs, they increase the cost of manufacturing the individual parts. A balance of the two types of costs must be achieved, with the objective of minimizing their sum (Kutz, 1986).

When tolerances are mentioned in terms of mechanical assemblies, one considers tolerance or assembly stack models used to determine the final assembly tolerance capability for a given set of component tolerances. Worst case or statistical equations are commonly used for optimization purposes, as discussed in detail in Chapter 3. However, this can be confusing, as a tolerance can exist on a component characteristic, but the component itself may not necessarily be capable of being assembled to form the assembly tolerance. This may be due to one or more features of the components being assembled preventing effective and repeatable placement. If this occurs then the product suffers not only from an assembly problem, but also a tolerance problem.

CA provides assembly capability predictions which can highlight potential assembly problems. We must also have a means of estimating the failure costs associated with component assembly processes in addition to the tolerance process capability predictions. These additional failure costs are independent of whether the tolerances assigned to an assembly stack or the single characteristic are capable. By identifying components with high assembly risks and potentially high failure costs, further design effort is highlighted and performed in order to identify the associated tolerances, for example clearance for the optimal fit and function of the components.

2.3.1 Design for assembly techniques

Early work looking at designing products for mechanized assembly started over 30 years ago (Boothroyd and Redford, 1968). Large cost savings were found to be made by careful consideration of the design of the product and its individual component parts for ease of assembly. Commercial DFA techniques are now used successfully by many companies in either workbook or software versions. The three most referred to methods are:

- Boothroyd–Dewhurst Design for Manufacture and Assembly (DFMA) (Boothroyd et al., 1994)
- Computer Sciences Corporation's (CSC) DFA/Manufacturing Analysis (MA) (CSC Manufacturing, 1995)
- Hitachi's Assembly Evaluation Method (AEM) (Shimada et al., 1992).

In fact, the use of DFA techniques is now mandatory in some companies, such as Ford and LucasVarity (now TRW) (Miles and Swift, 1998). These techniques offer the opportunity for a number of benefits, including:

- Reduced part count
- Systematic component costing and process selection
- Lower component and assembly costs
- Standardized components, assembly sequence and methods across product 'families' leading to improved reproducibility
- Faster product development and reduced time to market
- Lower level of engineering changes, modifications and concessions
- Fewer parts means: improved reliability, fewer stock costs, fewer invoices from fewer suppliers and possibly fewer quality problems.

A team-based application and systematic approach is essential as there are many subjective processes embedded, but many companies have found them to be pivotal techniques in designing cost-effective and competitive products (Miles and Swift, 1998).

An overview of each of the main commercial methods can be found in Huang (1996), but in general, a number of design for assembly guidelines can be highlighted (Leaney, 1996b):

- Reduce part count and types
- Modularize the design

- Strive to eliminate adjustments (especially blind adjustments)
- Design parts for ease of handling (from bulk)
- Design parts to be self-aligning and self-locating
- Ensure adequate access and unrestricted vision
- Design parts that cannot be installed incorrectly
- Use efficient fastening or fixing techniques
- Minimize handling and reorientations
- Maximize part symmetry
- Use good detail design for assembly
- Use gravity.

DFA minimizes part count by, for example, consolidating a number of features found on two components in a single component, but it cannot indicate the process capability for the design proposed. DFA techniques can only indicate that it may be costly to assemble the component or that the cost of the component may be relatively high, but still more cost effective than the two components it replaces. Production costs should not be the only measure of performance with which to select designs. The analysis should extend to the potential variability associated with the design when in production.

There is a strong correlation between assembly efficiency and reported defect levels for a number of Motorola products evaluated using the Boothroyd–Dewhurst DFMA technique (Branan, 1991). From the results of the study, it is claimed that DFA cuts assembly defects by 80% and, therefore, has a direct influence on manufacturing quality. This may be due to the fact that more efficient design solutions have relatively fewer component parts which naturally infers fewer quality problems. However, early life failures, which are caused by latent defects, are not necessarily highlighted by DFA techniques.

Quality or robustness is seen as a natural outcome of the product when effectively using DFA techniques, and is commonly listed as a potential benefit, but it is not evaluated explicitly by any of the commercial techniques. Too often, assembly is overlooked when assessing the robustness of a design and DFA techniques do not specifically address variability within the assembly processes. Undoubtedly, the better the assemblability, the better the product quality in terms of fewer parts and simpler assembly operations. Fewer parts lead to fewer breakdowns, fewer workstations, less time to assemble and fewer overheads. Simpler assembly operations imply that the product fits together more easily, leading to shorter lead times and less rework (Leaney, 1996b).

From the above arguments, it has been recognized that there is a need to predict variation at the assembly level. To facilitate an assembly analysis in the first instance, it is essential to understand the structure of the proposed product design and the assembly sequence diagram is useful in this respect.

2.3.2 Assembly sequence diagrams

Usually, the assembly sequence of the components that make up the product is examined late in the design process when manufacturing engineers are trying to

Castor assembly

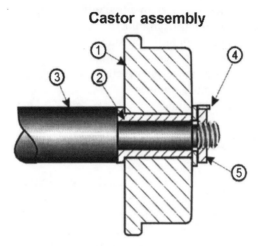

Assembly sequence diagram

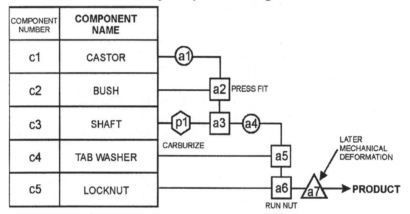

COMPONENT NUMBER	COMPONENT NAME
c1	CASTOR
c2	BUSH
c3	SHAFT
c4	TAB WASHER
c5	LOCKNUT

KEY TO ASSEMBLY SEQUENCE	
c	COMPONENT NUMBER
a	ASSEMBLY SEQUENCE NUMBER
p	POST MANUFACTURING PROCESS NUMBER
○	WORK HOLDER, e.g. handling prior to assembly
□	ASSEMBLY PROCESS, e.g. fitting, insertion
△	ADDITIONAL/POST ASSEMBLY PROCESS, e.g. riveting, welding, turn over
▽	ASSEMBLY IN MANUFACTURING PROCESS, e.g. inserts in moulding/casting
⬡	POST MANUFACTURING PROCESS, e.g. bulk and surface heat treatments/coatings

Figure 2.16 Assembly sequence diagram for a castor wheel

balance the assembly line. In turn, the choice of assembly sequence and the identification of potential subassemblies can affect or be affected by product testing options, market responsiveness and factory floor layout (Baldwin *et al.*, 1991), as well as engineering issues such as ease of servicing or ease of recycling. However, there are extensive implications for the successful generation of assembly sequences during the early stages of the design process. Cycle times can be reduced and rework is decreased (Barnes *et al.*, 1997).

Through an assembly sequence diagram for each component in the product, the assembly variability risks highlighted by an analysis are logically mapped. An assembly sequence declaration compels the designer to focus on each stage in the assembly and therefore makes the task of identification of potential problems much easier. Only CSC's DFA/MA method uses an explicit assembly sequencing method to augment the DFA process and the approach employed here is developed, with slight modifications, from this. Figure 2.16 shows an assembly sequence diagram for a simple castor wheel assembly. The development of assembly sequence diagrams for more complex assemblies does become a prohibitive task, but its generation should be one of the standard engineering design tasks.

2.4 Component Assembly Variability Risks Analysis

In the development of the assembly variability risks analysis, expert knowledge, data found in many engineering references and information drawn from the CSC DFA/MA practitioner's manual (CSC Manufacturing, 1995) were collated and issues related to variability converged on. Much of the knowledge for the additional assembly variability risks analysis was reviewed from the fabrication and joining data sheets called PRocess Information MAps (PRIMAs) as given in Swift and Booker (1997).

Product assembly is such a large area that focusing on the key operational issues, such as handling, fitting and joining, is justified at the design stage. Although much has been written about DFA, it was not appropriate to use the DFA indices directly for these assembly operations because they are based on relative cost, not potential variability. Where the knowledge and data cannot help, and where some of the more qualitative aspects of design commence, as well as expert knowledge, some Poka Yoke ideas were used, particularly for the component fitting chart. Poka Yoke is used to prevent an error being converted into a defect (Shingo, 1986). It is heavily involved in the development of any DFA process. Poka Yoke has two main steps: preventing the occurrence of a defect, and detecting a defect (Dale and McQuater, 1998). Some of the assembly variability risks charts reflect these Poka Yoke philosophies.

The measure of assembly variability, q_a, derived from the analysis should be used as a relative performance indicator for each design evaluated. The design with the least potential variability problems or least failure cost should be chosen for further development. The indices should not be taken as absolutes as assembly variability is difficult to measure and validate.

The component assembly variability risk, q_a, as determined by CA, attempts to better understand the affects of the assembly situation on variability by quantifying the risks that various operations inherently exhibit. A component's assembly situation typically involves the following process issues:

- Handling characteristics, h_p
- Fitting (placing and insertion) characteristics, f_p
- Additional assembly considerations (welding, soldering, etc.), a_p
- Whether the process is performed manually or automatically.

At the current stage of development, the risk index associated with component assembly variability is the product of the component's handling and fitting risks:

$$q_a = h_p \times f_p \qquad (2.9)$$

The additional assembly process risk, a_p, is considered in isolation after the handling and fitting operations, questioning the assembly situation of a component after initial placing, and which requires further processing, for example welding or adhesive bonding. Therefore, because additional assembly processes are performed after initial part placement, q_a simply defaults to a_p when it is considered in the assembly sequence:

$$q_a = a_p \qquad (2.10)$$

The additional assembly processes covered in the analysis are provided in Appendix VI, and are listed below:

- Miscellaneous operations (testing, reorientation, heating, cooling)
- Later mechanical deformation (riveting)
- Adhesive bonding
- Brazing and soldering
- Resistance welding
- Fusion welding.

A systematic assessment is made of every handling and fitting process involved in the product construction (manual or automated), including the effects of alignment, placement and fastening, through the assembly sequence diagram for the design. A number of charts are used for each aspect of the assembly analysis. Using the handling process risk chart (see Figure 2.17), the fitting process risk (Figure 2.18) and the additional process risk charts, the assembly situation for most components can be analysed, accruing penalties if the design has increased potential for variability. The measure of variability returned is the component assembly variability risk, q_a. Guidance notes accompany each chart for the designer.

The assembly indices are bounded in the same way as the component manufacturing variability risk indices, q_m. Starting with the notion that an ideal component assembly situation exists, this risk index is given a value of unity. The indices are then scaled within the same conditions described for t_p (used in the manufacturing variability risks analysis). For example, $1.7 \leq h_p \leq 2.2$ indicates that the handling process is within special control. High values of q_a indicate high-risk design to process solutions. The estimated values of q_a tend to be within the range 1 to 5, with only those components where the risk is exceptionally high coming out beyond 5. Where detailed knowledge and experience of a particular assembly situation is available,

h_p — HANDLING TECHNOLOGY				COMPONENT SENSITIVITY: DELICATE (1)	SENSITIVE TO TEMPERATURE CHANGE	SENSITIVE TO CONTAMINATION		NOT APPLICABLE (3)
						Chemical (2)	Mechanical	
MANUAL HANDLING & GRIPPING			Not thin/small (4)	1.3	1.2	1.5	1.1	1.0
MANUAL HANDLING & GRIPPING			Thin/small	1.7	1.3	2.0	1.2	1.0
MAGAZINE PROCESSES	No gripping			1.2	1.1	1.2	1.1	1.0
MAGAZINE PROCESSES	Gripping	Mechanical clamping	Not thin	1.4	1.1	1.2	1.3	1.0
MAGAZINE PROCESSES	Gripping	Mechanical clamping	Thin	1.9	1.3	1.2	1.4	1.0
MAGAZINE PROCESSES	Gripping	Vacuum/Magnetic contact		1.2	1.3	1.2	1.3	1.0
AUTOMATED PROCESSES	Parts feeder	No gripping		1.5	1.1	1.2	1.5	1.0
AUTOMATED PROCESSES	Parts feeder	Gripping: Mechanical clamping	Not thin	1.9	1.2	1.3	1.7	1.0
AUTOMATED PROCESSES	Parts feeder	Gripping: Mechanical clamping	Thin	2.2	1.3	1.3	1.8	1.0
AUTOMATED PROCESSES	Parts feeder	Gripping: Vacuum/magnetic contact		1.6	1.3	1.3	1.8	1.0
AUTOMATED PROCESSES	Conveyor (5)	No gripping		1.2	1.1	1.2	1.2	1.0
AUTOMATED PROCESSES	Conveyor (5)	Gripping: Mechanical clamping	Not thin	1.5	1.2	1.5	1.7	1.0
AUTOMATED PROCESSES	Conveyor (5)	Gripping: Mechanical clamping	Thin	2.2	1.3	1.5	1.8	1.0
AUTOMATED PROCESSES	Conveyor (5)	Gripping: Vacuum/magnetic contact		1.3	1.3	1.5	1.8	1.0

Notes

① Also includes high finish.
② Chemical contamination to operator/machine or product.
③ If component is not sensitive to the characteristics listed, $h_p = 1.0$.
④ A component is said to be thin if one of its major dimensions is ≤ 0.25 mm thick.
⑤ Includes mechanical and air assisted mechanisms.

Figure 2.17 Handling process risk, h_p

this expertise should be used to augment the analysis. Because the assembly variability knowledge collated is a mixture of quantitative and qualitative data, validation of the indices used in the charts is more difficult than the manufacturing variability risks analysis. Experts from the assembly industry were therefore used in their final assessment and allocation. However, despite the fact that the validation of the assembly capability measures are difficult to determine empirically, they are found to be of great benefit when comparing and evaluating a number of design principles.

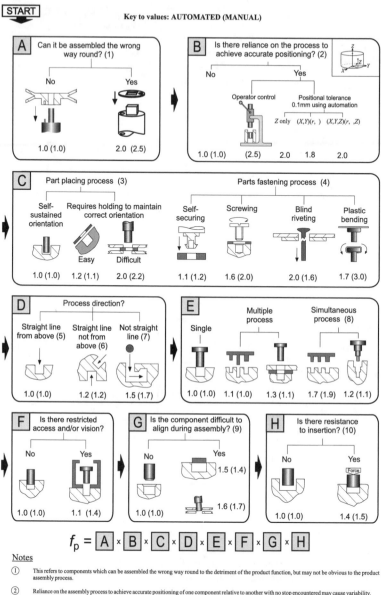

Figure 2.18 Fitting process risk, f_p

2.5 The effects of non-conformance

2.5.1 Design acceptability

FMEA can be used to provide a quantitative measure of the risk for a design. Because it can be applied hierarchically from system through subassembly and component levels down to individual dimensions and characteristics, it follows the progress of the design into detail. FMEA also lists potential failure modes and rates their Severity (S), Occurrence (O) and Detectability (D). It therefore provides a possible means for linking potential variability risks with consequent design acceptability and associated costs. Note that the ratings of Occurrence and Detectability are equated to probability levels.

To determine the costs of failing to design into a product the levels of process capability proper to the severity of consequential failure, it has been assumed that a fault is undetected and always results in a failure. Also, the estimate is not sensitive to changes in the product cost, volume and rework costs. The cost of customer dissatisfaction, over and above the accounted cost of rework or a returned product, is very difficult to assess. The model which follows is based on research with Lucas Automotive Electronics (Lucas PLC, 1994) and utilizes their specific ratings for FMEA Occurrence (O) and Severity (S) (see Figures 2.19 and 2.20 respectively). The analysis of failure consequence is a specialist area within each industry, as is the definition of the probability of occurrence (Smith, 1993). The alignment of the ratings chosen here with other existing FMEA ratings, for example those in Chrysler Corporation *et al.* (1995), may be considered when using CA with other business practices.

'Failure' in the context here means that product performance does not meet requirements and is related back in the design FMEA to some component/characteristic being out of specified limits – a fault. The probability of occurrence of failure (O) caused by a fault can be expressed as:

$$O = P_f \cdot d' \cdot P_{of} \tag{2.11}$$

where:

P_f = probability of a fault in the component/characteristic

d' = probability that the fault will escape detection

P_{of} = conditional probability that, given there is a fault,
the failure will actually occur.

Equation 2.11 recognizes that for a failure to occur, there must be fault, and that other events may need to combine with the fault to bring about a failure. The equation states that the probabilities of each of these factors occurring must be multiplied together to calculate the probability of failure.

In the special case when $d' = 1$ and $P_{of} = 1$, that is, when if a fault occurs it will not be detected and that failure is certain, occurrence equates directly to the probability of a fault. The probability of a fault in turn depends on the capability of the process used for the component/characteristic.

Occurrence (O)

Rating	Guide	ppm
1	Remote possibility of failure occurring	0.1
2	} Low possibility of occurrence	0.5
3		2
4	} Moderate possibility of occurrence	10
5		50
6	Significant number of failures possible	200
7	} High possibility of occurrence	1,000
8		5,000
9	Very high possibility of occurrence	20,000
10	Almost certain that many failures will occur	100,000

Figure 2.19 Typical FMEA Occurrence Ratings (O)

A standard for the minimum acceptable process capability index for any component/characteristic is normally set at $C_{pk} = 1.33$, and this standard will be used later to align costs of failure estimates. If the characteristics follow a Normal distribution, $C_{pk} = 1.33$ corresponds to a fault probability of:

$$P_f = 30\,\text{ppm} = 30 \times 10^{-6} \tag{2.12}$$

While 30 ppm may be acceptable as a maximum probability of occurrence for a failure of low severity, it is not acceptable as severity increases. An example table of FMEA Severity Ratings was shown in Figure 2.20. In the 'definite return to manufacturer' (a warranty return) or 'violation of statutory requirement' region ($S = 5$ or $S = 6$), the designer would seek ways to enhance the process capability or else utilize some inspection or test process. Reducing d' will reduce occurrence, as indicated by equation 2.11, but inspection or test is of limited efficiency.

If severity increases to, say, 'complete failure with probable severe injury and/or loss of life' ($S = 9$), the designer should reduce occurrence below the level afforded by two independent characteristics protecting against the fault. If each is designed to, then this implies a failure rate of:

$$P_f = (30 \times 10^{-6})^2 = 9 \times 10^{-10} \tag{2.13}$$

Again, this standard aligns with the costs of failure analysis below.

Severity (S)

Rating	Guide to effect on user
1 *	No effect or minimal effect on customer
2 *	Minor annoyance to customer
3 *	Annoyance to customer but no loss of major function
4 *	Possible return to manufacturer
5 *	Definite return to manufacturer
6	Failure leading to violation of statutory requirement
7	Failure leading to injury or a more safety critical related problem with secondary back-up
8	Safety problem – degradation of function with possible severe injury
9	Complete failure with probable severe injury and/or loss of life
10	Catastrophic failure with high probability of loss of life

* Note: These failures are non-safety critical

Figure 2.20 Typical FMEA Severity Ratings (S)

Figure 2.21 is a graph of Occurrence against Severity showing a boundary of the acceptable design based on these criteria for the case when $d' = 1$ and $P_{of} = 1$. The graph is scaled in terms of FMEA ratings for Occurrence, equivalent probabilities and parts-per-million (ppm).

In general, just as reducing d' reduces the probability of failure occurrence, so reducing the conditional probability of failure P_{of} moves the component/characteristic towards the acceptable design zone. However, this must be applied with considerable caution. The derivation of the acceptable maximum for failure occurrence at $S = 9$ and above is essentially an application of the conditional probability: the conditional probability of failure at one characteristic is the probability of failure at the other protecting characteristic. In this case, it would normally be more appropriate to design two characteristics than to rely on one very capable one. At such low probability levels, the chance of random unforeseen phenomena throwing the process out of control is significant.

On the other hand, consider the case of a 'secondary back-up' (external to the system) for a component/characteristic of severity $S = 7$. If the designer assumes the back-up is designed for $C_{pk} = 1.33$, then the analysis would reduce that for the internal back-up $S = 9$ case above. The acceptable design limits on Figure 2.21

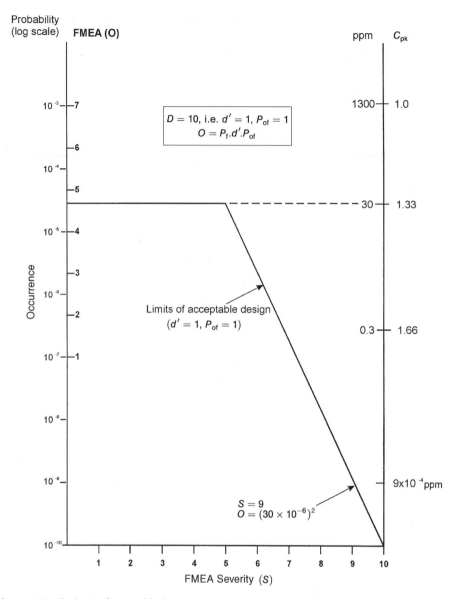

Figure 2.21 The limits of acceptable design

imply that it is prudent to allow a safety margin for external components not under the designer's control and scale up the conditional probability of failure.

In summary, for a component/characteristic it is possible to define an area of acceptable design on a graph of occurrence versus severity. The acceptability of the design can be enhanced somewhat by the addition of inspection and test operations. The requirements of process capability may be relaxed to a degree, as the conditional probability of failure reduces, but this should be subject to a generous safety margin.

It will now be demonstrated that the boundaries of acceptable design proposed above can be expressed in terms of the costs of failure.

2.5.2 Map of quality costs

From the quality–cost arguments made in Section 1.2, it is possible to plot points on the graph of Occurrence versus Severity and construct lines of equal failure cost (% isocosts). Figure 2.22 shows this graph, called a Conformability Map. Because of uncertainty in the estimates, only a broad band has been defined.

The boundary of acceptable design for a component or assembly characteristic in the zone $S \geq 6$ corresponds fairly closely to a failure cost line equivalent to 0.01% of the unit cost. The region of unacceptable design is bounded by the intersection of the horizontal line of $C_{pk} = 1$ and the 1% isocost. A process is not considered capable unless $C_{pk} \geq 1$ and a failure cost of less than 1% is thought to be acceptable.

Components/characteristics in the unacceptable design zone are virtually certain to cause expensive failures unless redesigned to an occurrence level acceptable for their failure severity rating (minimum $C_{pk} = 1.33$). In the intermediate zone, if acceptable design conformability cannot be achieved, then special control action will be required. If special action is needed then the component/characteristic is critical. However, it is the designer's responsibility to ensure that every effort is made to improve the design to eliminate the need for special control action.

The 0.01% line on Figure 2.22, implies that even in a well-designed product, there is an incurred failure cost. Just 100 dimensional characteristics on the limit of acceptable design are likely to incur a 1% cost of failure. The quality cost rises dramatically where design to process capability is inadequate. Just ten product characteristics on the 1% isocost line would give likely failure costs of 10%. (The 1% and 10% isocost lines on the Conformability Map are extrapolated from the conditions for the 0.01% and 1% isocost lines.) Clearly, the designer has a significant role in reducing the high cost of failure reported by many manufacturing organizations.

Variation is typically measured using process capability indices, as well as ppm failure, but it would be advantageous to scale the variability risks predicted by CA on the Conformability Map too. Scales showing C_{pk} and approximate variability risk measures, q_m and q_a, from CA equivalents have been added to the Conformability Map for use in the CA methodology. An empirical relationship exists between C_{pk} and q_m that allows us to do this, as discussed earlier. Because the assembly variability risk, q_a, is also scaled in the same way as q_m, it is assumed that the level of variability in assembly can also be associated to a notional process capability index. The Conformability Map, through the inclusion of q_m, q_a and FMEA Severity Rating (S) for a particular design, can be used to give the total risk potential in terms of the isocost at the design stage. The accumulated isocosts can then be translated into potential failure costs. In this way, a total cost of failure for a design can be evaluated.

The use of the Conformability Map in determining the potential costs of non-conformance or failure should be associated with the application of evaluating and comparing different design schemes in practice. The estimated failure costs act as a measure of performance by which to make a justified selection of a particular

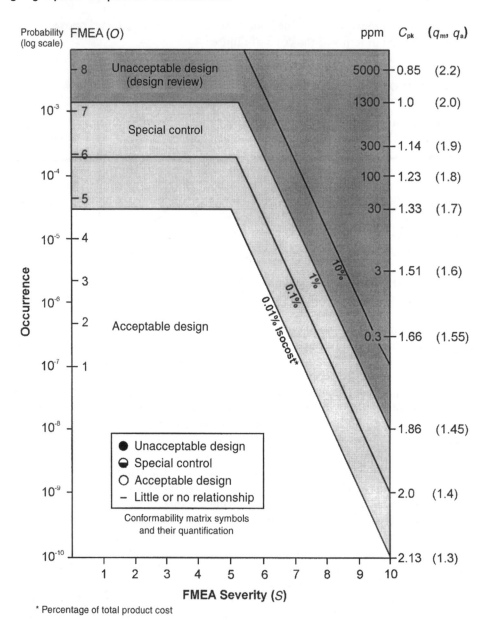

Figure 2.22 The Conformability Map

design scheme. Assuming that the costs are absolute values is not recommended due to the difficulty in obtaining accurate data for the quality–cost model (see later for guidance on the use and calculation of potential costs of failure for a design). It has been cited that ±5% is a good enough accuracy for the prediction of quality costs at the business assessment level (Bendell *et al.*, 1993), but the achievement of this at the design stage is a very difficult task.

The Conformability Map enables appropriate C_{pk} values to be selected and through the link with the component variability risks, q_m and q_a, it is possible to determine if a product design has characteristics that are unacceptable, and if so what the cost consequences are likely to be. The two modes of application are highlighted on Figure 2.23. Mode A shows that the quality loss associated with a characteristic at $C_{pk} = 1$ and FMEA Severity $(S) = 6$ could potentially be 8% of the total product

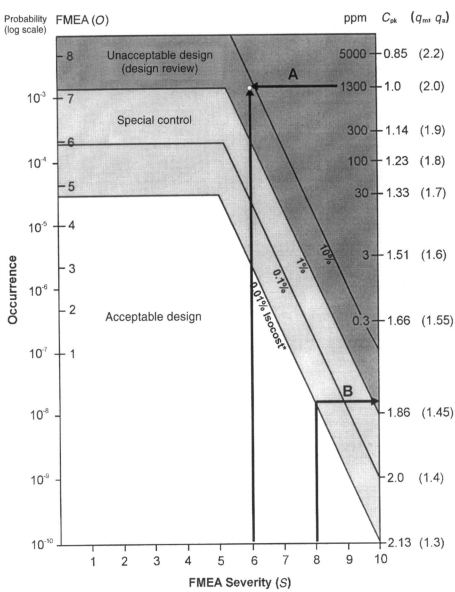

* Percentage of total product cost

Figure 2.23 Conformability Map applications

cost. Mode B indicates that a target $C_{pk} = 1.8$ is required for an acceptable design characteristic having an FMEA Severity Rating $(S) = 8$.

2.6 Objectives, application and guidance for an analysis

A short review of the CA process is given before proceeding with the applications of the technique to several industrial case studies. The three key stages of CA are shown in Figure 2.24 within the simplified process of assessing a design scheme.

- **Component Manufacturing Variability Risks Analysis** – As mentioned previously, the first of the three key stages in CA is the Component Manufacturing Variability Risks Analysis. When detailing a design, certain characteristics can be considered

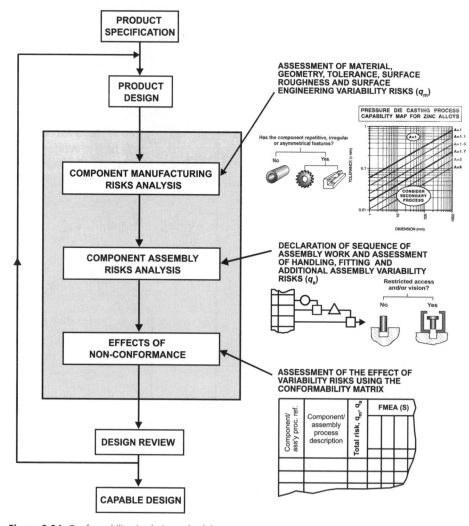

Figure 2.24 Conformability Analysis methodology

important. For example, tolerances on a dimension, a surface roughness requirement or some intricate geometry on a component. These important characteristics
need to be controlled and understood.

The 'quality' inherent in manufacturing a component is essentially dictated by
the designer, and can be accounted for by the level of knowledge in the following
design/manufacture interface issues:
 – Process precision and tolerance capability, t_p.
 – Material to process compatibility, m_p.
 – Component geometry to process limitations, g_p.
 – Surface roughness and detail capability, s_p.
 – Surface engineering process suitability, k_p.
From consideration of these issues, risk indices can be formulated using charts to
reflect the limits of variability set by the design which can be achieved by the
application of best practice in manufacturing. The analysis ultimately returns a
Component Manufacturing Variability Risk, q_m. The underlying notion of q_m is
that an ideal design exists with regard to tolerance where the risk index is unity,
indicating that variability is in control. Risk indices greater than unity exhibit a
greater potential for variability during manufacture. Central to the determination
of q_m is the use of the process capability maps showing the relationship between the
achievable tolerance and the characteristic dimension for a number of manufacturing processes and material combinations.

• **Component Assembly Variability Risks Analysis** – Too often, assembly is overlooked
when assessing the robustness of a design. DFA techniques offer the opportunity
for part count reduction through a structured analysis of the assembly sequence,
but they do not specifically address variability within assembly processes. A component's assembly situation involves the following process issues:
 – Handling characteristics, h_p.
 – Fitting (placing and insertion) characteristics, f_p.
 – Additional assembly considerations (welding, soldering, etc.), a_p.
 – Whether the process is performed manually or automatically.
Again, the notion is that an ideal component assembly design exists. Using expert
knowledge, charts for a handling process risk, fitting process risk and additional
processes question the assembly situation of the component, accruing penalties
if the design has increased potential for variability to return the Component
Assembly Variability Risk, q_a.

• **Effects of non-conformance** – The link between the component variability risks, q_m
and q_a, and FMEA Severity (S) is made through the Conformability Map.
Research into the effects of non-conformance and associated costs of failure has
found that an area of acceptable design can be defined for a component characteristic on a graph of Occurrence versus Severity. It is possible to plot points on this
graph and construct lines of equal failure cost. These isocosts in the non-safety
critical region (FMEA Severity Rating ≤ 5) come from a sample of businesses
and assume levels of cost at internal failure, returns from customer inspection or
test and warranty returns. The isocosts in the safety critical region (FMEA Severity
Rating >5) are based on allowances for failure investigations, legal actions and
product recall, but do not include elements for loss of current or future business
which can be considerable. The costs in the safety critical area have been more

difficult to assess and have a greater margin of error. In essence, as failures get more severe, they cost more, so the only approach available to a business is to reduce the probability of occurrence.

2.6.1 Objectives

CA is primarily a team-based product design technique that, through simple structured analysis, gives the information required by designers to achieve the following:

- Determine the potential process capability associated with the component and/or assembly characteristics
- Assess the acceptability of a design characteristic against the likely failure severity
- Estimate the costs of failure for the product
- Specify appropriate C_{pk} targets for in-house manufacturing or outside supply.

2.6.2 Application modes

The case studies that follow have mainly come from 'live' product development projects in industry. Whilst not all case studies require the methodology to predict an absolute capability, a common way of applying CA is by evaluating and comparing a number of design schemes and selecting the one with the most acceptable performance measure, either estimated C_{pk}, assembly risk or failure cost. In some cases, commercial confidence precludes the inclusion of detailed drawings of the components used in the analyses. CA has been used in industry in a number of different ways. Some of these are discussed below:

- **Variability prediction** – A key objective of the analysis is predicting, in the early stages of the product development process, the likely levels of out of tolerance variation when in production.
- **Tolerance stack analysis** – Tolerances on components that are assembled together to achieve an overall design tolerance across an assembly can be individually analysed, their potential variability predicted and their combined effect on the overall conformance determined. The analysis can be used to optimize the design through the explorations of alternative tolerances, processes and materials with the goal of minimizing the costs of non-conformance. This topic is discussed in depth in Chapter 3.
- **Evaluating and comparing designs** – CA may be used to evaluate alternative concepts or schemes that are generated to meet the specification requirements. It highlights the problem areas of the design, and calculates the failure cost of each, which acts as a measure of quality for the design. Since the vast majority of cost is built into a component in the early stages of the design process, it is advantageous to appraise the design as honestly and as soon as possible to justify choice in selecting a particular design scheme.
- **Requirements definition** – Where designers require tighter tolerances than normal, they must find out how this can be achieved, which secondary processes/process

developments are needed and what special control action is necessary to give the required level of capability. This must be validated in some way. The variability risks analysis described previously is useful in this connection. It provides systematic questioning of a design regarding the important factors that drive variability. It estimates quantities that can be related to potential C_{pk} values, and identifies those areas where redesign effort is best focused. The results serve as a good basis for component supplier dialogue, communicating the requirements that lead to a process capable design.

- **Generation mode** – The process capability knowledge used in CA to facilitate an analysis can be used in the early stages of product design to generate process capable dimensional tolerances and surface roughness values when the material and geometry effects are minimal. Similarly, the assembly charts can be used to generate an improved assembly by avoiding features known to exhibit high risk.

2.6.3 Analysis procedure

The procedure is shown in flow chart form in Figure 2.25. In order to obtain the best results, the following points should be clearly understood before starting to evaluate a component's manufacturing or assembly situation:

- Read through Chapter 2 of this book carefully before attempting a design analysis.
- It is important that the product development team be familiar with the main capabilities and characteristics of the manufacturing and assembly methods selected, as well as the materials considered, in order to obtain the full benefit from the methodology. Technical and economic knowledge for some common manufacturing processes can also be found in Swift and Booker (1997).
- The analysis requires the declaration of a sequence of assembly work, so familiarity with this supporting method is essential.
- When analysing a component or assembly process, complete all columns of the Variability Risks Results Table (see later for an example) and write additional notes and comments in the results table whenever possible. The table is a convenient means of recording the analysis for individual component manufacturing and assembly variability risks (q_m, q_a). It is recommended that the results table provided is used every time the analysis is applied to minimize possible errors.
- The Conformability Matrix (see later for an example) primarily drives assessment of the variability effects. The Conformability Matrix requires the declaration of FMEA Severity Ratings and descriptions of the likely failure mode(s). It is helpful in this respect to have the results from a design FMEA for the product.

2.6.4 Example – Component Manufacturing Variability Risks Analysis

Figure 2.26 shows the design details of a bracket, called a cover support leg. A critical characteristic is the distance from the hole centre to the opposite edge, dimension 'A'

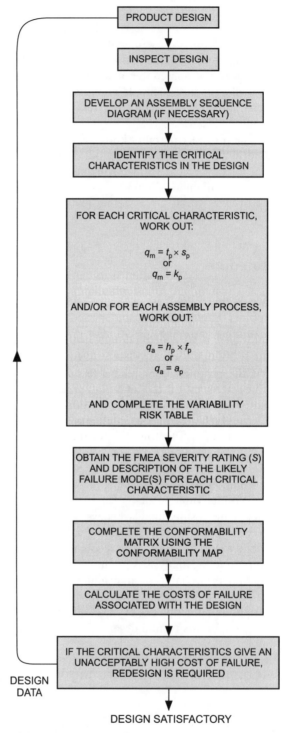

Figure 2.25 Conformability Analysis procedure flowchart

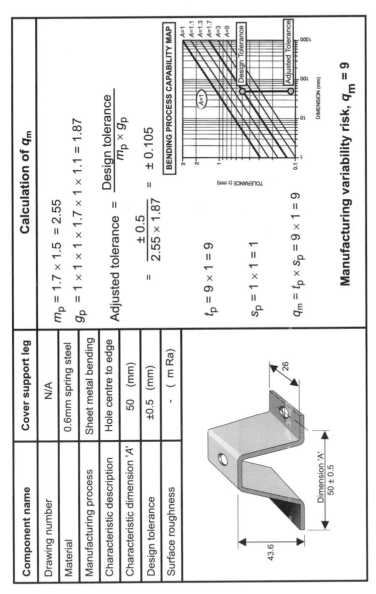

Figure 2.26 Component Manufacturing Variability Risks Analysis of a cover support leg

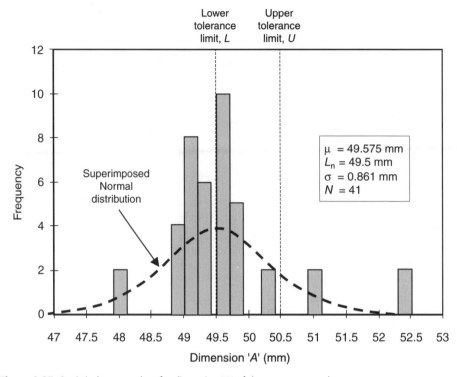

Figure 2.27 Statistical process data for dimension 'A' of the cover support leg

on the bracket, which must be 50 ± 0.5 mm to effectively operate in an automated assembly machine. The assembly machine is used to produce the final product, of which the support leg is a key component. Failure to achieve this specification would cause a major disruption to the production line due to factors including feeder jams, fastener insertion and securing.

Through an analysis using CA, high material and geometry variability risks were highlighted which could reduce the final capability of the characteristic. The Component Manufacturing Variability Risk, q_m, for the characteristic is calculated to be $q_m = 9$, this being determined from the adjusted tolerance using the process capability map for bending, as shown. This gives a predicted process capability index, $C_{pk} = 0.05$. In fact, the component was already in production when it was realized that the dimensions on most of the components were out of tolerance. A process capability analysis of the manufacturing process was conducted, and the results are shown in Figure 2.27. The calculated process capability index was found to be $C_{pk} = 0.03$, from equation 3 in Appendix II, which relates to approximately half the components falling out of specification, assuming a Normal distribution.

$$C_{pk} = \frac{|\mu - L_n|}{3\sigma} \qquad (2.14)$$

where:

$$\mu = \text{mean of distribution}$$

$$L_n = \text{nearest tolerance limit}$$

$$\sigma = \text{standard deviation.}$$

Therefore:

$$C_{pk} = \frac{|49.575 - 49.5|}{3 \times 0.861} = 0.03$$

As a result of poor capability, a high level of machine downtime was experienced, and the company which manufactured the assembly machine were involved in an expensive legal dispute with their customer. Although a third party manufactured the cover support leg, liability, it was claimed, lay with the final assembly machine manufacturer. This resulted in severe commercial problems for the company, from which it never fully recovered.

An analysis using CA at an early design stage would have highlighted the risks associated with using a spring steel and long unsupported sections in the design, which were the main tolerance reducing variabilities. With reference to the bending process capability map, the initial tolerance set at ± 0.5 mm was just within acceptable limits at a nominal dimension of 50 mm. However, the variability risks m_p and g_p decreased the probability of obtaining it substantially.

Completing a variability risks table

A variability risk table (as shown in Figure 2.28 for the cover support leg analysis above) is a more efficient and traceable way of presenting the results of the first part of the analysis. A blank variability risks results table is provided in Appendix VII. It catalogues all the important design information, such as the tolerance placed on the characteristic, the characteristic dimension itself, surface roughness value, and then allows the practitioner to input the results determined from the variability risks analysis, both manufacturing and assembly.

The documentation of the risks for each component and assembly operation follows the determination of the assembly sequence diagram, if appropriate, when the product consists of more than a single component. Every critical component characteristic or assembly stage is analysed through the use of the variability risks table. An assembly risk analysis will be performed during several of the case studies presented later.

2.6.5 Example – Component Assembly Variability Risks Analysis

Similar to the calculation of q_m, it is possible to identify the variability risks associated with the assembly of components, q_a. In the following example shown in Figure 2.29, a cover and housing (with a captive nut) are fastened by a fixing bolt, which in turn is secured by a tab washer.

For example, for assembly operation number 4 for the fixing bolt we can determine the risk indices h_p and f_p. For the handling process, assuming automatic assembly, the

Figure 2.28 Variability risks table for the cover support leg

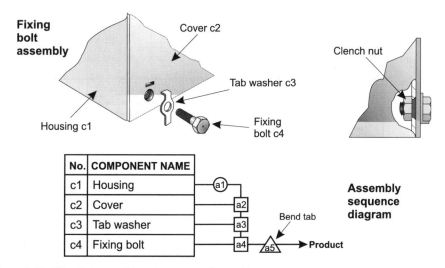

Figure 2.29 Fixing bolt assembly and sequence of assembly

fixing bolt is supplied in a feeder and does not have characteristics which complicate handling. Therefore, from the handling table in Figure 2.17, $h_p = 1.0$. Now assessing the fitting to process risk, f_p, from Figure 2.18:

A (cannot be assembled the wrong way)	A = 1.0
B (no positioning reliance to process)	B = 1.0
C (automatic screwing)	C = 1.6
D (straight line assembly from above)	D = 1.2
E (single process – one bolt in hole)	E = 1.1
F (no restricted access or vision)	F = 1.0
G (no alignment problems)	G = 1.0
H (no resistance to insertion)	H = 1.0

For the bolt fitting operation:

$$f_p = A \times B \times C \times D \times E \times F \times G \times H = 2.11$$

The Component Assembly Variability Risk is given by:

$$q_a = h_p \times f_p$$
$$q_a = 1.0 \times 2.11 = 2.11$$

Therefore,

$$q_a 4 = 2.11$$

However, once the cover, tab washer and bolt are in place an additional process is carried out on the washer to bend the tab. Thus from the additional assembly process

PRODUCT NAME: *Fixing Bolt Assembly*
PRODUCT CODE/ID: *FBA1*
PRODUCT QUANTITY: *100,000*

SHEET No.: *1 of 1*
ANALYSIS REF.: *FBA - CA 1*
ANALYSIS DATE: *1/1/99*
ENGINEER: *A.N.Other*

$q_m = t_p \cdot s_p$ (or $q_m = k_p$) $q_a = h_p \cdot f_p$ (or $q_a = a_p$)

Component/assembly process reference	Component/assembly process description	Material/process	Characteristic dimension (mm)	Design tolerance (mm)	Surface roughness (μm Ra)	Material process risk, m_p	Geometry process risk, g_p	Adjusted tolerance = $\dfrac{\text{Design tolerance}}{m_p \cdot g_p}$	Tolerance process risk, t_p	Surface roughness process risk, s_p	Surface eng. process risk, k_p	Handling process risk, h_p	A	B	C	D	E	F	G	H	Total fitting process risk, f_p	Additional process risk, a_p	Total risk (q_m or q_a)	Comments
a1	c1 → WH											1									1		1	ACCEPTABLE DESIGN
a2	c2 → a1											1	1	1	1.2	1	1	1		1	1.2		1.2	ACCEPTABLE DESIGN
a3	c3 → a2											1.2	1	1	1	1.2	1	1		1	1.2		1.44	ACCEPTABLE DESIGN
a4	c4 → a3											1	1	1	1.6	1.2	1.1	1		1	2.11		2.11	SPECIAL CONTROL
a5	Bend Washer Tab																					2.5	2.5	UNACCEPTABLE DESIGN

(Indices for fitting process risk, f_i)

Figure 2.30 Variability risks table for the fixing bolt assembly

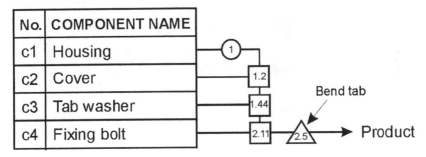

Figure 2.31 Revised sequence of assembly for the fixing bolt

table (see Appendix VI) for later mechanical deformation, the risk index, a_p, is:

A (automated process) $A = 1.0$
B (medium skill level) $B = 1.3$
C (poor access) $C = 1.6$
D (pressure used for deformation) $D = 1.2$
E (no heat applied) $E = 1.0$
F (one stage operation) $F = 1.0$

For the bending the tab washer operation, the Additional Assembly Variability Risk is given by:

$$a_p = A \times B \times C \times D \times E \times F$$

$$a_p = 2.5$$

Therefore,

$$q_a 5 = 2.5$$

A completed variability risks result table for the fixing bolt assembly is shown in Figure 2.30, highlighting the assembly variability risks only. The risk indices can then be entered on the assembly sequence diagram as shown in Figure 2.31. It is evident that insertion of the fixing bolt and bending of the tab washer operations will potentially be problematic on assembly, scoring assembly risk values greater than 2.

The previous sections have illustrated the use of the various process risk charts and tables to obtain variability indices associated with the manufacture and assembly of products. It is important to be systematic with the application of the methodology and the recording of any product analyses, especially as products often contain many parts. It is also important to first declare a sequence of assembly for the individual components before proceeding with an analysis.

2.6.6 Completing the Conformability Matrix

The final part of the analysis is based around the completion of a Conformability Matrix relating variability risk indices for component manufacturing/assembly

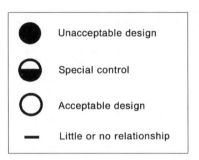

Figure 2.32 Conformability matrix symbols and their quantification

processes to potential failure modes, their severity and the costs of failure. A blank Conformability Matrix is provided in Appendix VII. The final results of an analysis are best displayed in the Conformability Matrix to provide a traceable record of the costs of failure and how these costs are related to the conformance problems through the design decisions made. The symbols, shown in Figure 2.32, are placed in the nodes of the Conformability Matrix, and represent the levels of design acceptability obtained with reference to the Conformability Map given in Figure 2.22.

The link with FMEA brings into play the additional dimension of potential variability into the assessment of the failure modes and the effects on the customer. The Conformability Matrix also highlights those 'bought-in' components and/or assemblies that have been analysed and found to have conformance problems and require further communication with the supplier. This will ultimately improve the supplier development process by highlighting problems up front.

Once the variability risks, q_m and q_a, have been calculated, the link with the particular failure mode(s) from an FMEA for each critical characteristic is made. However, determining this link, if not already evident, can be the most subjective part of the analysis and should ideally be a team-based activity. There may be many component characteristics and failure modes in a product and the matrix must be used to methodically work through this part of the analysis. Past failure data on similar products may be useful in this respect, highlighting those areas of the product that are most affected by variation. Variation in fit, performance or service life is of particular interest since controlling these kinds of variation is most closely allied with quality and reliability (Nelson, 1996).

For each q_m and q_a risk value and the Severity Rating (S), a level of design acceptability is determined from where these values intersect on the Conformability Map. The symbols, relating to the levels of design acceptability, are then placed in the nodes of the Conformability Matrix for each variability risk which the failure mode is directly dependent on for the failure to occur. Once the level of design acceptability has been determined, it can then be written on the Conformability Matrix in the 'Comments' section. C_{pk} values predicted or comments for suppliers can be added too, although predicted C_{pk} values can also be written in the variability risks results table.

PRODUCT NAME	Cover Support Leg	SHEET No.	1 of 1
PRODUCT CODE/ID	CSL1	ANALYSIS REF.	CSL – CA1
PRODUCT QUANTITY	50,000	ANALYSIS DATE	1/1/99
		ENGINEER	A.N.Other

Failure Mode Description and FMEA Severity Rating (S)

Component/assembly process reference	Component/ assembly process description	Total risk (*q*ₘ or *q*ₛ)	Assembly Line Disrupted											Comments (including action for suppliers)
			9											
C1	Cover Support Leg Dimension A	9	●											UNACCEPTABLE DESIGN
Total Failure Mode Isocost %			10											**TOTAL FAILURE COST**
Total Failure Mode Cost			£30K*											£30K

* Number of units = 50 000
Product cost (*Pc*) = £5.93
Total failure mode isocost (%) = 10

Therefore, total failure mode cost = $\dfrac{10 \times 50\,000 \times 5.93}{100}$ = £29 650

Figure 2.33 Conformability matrix example

Example – determining the failure costs for product design

We will now consider calculating the potential costs of failure in more detail for the cover support leg shown earlier. The process for calculating the failure costs for a component is as follows:

- Determine the value of q_m or q_a
- Obtain an FMEA Severity Rating (S)
- Estimate the number of components to be produced (N)
- Estimate the component cost (Pc).

For example, the characteristic dimension 'A' on the cover support leg was critical to the success of the automated assembly process, the potential failure mode being a major disruption to the production line. An FMEA Severity Rating $(S) = 8$ is allocated. See a Process FMEA Severity Ratings table as provided in Chrysler Corporation *et al.* (1995) for guidance on process orientated failures. The component cost, $Pc = £5.93$ and the number planned to be produced *per annum*, $N = 50\,000$.

The characteristic was analysed using CA and q_m was found to be 9. The values of $q_m = 9$ and $S = 8$ are found to intersect on the Conformability Map above the 10% isocost line. (If they had intersected between two isocost lines, the final isocost value is found by interpolation.) If there is more than one critical characteristic on the component, then the isocosts are added to give a total isocost to be used in equation 2.15. The total failure cost is determined from:

$$\text{Total failure cost} = \frac{\text{isocost}(\%) \times N \times Pc}{100}$$

$$\text{Total failure cost} = \frac{10 \times 50\,000 \times 5.93}{100} = £29\,650 \tag{2.15}$$

This figure is of course an estimate of lost profit and may even be conservative, but it clearly shows that the designer has a significant role in reducing the high costs of failure reported by many manufacturing companies. The results are repeated in the Conformability Matrix in Figure 2.33.

2.7 Case studies

2.7.1 Electronic power assisted steering hub design

Under this heading, a flexible hub design for an automotive steering unit is analysed. The application of CA resulted from the requirement to explain to a customer how dimensional characteristics on the product, identified as safety critical, could be produced capably. A key component in this respect is the hub. The component is made by injection moulding, the material being unfilled polybutylene terephthalate (PBT) plastic. The moulding process was selected for its ability to integrate a number of functional elements into a single piece and reduce assembly costs. The design outline of the hub is shown in Figure 2.34(a) together with an optical plate that is mounted on it.

Figure 2.34(b) shows a line from the design FMEA related to the plastic moulded hub. It gives the component function, the potential failure mode, the potential effects and potential causes of failure. In addition, the columns of current controls, Occurrence (O), Severity (S), Detectability (D) ratings and associated Risk Priority Number (RPN) have been completed. A high severity rating was given ($S = 8$) since faulty positional readings due to out of tolerance variation could cause loss of car control and driver injury. The Detectability was rated at $D = 8$ because complex inspection processes would be required. The possibility of Occurrence was provisionally estimated at $O = 4$, giving an RPN equal to 256. The design FMEA specifies compatible dimensioning as the current control to avoid failure, and it was this aspect of the design that needed to be explored further. A quotation from a supplier had been received for volume production of the component.

Some detail on the analysis of the hub design is given below. The hub performed several functions in the controller and therefore carried several critical characteristics. The positional tolerance of the recesses to accommodate a system of location pegs needed to be close. Faces on the hub for mounting the optical plate required precise positioning to provide the necessary spacing between two optical grids (one mounted on the hub and the other carried on a torsion shaft). The depth of the moulded recesses needed to be controlled as they were part of a tolerance chain. It is important to note that the recesses and faces were in different planes and the depth was across a mould parting line.

The positional tolerance on a 10 mm dimension was ± 0.1 mm (tol_1), providing an angular position of $0.6°$. Additionally, the widths of the recesses needed to be held to 4 ± 0.08 mm (tol_2), the dimensional tolerance on the faces to 1 ± 0.07 mm (tol_3) and the depth to 10 ± 0.12 mm (tol_4). Also, the thin sections of the hub gave two geometrical concerns as these vanes were on the limits of plastic flow and distortion was likely on cooling.

The results of the analyses carried out by the business on the hub are given in the variability risks table shown in Figure 2.34(c). The four critical characteristics described above were examined. The positional tolerance (tol_1) set across the thin vanes resulted in high geometry to process risk (g_p) and gave $q_m = 9$. This equates to an out of control C_{pk}, as does the q_m value for the recess depth across the mould parting line (tol_4) which came to 8. The q_m scores for the characteristics tol_2 and tol_3 suggest initially that the process will be in control giving estimated C_{pk} values of 1.33 and 1.75 respectively.

Following the completion of the variability risks table, a Conformability Matrix was produced. This was used to relate the failure modes and their severity coming out of the design FMEA to the results of the Component Manufacturing Variability Risk Analysis. The portion of the matrix concerned with the moulded hub can be found in Figure 2.34(d) and was completed using the Conformability Map.

It is evident that the two characteristics described earlier as being out of control, tol_1 and tol_4, give costs of failure greater than 10%. Also, the characteristics tol_2 and tol_3 which may have been regarded as having acceptable C_{pk} values are shown to have costs of failure of greater than 10% and 0.2% respectively, due to the high Severity Rating (S) = 8 for the potential failure modes in question. Note that two additional failure modes are also illustrated.

(a) Hub Design

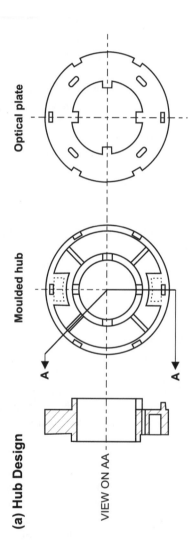

Optical plate

Moulded hub

VIEW ON AA

(b) Design FMEA

Product function	Potential failure mode	Potential effects of failure	Potential causes of failure	Current controls	Existing conditions			
					Occurrence (O)	Severity (S)	Detectability (D)	RPN = O×S×D
LOCATE AND SUPPORT OPTICAL PLATE	MISALIGNMENT OF OPTICAL PLATE	SPURIOUS POSITIONAL SIGNALS	FACES OUT OF TOLERANCE	COMPATIBLE DIMENSION OF MATING PARTS	4	8	8	256

(c) Variability risks

$q_m = t_p \cdot s_p$ (or $q_m = k_p$) $q_a = h_p \cdot f_p$ (or $q_a = a_p$)

Component/assembly process reference	Component/assembly process description	Material process	Characteristic dimension (mm)	Design tolerance (mm)	Surface roughness (μm Ra)	Material process risk, m_p	Geometry process risk, g_p	Adjusted tolerance $= \dfrac{\text{Design tolerance}}{m_p \cdot g_p}$	Tolerance process risk, t_p	Surface roughness process risk, s_p	Surface eng. process risk, k_p	Handling process risk, h_p	Indices for fitting process risk, f_p (A B C D E F G H — NOT APPLICABLE TO THIS ANALYSIS)	Total fitting process risk, f_p	Additional process risk, a_p	Total risk (q_m or q_a)	Comments
c1	HUB - tol1	INJ. MOULDED UNFILLED PBT	10	±0.1	1.6	1	3.18	±0.031	9	1						9	UNACCEPTABLE DESIGN
	- tol2		4	±0.08	1.6	1	1.1	±0.075	1.7	1						1.7	$C_{pk} \approx 1.33$
	- tol3		1	±0.07	1.6	1	1.1	±0.064	1.5	1						1.5	$C_{pk} \approx 1.75$
	- tol4		10	±0.12	1.6	1	1.87	±0.064	8	1						8	UNACCEPTABLE DESIGN

(d) Conformability matrix

Component/assembly process reference	Component/assembly process description	Total risk (q_m or q_a)	Failure Mode Description and FMEA Severity Rating (S)			Comments (including action for suppliers)
			MISALIGNMENT OF OPTICAL PLATES (8)	RELATIVE MOVEMENT OF PLATES (8)	CONTACT OF PLATES (8)	
c1	HUB - tol1	9	●	—	—	UNACCEPTABLE DESIGN
	- tol2	1.7	—	●	—	UNACCEPTABLE DESIGN
	- tol3	1.5	—	—	◐	SPECIAL CONTROL
	- tol4	8	—	—	●	UNACCEPTABLE DESIGN
Total Failure Mode Isocost (%)			10	10	10.05	TOTAL FAILURE COST 30.05%
Total Failure Mode Cost						

Figure 2.34 Hub analysis results

The analysis indicated that the conformance problems associated with the hub design had a cost of failure of more than 30%. This would represent at the annual production quantity required and target selling price, a loss to the business of several million pounds. As a result of the study the business had further detailed discussions with their suppliers and not surprisingly it turned out that the supplier would only be prepared to stand by its original quotation provided the tolerances on the hub, discussed above, were opened up considerably (more than 50%). Subsequently, this result supported the adoption of another more capable design scheme.

2.7.2 Solenoid security cover

This case study concerns the initial design and redesign of a security cover assembly for a solenoid. The analysis only focuses on those critical aspects of the assembly of the product that must be addressed to meet the requirement that the electronics inside the unit are sealed from the outside environment. An FMEA Severity Rating (S) for the assembly was determined as $S = 5$, a warranty return if failure is experienced.

Cover assembly initial design

The initial design of the cover assembly as shown in Figure 2.35 uses an O-ring to seal the electronics against any contamination. Concerns were raised about three mains aspects of the assembly, these being:

- The compression of the O-ring may work against the needs of an adhesive cure on final assembly with an end unit.
- There is a risk that the O-ring will not maintain its proper orientation in the cover recess during subsequent assembly processes, and therefore may not be correctly positioned on final assembly. Restricted vision of the inside of the cover is the key problem here.
- The wire cable may present problems using either manual or automated assembly.

The analysis in Figure 2.35 shows that there is a high risk of non-conformance for the insertion of the frame into the cover, the process relying on the position of the O-ring being maintained (operation a8). The situation is complicated by the restriction of vision during O-ring placement, and this is reflected in the analysis. Using the Conformability Map, it is possible to calculate the potential failure costs for this design scheme in meeting the sealing integrity requirement, as documented in the Conformability Matrix. The final failure cost is calculated to be £805 000. This potential failure cost for this single failure mode is far too high, representing over 10% of the total product cost. A more reasonable target value would be less than 1%. An alternative design scheme should be developed, focusing on reducing the risks of the final assembly operation to reduce the potential for non-conformance as highlighted by the analysis.

Cover assembly redesign

Unfortunately, the design of the wire could not be changed to a more simple arrangement, for example using a spade connector integrated with a recess for the O-ring. The wire is part of the customer's requirements and will inevitably present problems using

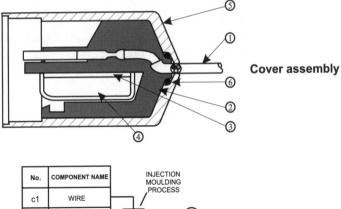

Cover assembly

Assembly sequence diagram

No.	COMPONENT NAME
c1	WIRE
c2	FRAME
c3	HEADER
c4	CAN
c5	COVER
c6	O-RING

Component/assembly process reference	Component assembly process description	Total risk (q_m or q_a)	Failure Mode Description and FMEA Severity Rating (S)					Comments (including action for suppliers)
			SEALING INTEGRITY					
			5					
c2 ¦ a1	FRAME MOULDING	1.5	◯					ACCEPTABLE DESIGN
¦ a4	CAN TO HEADER	2.1	—					
¦ a5	CAN/HEADER TO FRAME	2.4	—					
¦ a7	O-RING INTO COVER	1.94	◑					SPECIAL CONTROL
¦ a8	FRAME INTO COVER	5.1	●					UNACCEPTABLE DESIGN
¦								
¦								
Total Failure Mode Isocost %		10.51						**TOTAL FAILURE COST**
Total Failure Mode Cost		£805K*						£805K

*Number of units = 1 000 000

Product cost (Pc) = £7.66

Total failure mode isocost (%) = 0.01 + 0.5 + 10 = 10.51

Therefore, total failure mode cost = $\dfrac{10.51 \times 1\,000\,000 \times 7.66}{100}$ = £805 066

Conformability matrix

Figure 2.35 Cover assembly design analysis

either manual or automatic assembly operations. Looking to the O-ring, a better design would be to eliminate it altogether and integrate the seal with the wire as shown in Figure 2.36. The wire is then positively located with the seal in the cover hole. Clearly, the risks associated with the cover assembly have been reduced

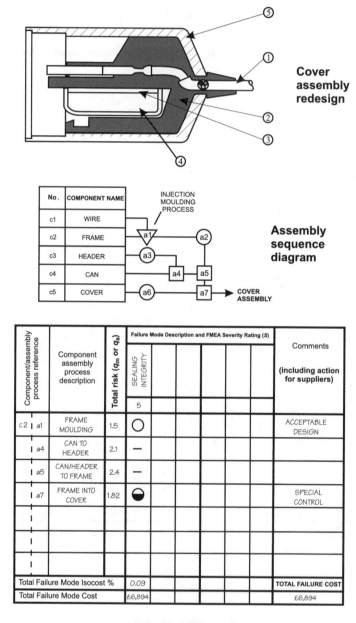

Figure 2.36 Cover assembly redesign analysis

following the elimination of one positionally unstable component and its integration with another. Again, with an FMEA $(S) = 5$, and referring to the Conformability Map, isocosts for each assembly variability risk can be evaluated and the total failure cost is calculated to be £7000.

Comparing this value with the initial design's high potential cost of failure, it is evident that a major design fault in the cover has been eliminated although the assembly process must remain within special control. Subsequently, the redesign solution was chosen for further design development.

2.7.3 Telescopic lever assembly

Consider the telescopic lever assembly, Design A in Figure 2.37, which is part of a stretcher, and hence safety critical. The assembly has an FMEA Severity Rating $(S) = 8$, and is used in a product having a cost of £150. It is estimated that 5000 units are produced *per annum*. The assembly is subjected to bending in operation, with the maximum bending stresses occurring at a point on the main tube corresponding to the stop ring recess. In order to provide additional support, a reinforcing tube is positioned as shown. It is crucial that the tube is placed where it is, since fracture of the telescopic lever may result in injury to users and third parties. Figure 2.37 shows part of the results from the Conformability Matrix for Design A. At each node in the matrix, consideration has been given to the effect of the component or assembly variability risk, represented by q_m or q_a, on the failure mode in question determined from an FMEA.

While the design is satisfactory from a design for strength point of view, the analysis highlights a number of areas where potential variability and failure severity combine to make the risks unacceptably high. For example, there are no design features which ensure the positioning of the reinforcement tube in the assembly (assembly process a3 in Figure 2.37). There are also several critical component tolerances which need to be controlled if its position is to be maintained in service, such as the inner diameter of the main tube, c1, the outside diameter of the reinforcement tube, c2, and the application of the adhesive, a2.

The conclusion from the analysis is that the assembly should be redesigned. This is further justified by calculating the potential costs of failure for the assembly. If this design of telescopic lever assembly fractured in service, user injury, high losses, including legal costs, could be incurred. A cost of failure of £257 000 was calculated from the analysis, which is far too high representing more than 36% of annual revenue from the product. This figure was calculated by summing the isocosts for each characteristic/assembly process whose variability risks potentially contribute to each failure mode, and then multiplying the total failure mode isocost (%) by the product cost and number of items produced. The calculation of the costs for the second failure mode type (reinforcing tube moves out of position) on Design A is shown in detail in Figure 2.37.

Failure of this design in service did in fact result in user injury. High losses of the order of those calculated above for the particular failure mode, including legal costs, were incurred. A number of alternative designs are possible, and one which does not involve the above problems is included with its Conformability Matrix in Figure 2.38.

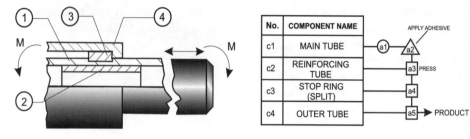

DESIGN A   **ASSEMBLY SEQUENCE DIAGRAM**

Component/assembly process reference	Component/ assembly process description	Total risk (q_m or q_a)	REINFORCING TUBE FITTED IN THE WRONG PLACE	REINFORCING TUBE MOVES OUT OF POSITION	DEPTH GEOMETRY OF STOP RING RECESS	STOP RING MOVES OUT OF POSITION			Comments (including action for suppliers)
			8	8	8	8			
c1	MAIN TUBE I.D.	1.6	–	●	–	–			UNACCEPTABLE DESIGN
c1'	MAIN TUBE RECESS	1.6	–	–	●	–			UNACCEPTABLE DESIGN
a1	c1 → WH	1	○	–	–	–			ACCEPTABLE DESIGN
a2	APPLY GLUE TO TUBE BORE	2.4	–	●	–	–			UNACCEPTABLE DESIGN
c2	REINFORCING TUBE	1.5	–	◐	–	–			SPECIAL CONTROL
a3	c2 → a2	4.5	●	–	–	–			UNACCEPTABLE DESIGN
c3	STOP RING (SPLIT)	1.3	–	–	–	○			ACCEPTABLE DESIGN
a4	c3 → a3	2	–	–	–	●			UNACCEPTABLE DESIGN
c4	OUTER TUBE	1	–	–	–	–			
a5	c4 → a4	1	–	–	–	–			
Total Failure Mode Isocost (%)			10.01	12.05	2	10.01			**TOTAL FAILURE COST**
Total Failure Mode Cost			£75.8K	£90.4K*	£15K	£75.8K			£257K

*Sample calculation of failure costs:

Number of units = 5000

Product cost (Pc) = £150

Total failure mode isocost (%) = 2 + 10 + 0.05 = 12.05

Therefore, total failure cost = $\dfrac{12.05 \times 5000 \times 150}{100}$ = £90 375

CONFORMABILITY MATRIX FOR DESIGN A

Figure 2.37 Telescopic lever assembly analysis for Design A

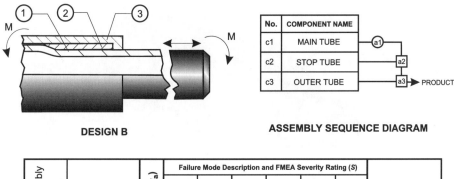

DESIGN B

ASSEMBLY SEQUENCE DIAGRAM

No.	COMPONENT NAME
c1	MAIN TUBE
c2	STOP TUBE
c3	OUTER TUBE

Component/assembly process reference		Component/ assembly process description	Total risk (q_m or q_a)	Failure Mode Description and FMEA Severity Rating (S)					Comments (including action for suppliers)
				STOP TUBE NOT FITTED					
				8					
c1		MAIN TUBE	1.5	–					
	a1	c1 → WH	1	◯					ACCEPTABLE DESIGN
c2		STOP TUBE	1.3	–					
	a2	c2 → a1	1.3	◯					
c3		OUTER TUBE	1	–					ACCEPTABLE DESIGN
	a3	c3 → a2	1	–					
Total Failure Mode Isocost (%)			0.02						TOTAL FAILURE COST
Total Failure Mode Cost			£150						£150

CONFORMABILITY MATRIX FOR DESIGN B

Figure 2.38 Telescopic lever assembly analysis for Design B

The costs of failure for this Design B were subsequently reduced to a negligible amount.

The above example demonstrates the use of CA in supporting the identification of manufacturing and assembly problems before production commences, but more importantly safety problems can be systematically identified and the potential for harm eliminated before the product is in use by the customer. Given the huge losses that are associated with safety critical products when they fail, considerations of this type must be on the agenda of all manufacturing companies, especially when the product has a high degree of interaction with the user.

2.7.4 Solenoid end assembly

The following case study determines the manufacturing and assembly variability risks for a solenoid end assembly design, shown in Figure 2.39, and projects the potential

Figure 2.39 Solenoid end assembly

costs of failure associated with the capability of an assembly tolerance stack. The solenoid is to operate as a fuel cut-off device in a vehicle, operated when a signal is received from the ignition. The signal allows current to flow to the inductor coil which then withdraws the plunger seal and allows the fuel to flow. The solenoid assembly is to be screwed into an engine block at the fuel port in a counterbored hole. An important requirement is that the plunger displacement from the engine block face through the solenoid tolerance stack to the plunger seal face must be within a tolerance of ± 0.2 mm. If this requirement is not met, fuel flow restriction could occur, this being the main failure mode. The product will be in the warranty return category as it has little effect on user safety if it fails in service, which relates to an FMEA Severity Rating $(S) = 5$. The product cost is £7.66 and it is estimated that one million units will be manufactured in total.

Solenoid end assembly initial design

The initial design is analysed using CA at a component level for their combined ability to achieve the important customer requirement, this being the tolerance of ± 0.2 mm for the plunger displacement. Only those characteristics involved in the tolerance stack are analysed. The 'worst case' tolerance stack model is used as directed by the customer. This model assumes that each component tolerance is at its maximum or minimum limit and that the sum of these equals the assembly tolerance, given by equation 2.16 (see Chapter 3 for a detailed discussion on tolerance stack models):

$$\sum_{i=1}^{n} t_i \leq t_a \tag{2.16}$$

where:

$t_i =$ bilateral tolerance for ith component characteristic

$t_a =$ bilateral tolerance for assembly stack.

Figure 2.40 shows the initial detailed design including the tolerances required on each component in the stack to achieve the ± 0.2 mm assembly tolerance, t_a (not included is the dimensional tolerance on the fuel port block of 12 ± 0.05 mm which is set by the

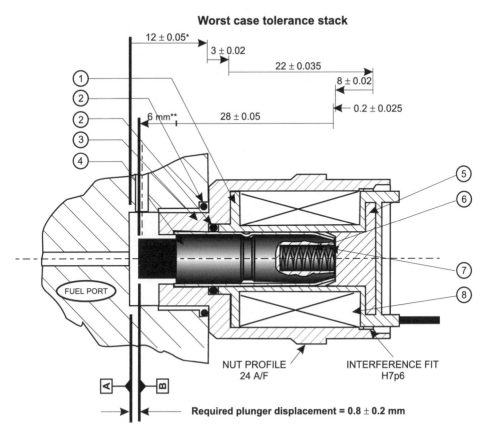

Worst case tolerance stack

PART No.	DESCRIPTION	MATERIAL	PROCESS
1	BOBBIN	PBT(30% FILL)	INJECTION MOULDED WITH ITEM 5
2	O-RING	RUBBER	–
3	BODY	COLD FORMING STEEL	IMPACT EXTRUSION
4	PLUNGER	STEEL/SILICONE RUBBER	M/C (MOULDED SILICONE RUBBER END SEAL)
5	MAGNETIC POLE	COLD FORMING STEEL	IMPACT EXTRUSION
6	SPRING	STEEL	–
7	TUBE	BRASS	ROLLED/DEEP DRAWN
8	COIL	–	–

* Tolerance fixed by supplier
** The 6 mm end seal dimension will be used in the analysis, not 28 mm

Figure 2.40 Solenoid end assembly initial design

supplier). Also shown is a table describing the process used to manufacture each component and an assembly sequence diagram is given in Figure 2.41. Referring back to Figure 2.40, the tolerance stack starts at face A on the fuel port and accumulates through the individual components to face B on the plunger seal.

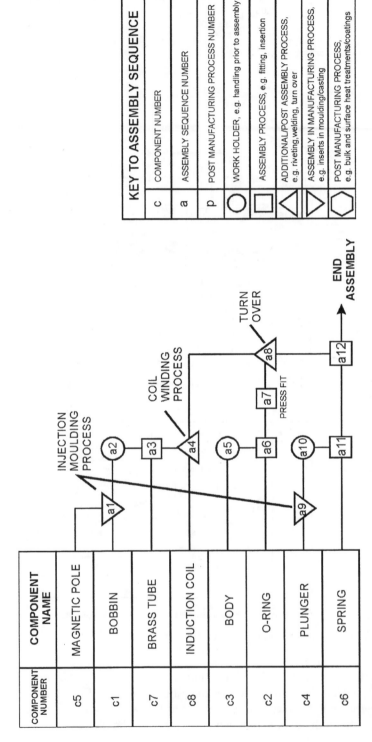

Figure 2.41 Assembly sequence diagram for the solenoid end assembly design

The body is impact extruded from a cold forming steel. The characteristic dimension to be analysed in the tolerance stack is the base thickness of 3 mm (on a $\varnothing 20$ mm bore) and this dimension has been assigned a tolerance of ± 0.02 mm.

Following the tolerance stack through the end assembly, the bobbin dimension of 22 mm from the outside face to the back face of the magnetic pole is analysed next. This characteristic dimension does not include the tolerance on the impact extruded pole. The pole is to be moulded into the bobbin and the pole face is considered to be part of a mould related dimension. The bobbin is injection moulded using 30% filled polybutylene terephthalate (PBT). The tolerance assigned to the bobbin dimension is ± 0.035 mm.

Component/assembly process reference	Component/assembly process descrption	Material/process	Characteristic dimension (mm)	Design tolerance (mm)	Surface roughness (μm Ra)	Material process risk, m_p	Geometry process risk, g_p	Adjusted tolerance = Design tolerance / $m_p.g_p$	Tolerance process risk, t_p	Surface roughness process risk, s_p	Surface eng. Process risk, k_p	Handling process risk, h_p	A	B	C	D	E	F	G	H	Total fitting process risk, f_p	Additional process risk, a_p	Total risk (q_m or q_a)	Comments
c5	MAGNETIC POLE	IMPACT EXTRUDED COLD FORMING STEEL	8	±0.02	0.8	1.2	1.7	±0.01	9	1													9	UNACCEPTABLE DESIGN
c1 a1	BOBBIN	PBT (30% FILL) INJECTION MOULD WITH INSERT c5	22	±0.035	−	1	1	±0.035	9	1													9	UNACCEPTABLE DESIGN
a1 ▶ WH												1									1		1	ACCEPTABLE DESIGN
c7	TUBE	ROLLED/ DEEP DRAWN BRASS	0.2	±0.025	1.6	1	1	±0.025	1	1													1	ACCEPTABLE DESIGN
a2 c7 ▶ a2	TUBE INSERTED INTO BOBBIN I/D											1	1	1	1.1	1	1	1	1	1.4	1.54		1.54	ACCEPTABLE DESIGN
a4	COIL WINDING PROCESS																							
c3	BODY	IMPACT EXTRUDED COLD FORMING STEEL	3 (Ø20)	±0.02	0.8	1.2	1.1	±0.015	9	1													9	UNACCEPTABLE DESIGN
a5 c3 ▶ WH												1									1		1	ACCEPTABLE DESIGN
c2	O-RING	RUBBER																						
a6 c2 ▶ a5	O-RING INSERTED INTO BODY											1.5	1	1	1	1	1	1.1	1	1	1.1		1.65	ACCEPTABLE DESIGN
a7 a3 ▶ a5	COIL ASS'Y FITTED INTO BODY ASS'Y											1	1	1	1	1	1.2	1.1	1	1.4	1.85		1.85	SPECIAL CONTROL
a8	TURN OVER																					1.3	1.3	ACCEPTABLE DESIGN
c4 a9	PLUNGER	RUBBER END SEAL MOULDED ONTO M/C PLUNGER	6 (28)	±0.05	−	1	1	±0.05	4	1													4	UNACCEPTABLE DESIGN
a10 a9 ▶ WH												1									1		1	ACCEPTABLE DESIGN
c6	SPRING	STEEL																						
a11 c6 ▶ a10	SPRING INSERTED INTO PLUNGER											1.5	1	1	1	1	1	1	1	1	1		1.5	ACCEPTABLE DESIGN
a12 A11 ▶ a8	PLUNGER ASSEMBLY INSERTED											1	1	1	1.1	1.2	1	1	1	1.4	1.85		1.85	SPECIAL CONTROL

Figure 2.42 Variability risks analysis for the solenoid end assembly initial design

The pole has a characteristic dimension of 8 mm from the rear of the bobbin to the recess face and has a tolerance assigned to it of ±0.02 mm. From the pole recess face, the tube base tolerance is the last component to make up the tolerance stack. The brass tube has been given a dimensional tolerance on its base of 0.2 ± 0.025 mm. Note, the dimensional tolerance on the plunger is 28 ± 0.05 mm, but the analysis will concentrate on the silicone rubber seal length of 6 mm because this is moulded onto the plunger and again is a mould related dimension.

Figure 2.42 shows the variability risks analysis based on the tolerances assigned to meet the ±0.2 mm tolerance for the assembly. Given that an FMEA Severity Rating $(S) = 5$ has been determined, which relates to a 'definite return to manufacturer', both impact extruded components are in the unacceptable design region, as well as the bobbin and plunger end seal as shown on the Conformability Matrix in Figure 2.43. The tolerance for the brass tube base thickness has no risk and is an acceptable design.

Component/assembly process reference	Component/ assembly process description	Total risk (q_m or q_a)	Failure Mode Description and FMEA Severity Rating (S)						Comments (including action for suppliers)
			FUEL FLOW RESTRICTION						
			5						
c1	BOBBIN LENGTH	9	●						UNACCEPTABLE DESIGN
c3	BODY THICKNESS	9	●						UNACCEPTABLE DESIGN
c4	PLUNGER END SEAL LENGTH	4	●						UNACCEPTABLE DESIGN
c5	MAGNETIC POLE LENGTH	9	●						UNACCEPTABLE DESIGN
c7	TUBE	1	○						ACCEPTABLE DESIGN
Total Failure Mode Isocost (%)			40.01						TOTAL FAILURE COST
Total Failure Mode Cost			£3.07M*						£3.07M

* Number of units = 1 000 000

Product cost $(Pc) = £7.66$

Total failure mode isocost (%) $= 10 + 10 + 0.01 + 10 + 10 = 40.01$

Therefore, total failure mode cost $= \dfrac{40.01 \times 1\,000\,000 \times 7.66}{100} = £3\,064\,766$

Figure 2.43 Conformability matrix for the solenoid end assembly initial design

The associated cost of failure for the solenoid end assembly is calculated to be over £3 million for a product cost of £7.66 and production volume of one million units. This figure is for the tolerance stack failure mode alone as this is most important to the customer. Although the assembly variability risks are analysed, they are not taken into account in the final costs of failure. In conclusion, the process capabilities

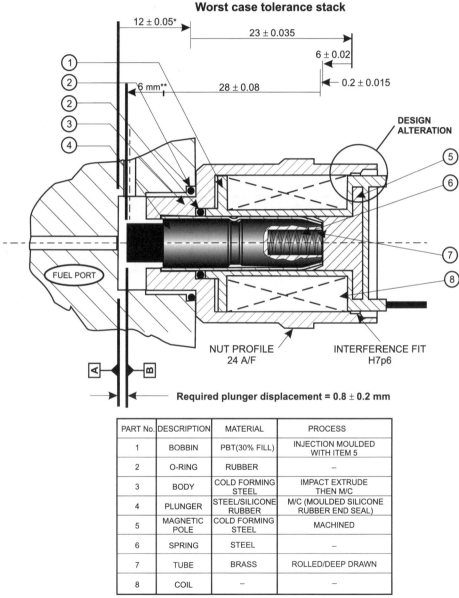

PART No.	DESCRIPTION	MATERIAL	PROCESS
1	BOBBIN	PBT(30% FILL)	INJECTION MOULDED WITH ITEM 5
2	O-RING	RUBBER	–
3	BODY	COLD FORMING STEEL	IMPACT EXTRUDE THEN M/C
4	PLUNGER	STEEL/SILICONE RUBBER	M/C (MOULDED SILICONE RUBBER END SEAL)
5	MAGNETIC POLE	COLD FORMING STEEL	MACHINED
6	SPRING	STEEL	–
7	TUBE	BRASS	ROLLED/DEEP DRAWN
8	COIL	–	–

* Tolerance fixed by supplier
** The 6 mm end seal dimension will be used in the analysis, not 28 mm

Figure 2.44 Solenoid end assembly redesign

of several characteristics in this tolerance stack are inadequate and will not meet the customer's requirements consistently.

Solenoid end assembly redesign

This is similar to the initial design, but involving turning as a secondary process on the body to improve a key tolerance capability as indicated in Figure 2.44. The body is still impact extruded, but the face which mates with the fuel port block is machined, together with a shoulder on the inside diameter. The front face of the pole, now fully machined and assuming no component cost increase, is assembled up to the machined shoulder on the body. Only the tolerances on the pole length, tube and plunger end seal remain in the stack, reducing the number of components to five.

Component/assembly process reference	Component/assembly process description	Material/process	Char. dim. (mm)	Design tol. (mm)	Surf. rough. (µm Ra)	m_p	g_p	Adj. tol. = $\frac{\text{tol.}}{m_p.g_p}$	t_p	s_p	k_p	h_p	A	B	C	D	E	F	G	H	f_p total	a_p	Total risk (q_m or q_a)	Comments
c5	MAGNETIC POLE	M/C FREE CUTTING STEEL	6	±0.02	0.8	1.2	1	±0.017	1.07	1													1.07	ACCEPTABLE DESIGN
c1 a1	BOBBIN	PBT (30% FILL) INJECTION MOULD WITH INSERT c5																						
a1 → WH												1									1		1	ACCEPTABLE DESIGN
c7	TUBE	ROLLED/DEEP DRAWN BRASS	0.2	±0.015	1.6	1	1	±0.015	1.06	1													1.06	ACCEPTABLE DESIGN
a2 c7 → a2	TUBE INSERTED INTO BOBBIN I/D											1	1	1	1	1.1	1	1	1	1	1.4	1.54	1.54	ACCEPTABLE DESIGN
a4		COIL WINDING PROCESS																						
c3	BODY	IMPACT EXTRUDED COLD FORMING STEEL THEN M/C	25	±0.035	0.8	1.2	1	±0.029	1.09	1													1.09	ACCEPTABLE DESIGN
a5 c3 → WH												1									1		1	ACCEPTABLE DESIGN
c2	O-RING	RUBBER																						
a6 c2 → a5	O-RING INSERTED INTO BODY											1.5	1	1	1	1	1	1.1	1	1	1.1		1.65	ACCEPTABLE DESIGN
a7 a3 → a5	COIL ASS'Y FITTED INTO BODY ASS'Y											1	1	1	1	1.2	1.1	1	1.4	1.85		1.85	SPECIAL CONTROL
a8		TURN OVER																				1.3	1.3	ACCEPTABLE DESIGN
c4 a9	PLUNGER	RUBBER END SEAL MOULDED ONTO M/C PLUNGER	6 (26)	±0.08	–	1	1	±0.08	1.7	1													1.7	SPECIAL CONTROL
a10 a9 → WH												1									1		1	ACCEPTABLE DESIGN
c6	SPRING	STEEL																						
a11 c6 → a10	SPRING INSERTED INTO PLUNGER											1.5	1	1	1	1	1	1	1	1	1		1.5	ACCEPTABLE DESIGN
a12 A11 → a8	PLUNGER ASSEMBLY INSERTED											1	1	1	1.1	1.2	1	1	1.4	1.85		1.85	SPECIAL CONTROL

Figure 2.45 Variability risks analysis for the solenoid end assembly redesign

The variability risks table for the redesign is shown in Figure 2.45 and the Conformability Matrix in Figure 2.46. Clearly, machining the critical faces on the impact extruded components has reduced the risks associated with conforming to the ±0.2 mm tolerance for the plunger displacement.

The associated potential cost of failure has reduced significantly to a little over £3000. However, there is an additional cost associated with the extra machining process which adds to the overall product cost. Since it is likely a secondary process involving machining will take place on the body thread anyway, the case for turning these critical faces may be further justified. Although machining these faces will raise the cost of the component slightly, this must be secondary to satisfying the overriding customer requirement of meeting the plunger displacement tolerance.

As highlighted by the difference in the potential failure costs, the redesign scheme must be chosen for further design development. Of course, other design schemes

Component/assembly process reference	Component/ assembly process description	Total risk (q_m or q_a)	Failure Mode Description and FMEA Severity Rating (S)						Comments (including action for suppliers)
			FUEL FLOW RESTRICTION						
			5						
c3	BODY THICKNESS	1.09	○						ACCEPTABLE DESIGN
c4	PLUNGER END SEAL LENGTH	1.7	◑						SPECIAL CONTROL
c5	MAGNETIC POLE LENGTH	1.07	○						ACCEPTABLE DESIGN
c7	TUBE	1.06	○						ACCEPTABLE DESIGN
Total Failure Mode Isocost (%)		0.04							TOTAL FAILURE COST
Total Failure Mode Cost		£3K*							£3K

*Number of units = 1 000 000
Product cost (Pc) = £7.66
Total failure mode isocost (%) = 0.01 + 0.01 + 0.01 + 0.01 = 0.04

Therefore, total failure mode cost = $\dfrac{0.04 \times 1\,000\,000 \times 7.66}{100}$ = £3064

Figure 2.46 Conformability matrix for the solenoid end assembly redesign

could also be explored, but the initial design shown here is of inherently poor quality and therefore must be rejected.

2.8 Summary

Decisions made during the design stage of the product development process account for a large proportion of the problems that incur failure costs in production and service. It is possible to relate these failure costs back to the original design intent where variability, and the lack of understanding of variability, is a key failure costs driver. The correct choice of tolerance on a dimensional characteristic can be crucial for the correct functioning of the product in service and tolerance selection can have a large contribution to the overall costs of the product, both production and quality loss.

Process capability indices are not generally specified by designers and subsequently the impact of design decisions made on the production department cannot be fully understood because tolerances alone do not contain enough information. Variability in component manufacture has proved difficult to predict in the early stages of the design process and there are many influencing factors that the designer may not necessarily be able to anticipate. The material and geometrical configuration of the design, and the compatibility with the manufacturing process are the main variability drivers. Although design rules and general manufacturing capability information are available, they are rarely presented in a useful or practical form, especially when innovative design is required. There is a need to set realistic tolerances and anticipate the variability associated with the design to help reduce failure costs later in the product's life-cycle.

The CA methodology is useful in this respect. It is comprised of three sections: the Component Manufacturing Variability Risks Analysis, the Component Assembly Variability Risks Analysis and the determination of the Effects of Non-conformance through the Conformability Map.

The Component Manufacturing Variability Risks Analysis presented, models the important design/manufacture interface issues which reflect the likely process capability that can be achieved. Included is the assessment of tolerance, geometry, material and surface roughness variability in component manufacture. Quantitative and qualitative manufacturing knowledge is used to support various aspects of the analysis and is taken from a wide range of sources. The concept of an ideal design allows the analysis to generate risk indices, where values greater than unity have a potential for increased variation in production. A simple cost–tolerance relationship is used in the Process Capability Maps, developed for over 60 manufacturing process/material combinations. The maps are subsequently employed to determine the process capability estimates for the component characteristics analysed. Through empirical studies, a close correlation between the process capability estimates using the Component Manufacturing Variability Risks Analysis and shop-floor process capability has been observed.

Most literature tends only to focus on tolerance stack analysis when assessing the capability of assemblies. The variability of the actual assembly operations is rarely considered and does not rely solely on the tolerances accumulating throughout the assembly, but on the feasibility and inherent technical capability of the assembly operations performed, manually or automatically. Developers and practitioners of DFA techniques reason that an assembly with a high assembly efficiency is a better quality product. The natural outcome of having a high assembly efficiency leads to fewer parts in the assembly and, therefore, fewer quality problems to tackle in production. The outcome is not due to any specific analysis process in the DFA technique to address variability, and there still exists a need for analysing the assembly capability of designs, rather than a production cost driven approach. A useful technique for facilitating an assembly risks analysis is the declaration of a sequence of assembly for the components. Through such a diagram, each component in the assembly and, therefore, the potential areas for assembly risk can be logically mapped through the product design.

The Component Assembly Variability Risks Analysis has the purpose of better understanding the effects of a component's assembly situation on variability, by quantifying the risks that various assembly operations inherently exhibit. The analysis processes are supported by expert knowledge and presented in charts. Again, the theory is that an ideal component assembly situation exists where the assembly risk is unity. Using the charts to reflect the handling process risk, fitting process risk and the risks associated with additional assembly and joining processes, the assembly situation of the component is questioned, accruing penalties at each stage if the design has increased potential for variability.

Current quality–cost models are useful for identifying general trends in a long-term improvement programme, but are of limited use in the identification of the failure costs associated with actual design decisions. A link between the costs that can be typically expected in practice due to failure or non-conformance of the product in production or service, and the probability of fault occurrence, is made using FMEA through the Conformability Map. The underlying concept assumes that as failures become more severe, they are going to cost more when they fail. The quality–cost model embedded in the Conformability Map allows the designer to assess the level of acceptability, special control or unacceptability for non-safety critical and safety critical component characteristics in the design by determination of the process capability measures from the previous two stages of the analysis. The Conformability Map also allows failure isocosts (percentages of total product cost), and, therefore, the total failure cost to be estimated with knowledge of the likely product cost and production volume. The nature of the underlying cost models limits the accuracy of the failure cost estimates at an absolute level and so they become useful in evaluating and comparing design schemes for their potential quality loss. The model can alternatively be employed in setting capability targets for characteristics to incur allowable failure costs dependent on the failure severity of the product.

Through performing an analysis using CA, many modes of application have been highlighted. This has resulted from the way that the CA design performance measures allow a non-judgemental 'language' to develop between the design team. It has also been found not to inhibit the design process, but provide a structured analysis with which to trace design decisions. The knowledge embedded within CA also allows the designer to generate process capable solutions and open up discussion with suppliers. The analysis is currently facilitated through the use of a paper-based assessment. This has many benefits, including improved team working, and provides

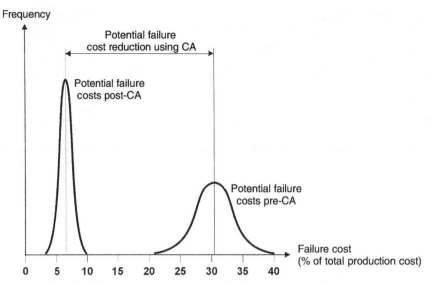

Figure 2.47 Influence of the team-based application of CA on several product introduction projects

a more unconstrained approach than if the analysis were computer based. It also allows the knowledge to be readily visible and available at anytime for the designer to scrutinize and manipulate if they chose to do so.

The potential benefits of using CA in the early stages of the product development process have been found to be:

- Early awareness of potential design problems through a systematic analysis
- Produces more process capable designs with regard to their manufacture and assembly
- Reduces internal/external failure costs
- Reduces lead times
- Focused discussions with suppliers.

Finally, the main benefit as far as competitive business performance is concerned is the potential for reduction in failure costs. Studies using CA very early in the development process of a number of projects have indicated that the potential failure costs were all reduced through an analysis. This is shown in Figure 2.47, where this potential failure cost reduction is shown as the difference between *pre-CA* and *post-CA* application by the teams analysing the product designs.

Designing capable assembly stacks

3.1 Introduction

The analysis of process capable tolerances on individual component dimensions at the design stage has already been explored in Chapter 2. An important extension to this work is the identification and assignment of capable tolerances on individual component dimensions within an assembly stack forming some specified assembly tolerance, where typically the number of component tolerances in the assembly stack is two or more*. This engineering task is called *tolerance allocation* and is a design function performed before any parts have been produced or tooling ordered. It involves three main activities. First, deciding what tolerance limits to place on the critical clearances/fits for the assembly based on performance/functional requirements. Next, creating an assembly model to identify what dimensions characterize the final assembly dimension. And finally, deciding how much of the assembly tolerance to assign to each of the components in the assembly (Chase and Parkinson, 1991). *Tolerance analysis*, on the other hand, is used for determining the assembly tolerance from knowledge of the tolerances on each component.

Tolerance analysis does not readily lend itself to a 'designing for quality' philosophy because the required functional parameter of the assembly tolerance is not required as an input. However, by setting a target tolerance on the assembly from the customer specification and for functional performance, as in tolerance allocation, we can establish the component tolerances in order to keep the failure cost of the assembly to an acceptable level. This way, there is no need to calculate a failure cost associated with the assembly tolerance as would be performed using CA; an acceptably low failure cost is actually built in at the design stage using manufacturing knowledge to optimize the capability of the individual components. In essence, the philosophy is for customer, for function and for conformance.

Today's high technology products and growing international competition require knowledgeable design decisions based on realistic models that include manufacturability requirements. A suitable and coherent tolerance allocation methodology

---

* See Appendix VIII for an approach to solve clearance/interference problems with two tolerances.

can be an effective interface between the customer, design and manufacturing. However, there are several issues with regard to tolerance allocation that must be addressed (Chase and Greenwood, 1988):

- Engineering design and manufacturing need to communicate their needs effectively
- The choice of tolerance model must be both realistic and applicable as a design tool
- The role of advanced statistical and optimization methods in the tolerance model
- Sufficient data on process distributions and costs must be collated to characterize manufacturing processes for advanced tolerance models.

The effective use of the capability data and knowledge as part of CA is beneficial in the design of capable assembly stacks providing the necessary information, which has been lacking previously. The aim of this chapter is to present a methodology for the allocation of capable component tolerances within assembly stack problems (one dimensional) and optimize these with respect to cost/risk criteria in obtaining a functional assembly tolerance through their synthesis. The methodology and the demonstration software presented to aid an analysis, called *CAPRAtol*, forms part of the *CAPRA* methodology (**CA**pabilty and **PR**obabilistic Design **A**nalysis).

3.2 Background

Proper assignment of tolerances is one of the least well-understood engineering tasks (Gerth, 1997). Assignment decisions are often based on insufficient data or incomplete models (Wu *et al.*, 1988). The precise assignment of the component tolerances for this combined effect is multifarious and is dictated by a number of factors, including:

- Number of components in the stack
- Functional performance of the assembly tolerance
- Level of capability assigned to each component tolerance
- Component assemblability
- Manufacturing processes available
- Accuracy of process capability data
- Assumed characteristic distributions and degree of skew and shift
- Cost models used
- Allowable costs (both production and quality loss)
- Tolerance stack model used
- Optimization method.

Some of the above points are worth expanding on. Tolerances exist to maintain product performance levels. Engineers know that tolerance stacking or accumulation in assemblies controls the critical clearances and interferences in a design, such as lubrication paths or bearing mounts and that these affect performance (Vasseur *et al.*, 1992). Tolerances also influence the selection of manufacturing processes and determine the assemblability of the final product (Chase and Greenwood, 1988). The first concern in allocating tolerances should be then to guarantee the proper functioning of the product, and therefore to satisfy technical constraints. In general design practice, the final assembly specifications are usually derived from customer requirements (Lin *et al.*, 1997). The functional assembly tolerance is a specification of the design and maintains integrity with its mating assemblies only when this tolerance is realized within a suitable level of capability.

The random manner by which the inherent inaccuracies within the process are generated produces a pattern of variation for the dimension that resembles the Normal distribution, as discussed in Chapter 2. As a first supposition then in the optimization of a tolerance stack with '*n*' number of components, it is assumed that each component follows a Normal distribution, therefore giving an assembly tolerance with a Normal distribution. It is also a good approximation that if the number of components in the stack is greater than 5, then the final assembly characteristic will form a Normal distribution regardless of the individual component distributions due to the *central limit theorem* (Mischke, 1980).

Shift or drift is a critical factor in an assembly model as is the determination of the variability for each component tolerance. It has been estimated that over a very large number of batches produced, the mean of a tolerance distribution could expect to drift about 1.5 times the standard deviation due to tool wear, differences in raw material or change of suppliers (Evans, 1975; Harry and Stewart, 1988). The degree of shift inherent within the process or the shift over time should also be accounted for in the tolerance stack model as its omission can severely affect the precision of the results. Figure 3.1(a) shows the effect that a dominant component distribution, prone to shift, in a stack of '*n*' components has on the overall assembly distribution. It is more than likely for this case that the assembly tolerance distribution will also be shifted. As dominance reduces in any one component, the probability of the assembly distribution being shifted from the target is much lower and tends to average out to a Normal centred distribution as shown in Figure 3.1(b). This is also true for skew or kurtosis of the distribution (Chase and Parkinson, 1991).

Designers seldom have sufficient data by which to specify the variability of the manufacturing processes. In practice, most designers do not worry about the true behaviour of the process and compensate for the lack of knowledge with large process capability indices as discussed in Chapter 2. The precise process capability of a process cannot be determined before statistical control of the actual process has been established. Therefore, during the product design phase, the designer must use the best available process capability data for similar processes (Battin, 1988).

A good tolerance allocation model should maximize component tolerances for a specified assembly tolerance in order to keep production costs low (Wu *et al.*, 1988). Any concessions to this should be in meeting the functional assembly tolerance in order to keep the failure costs low. This can only be achieved by optimization of the component tolerances by important technical and/or economic criteria. Design optimization techniques are useful in minimizing an objective function such as production cost or quality loss (Chase and Parkinson, 1991), the optimal statistical tolerances being those that minimize this aggregate quantity of production cost plus quality loss (Vasseur *et al.*, 1992).

Optimization of the tolerances assigned to the tolerance stack to satisfy the assembly tolerance is required because of the risk of assigning impractical and costly values. The tolerance stack models, presented later, do not optimize the tolerances and additional methods are required. The actual optimization operations usually take the form of minimization or maximization of the results from the cost

(a) One dominant component distribution – shifted

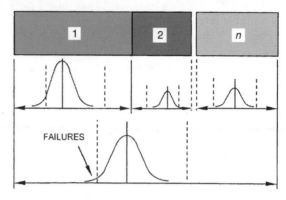

(b) No dominant component distributions

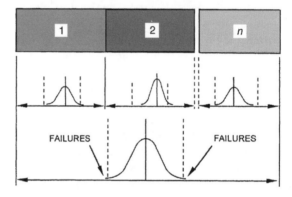

Figure 3.1 Effect of component distribution shift scenarios on the final assembly distribution

models through computer coded algorithms. Common optimization methods given in the literature include: Lagrange multipliers, linear and non-linear programming, geometric programming and genetic algorithms (Chase and Parkinson, 1991; Lin *et al.*, 1997; Wu *et al.*, 1988).

The combination of the cost model and the optimization method will then give an augmented model from which the allocation of the component tolerances are optimized for competitive results. Optimization methods have also been extended to include procedures that select the most cost-effective manufacturing process for each component tolerance in the assembly stack (Chase and Parkinson, 1991).

Research looking into tolerance allocation in assembly stacks is by no means new. A current theme is towards an optimization approach using complex routines and/or cost models (Lin *et al.*, 1997; Jeang, 1995). Advanced methods are also available, such as *Monte Carlo Simulation* and *Method of Moments* (Chase and Parkinson, 1991; Wu *et al.*, 1988). The approach presented here is based on empirical process capability measures using simple tolerance models, cost analogies and optimization procedures. Using the methodology described, a number of design schemes can be quickly compared and the most capable, relatively, selected as the final design solution.

The optimization of the tolerances allocated will be based on achieving the lowest assembly standard deviation for the largest possible component tolerances. By aiming for tolerances with low standard deviations also makes the tolerances robust against any further unseen variations whilst at the design stage. A low standard deviation translates into a lower probability of encountering assembly problems, which in turn means higher manufacturing confidence, lower costs, shorter cycle time, and perhaps the most important, enhanced customer satisfaction (Harry and Stewart, 1988). It is important to predict the probability of successful assembly of the parts so that the tolerance specifications can be re-evaluated and modified if necessary in order to increase the probability of success and lower the associated production costs (Lee et al., 1997).

3.3 Tolerance stack models

Many references can be found reporting on the mathematical/empirical models used to relate individual tolerances in an assembly stack to the functional assembly tolerance. See the following references for a discussion of some of the various models developed (Chase and Parkinson, 1991; Gilson, 1951; Harry and Stewart, 1988; Henzold, 1995; Vasseur et al., 1992; Wu et al., 1988; Zhang, 1997). The two most well-known models are highlighted below. In all cases, the linear one-dimensional situation is examined for simplicity.

In general, tolerance stack models are based on either the *worst case* or *statistical* approaches, including those given in the references above. The worst case model (see equation 3.1) assumes that each component dimension is at its maximum or minimum limit and that the sum of these equals the assembly tolerance (initially this model was presented in Chapter 2). The tolerance stack equations are given in terms of bilateral tolerances on each component dimension, which is a common format when analysing tolerances in practice. The worst case model is:

$$\sum_{i=1}^{n} t_i \leq t_a \tag{3.1}$$

where:

t_i = bilateral tolerance for ith component characteristic

t_a = bilateral tolerance for assembly stack.

The statistical model makes use of the fact that the probability of all the components being at the extremes of their tolerance range is very low (see equation 3.2). The statistical model is given by:

$$z_a \left[\sum_{i=1}^{n} \left(\frac{t_i}{z_i} \right)^2 \right]^{0.5} \leq t_a \tag{3.2}$$

where:

z_a = assembly tolerance standard deviation multiplier

z_i = ith component tolerance standard deviation multiplier.

Equation 3.2 is essentially the root of the sum of the squares of the standard deviations of the tolerances in the stack, which equals the standard deviation of the assembly tolerance, hence its other name, *Root Sum Square* or RSS model. This can be represented by:

$$\left[\sum_{i=1}^{n} \sigma_i^2 \right]^{0.5} \leq \sigma_a \qquad (3.3)$$

where:

σ_a = assembly tolerance standard deviation

σ_i = ith component tolerance standard deviation.

The statistical model is potentially unappealing to designers because a defective assembly can result even if all components are within specification, although the probability of this occurring may be low. The worst case model is, therefore, more popular as a safeguard (Gerth, 1997), although it has been argued that it results in tighter tolerances that are often ignored by manufacturing when the design goes into production.

From the above considerations and models, we will now develop the relationships used in the *CAPRAtol* methodology. The tolerance model developed in addressing the assembly stack problem is based on the statistical model in equation 3.2, which is generally an accurate predictor (Wu *et al.*, 1988).

3.4 A methodology for assembly stack analysis

3.4.1 Application of the process capability estimates from CA

In Chapter 2, the Component Manufacturing Variability Risk, q_m, was effectively used to predict the process capability measures, C_{pk} and C_p, for individual component tolerances. This risk index therefore becomes useful in the allocation of capable tolerances and analysis of their distributions in the assembly stack problem. The key element of the CA methodology for determining the tolerance risk is the use of the process capability maps for the manufacturing processes in question. Figure 3.2 shows a process capability map for turning/boring and the equations modelling the risk contours. The contours indicate the level of risk associated with the achievement of a tolerance on a dimensional characteristic. The valid tolerance range of a production operation represents the accuracy improving capability of a production operation. Within this valid range, a tighter tolerance or higher accuracy demand leads to higher manufacturing costs, and a looser tolerance or lower accuracy demand leads to lower manufacturing costs (Dong, 1997).

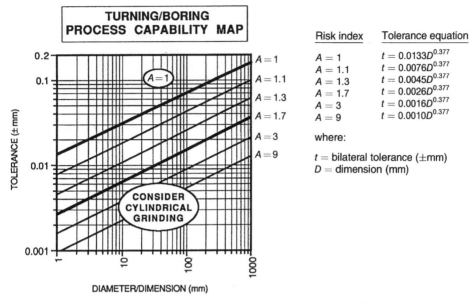

Risk index	Tolerance equation
$A = 1$	$t = 0.0133D^{0.377}$
$A = 1.1$	$t = 0.0076D^{0.377}$
$A = 1.3$	$t = 0.0045D^{0.377}$
$A = 1.7$	$t = 0.0026D^{0.377}$
$A = 3$	$t = 0.0016D^{0.377}$
$A = 9$	$t = 0.0010D^{0.377}$

where:

t = bilateral tolerance ($\pm$mm)
D = dimension (mm)

Figure 3.2 Equations describing levels of processing risk for turning/boring

By the nature of their modelling then, the maps are a relative cost/risk indicator with regards to the tolerance specified, and could be useful when optimizing tolerances. This omits the necessity of cost equations with difficult to ascertain constants and complicated optimization techniques to simplify the synthesis of tolerance allocation. However, economic and technical indicators for the initial selection of the manufacturing process to be used would need to be established at the design stage. A methodology for selecting manufacturing processes and costing designs can be found in Swift and Booker (1997).

The two most commonly used process capability indices used are C_{pk} and C_p, as discussed in Appendix II. These are considered to be adequate for design estimates and are easily tractable mathematically. As already discussed in Chapter 2, empirical relationships exist which relate q_m to C_{pk} and q_m to C_p. The equations that model these relationships are given below, together with the statistical definitions of C_{pk} and C_p respectively:

$$C_{pk} \approx \frac{4}{q_m^2} \qquad (3.4)$$

and

$$C_{pk} = \frac{|\mu - L_n|}{3\sigma} \qquad (3.5)$$

$$C_p \approx \frac{4}{q_m^{4/3}} \qquad (3.6)$$

and

$$C_p = \frac{t}{3\sigma} \tag{3.7}$$

where:

μ = mean

L_n = nearest tolerance limit

q_m = component manufacturing variability risk

σ = standard deviation

t = bilateral tolerance.

It is difficult to determine $|\mu - L_n|/3\sigma$ in equation 3.5 without statistical process data, which is unknown at the design stage unless similar characteristics have been manufactured and inspected for past products. C_{pk} for a shifted distribution tends to C_p for a centred or symmetrical distribution about its tolerance range. Therefore, assuming C_{pk} approaches C_p, and combining equations 3.4 and 3.7 gives:

$$\frac{4}{q_m^2} = \frac{t}{3\sigma} \tag{3.8}$$

In the specific case to the ith component dimension:

$$\sigma_i' = \frac{t_i q_{mi}^2}{12} \tag{3.9}$$

The $'$ relates to the fact that this is not the true standard deviation, but an estimate to measure the potential shift in the distribution.

The standard deviation multiplier, z, is the ratio of the tolerance and standard deviation, for one half of the distribution in this case:

$$z = \pm \frac{t}{\sigma} \tag{3.10}$$

Therefore, combining equations 3.9 and 3.10 gives an estimate for z_i in equation 3.2, and for bilateral tolerances for the specific case of the ith component tolerance where shift is anticipated, this gives:

$$z_i' = \pm \frac{12}{q_{mi}^2} \tag{3.11}$$

It follows for bilateral tolerances with no shift anticipated, from equation 3.6, that:

$$\sigma_i = \frac{t_i q_{mi}^{4/3}}{12} \tag{3.12}$$

and

$$z_i = \pm \frac{12}{q_{mi}^{4/3}} \tag{3.13}$$

Equation 3.12 is the best estimate for the standard deviation of the distribution as determined by CA.

3.4.2 Model for centred distributions

We now have a means of predicting the standard deviation multiplier z_i which can be used in equation 3.2. However, z_a, the assembly tolerance standard deviation multiplier, must be estimated before this equation can be satisfied. This is achieved by setting a capability requirement. The level of capability required typically by industry is $C_p = 2$ (Harry and Stewart, 1988; O'Connor, 1991) which equates to 0.002 parts-per-million (ppm) (see Appendix II for a relationship between C_p, C_{pk} and ppm). Note, this value is with no failure severity taken into account, but is a 'blanket' target value difficult to realize in practice. It follows then for the overall assembly process capability for a bilateral assembly tolerance can be given from equation 3.7 as:

$$C_{pa} = \frac{t_a}{3\sigma} \tag{3.14}$$

and

$$z_a = \pm 3C_{pa} \tag{3.15}$$

Substituting equations 3.13 and 3.15 into equation 3.2 gives the *CAPRAtol* tolerance stack model for bilateral tolerances with no shift assumed:

$$3C_{pa}\left[\sum_{i=1}^{n}\left(\frac{t_i q_{mi}^{4/3}}{12}\right)^2\right]^{0.5} \leq t_a \tag{3.16}$$

where:

t_i = bilateral tolerance for ith component characteristic

t_a = bilateral tolerance for assembly

q_{mi} = component manufacturing variability risk for ith characteristic

C_{pa} = required process capability for the assembly tolerance (centred).

3.4.3 Model for shifted distributions

Other factors can further enhance equation 3.16, such as factors to account for the type of distributions anticipated and the shift in the component distributions from the target. When components with shifted distributions are assembled, a large percentage of rejects would result if a model which does not effectively handle non-symmetrical distributions is used (Chase and Greenwood, 1988; Lin et al., 1997), particularly when one component distribution is dominant in its variance contribution as shown in Figure 3.1(a). For example, plotting C_p against C_{pk} for the components analysed in the empirical study in Chapter 2 gives Figure 3.3. This indicates that the distributions have an offset approaching ± 1.5 standard deviations, where in fact $C_{pk} = C_p - 0.5$ (Harry, 1987), but on average an offset determined by the equation:

$$C_{pk} = 0.93C_p - 0.19 \tag{3.17}$$

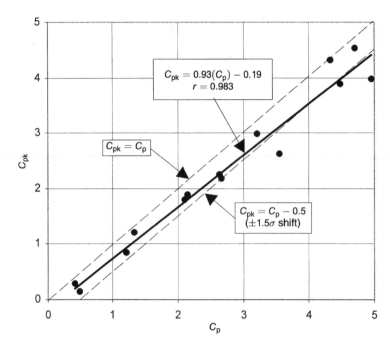

Figure 3.3 Empirical relationship between C_p and C_{pk} showing degree of process shift expected for the components analysed

Therefore, C_{pk} as predicted by q_m can be used to measure the potential shift in component tolerance distributions due to its formulation.

To determine z_a used in equation 3.2, the CA methodology is used in determining a process capability target for the assembly tolerance. Figure 3.4 shows the Conformability Map used in CA to assess the failure costs associated with a particular design characteristic and FMEA Severity Rating (S). Remember, in the acceptable design region a relationship exists between C_{pk} and FMEA (S). For example, at FMEA (S) = 6, you would expect a process capability index of $C_{pk} \approx 1.5$ (or approximately 3 ppm) to be met in order to keep the failure costs to a minimum of 0.01% of the product cost. This is the typical level of capability set by Motorola (Harry and Stewart, 1988), where $C_{pk} = 1.5$ (or ± 1.5 standard deviation shift for $C_p = 2$).

From equation 3.15, an estimate for a shifted version of z_a is given by:

$$z'_a = \pm 3 C_{pka} \tag{3.18}$$

Since the poorest performance of the assembly distribution would occur when shifted, C_{pk} values rather than C_p values are better design targets. The C_{pk} value can be used as a target for the assembly based on the severity of application and minimum failure cost of 0.01% of the total product cost as determined by the Conformability Map. If only capable solutions are to be generated, which have a minimum process capability index of $C_{pk} = 1.33$ (or 30 ppm) for both component and assembly distributions, then the number of components in the assembly stack can be as low as three using the proposed statistical model (Harry and Stewart, 1988). The overall requirement is

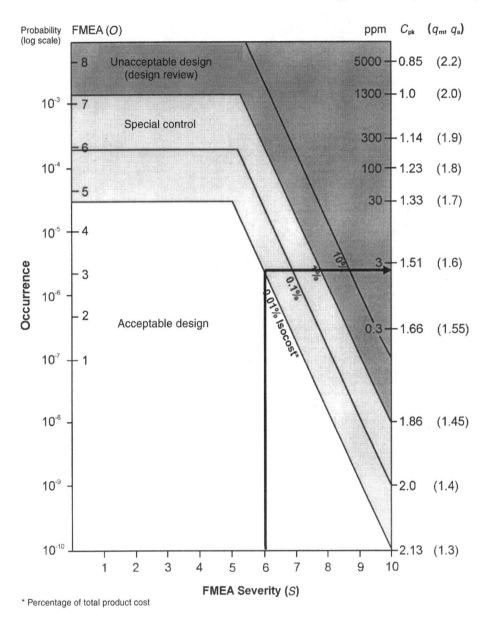

Figure 3.4 Conformability Map

that the predicted C_{pk} must be greater or equal to that determined from its FMEA Severity Rating (S) to be capable.

The bilateral tolerance stack model including a factor for shifted component distributions is given below. It is derived by substituting equations 3.11 and 3.18 into equation 3.2. This equation is similar to that derived in Harry and Stewart (1988), but using the q_m estimates for C_{pk} and a target C_{pk} for the assembly tolerance from the Conformability Map based on the FMEA Severity Rating (S):

$$3C_{pka}\left[\sum_{i=1}^{n}\left(\frac{t_i q_{mi}^2}{12}\right)^2\right]^{0.5} \leq t_a \tag{3.19}$$

where:

t_i = bilateral tolerance for ith component characteristic

t_a = bilateral tolerance for assembly

q_{mi} = component manufacturing variability risk for ith characteristic

C_{pka} = required process capability for the assembly tolerance (shifted).

3.5 Application issues

A flow chart for the tolerance stack methodology *CAPRAtol* is shown in Figure 3.5. Elements of FMEA, CA, process selection methodology, assembly sequence diagrams (through DFA techniques or CA) and, of course, adequate tolerance stack models, should be used in order to provide a complete solution to the assembly stack problem. Additionally, an understanding of geometric tolerancing, process capability indices and selection of key characteristics is useful (Leaney, 1996a).

Initially, it is recommended that an assembly sequence for the tolerance stack design be developed and that any customer specifications be noted, typically the final assembly tolerance, potential failure mode(s) and the FMEA Severity Rating (S). Once an assembly tolerance has been assigned and the level of capability determined from the Conformability Map, design tolerances for each component in the stack can be assigned. These are then optimized between acceptable risk values initially, which are in the capable region of the process (and 'acceptable' production cost) by running a simple routine. Of course, the use of computers in this respect greatly speeds up the time for a solution; however, a paper-based analysis is also possible as shown later. If the assembly tolerance cannot be met by optimization, then more capable processes will need to be investigated by the designer. These candidate processes could be a secondary operation for increased accuracy, for example turning/boring followed by cylindrical grinding, or a completely different process based on some technical, economic or business requirement. The importance of having adequate manufacturing knowledge and component costing processes is further emphasized.

If optimization is achieved using the processes chosen, then the standard deviation estimated for each component tolerance can be compared to the required assembly standard deviation to see if overall capability on the assembly tolerance has been achieved. If excessive variability is estimated at this stage on one or two characteristics, then redesign will need to be performed. Guidance for redesign can be simplified by using *sensitivity analysis*, used to estimate the percentage contribution of the variance of each component tolerance to the assembly variance, where the variance equals σ^2. It follows that from equation 3.3:

$$1 = \left(\frac{\sigma_1^2}{\sigma_a^2}\right) + \left(\frac{\sigma_2^2}{\sigma_a^2}\right) + \cdots + \left(\frac{\sigma_n^2}{\sigma_a^2}\right) \tag{3.20}$$

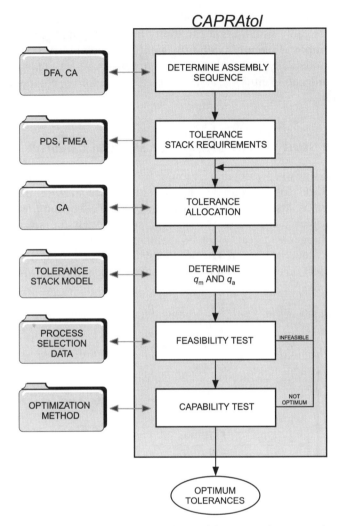

Figure 3.5 Elements of the *CAPRAtol* methodology (adapted from Lee and Woo, 1990)

Each term when multiplied by 100 gives the percentage contribution of each component's variance, $1, 2, \ldots, n$. From this and other information in Pareto chart form (such as the standard deviation multiplier, z), the designer can quickly focus on the problem components in terms of dominant variation, lack of capability, need for SPC or supplier dialogue. As a design evaluation package, *CAPRAtol* is able to quickly assess whether a design is going to be capable, and if not what components require redesigning.

In addition to understanding the statistical tolerance stack models and the FMEA process in developing a process capable solution, the designer should also address the physical assembly aspects of the tolerance stack problem. Any additional failure costs determined using CA are independent of whether the tolerances assigned to the assembly stack are capable or not. As presented in Chapter 2, the Component

Assembly Variability Risks Analysis is key to better understanding the effects of a component's assembly situation on variability by quantifying the risks that various assembly operations inherently exhibit. By identifying components with high assembly risks and potentially high failure costs, further design effort is highlighted and performed in order to identify the associated tolerances for the component's optimal fit and function.

3.6 Case study – revisiting the solenoid design

A familiar case study is presented next to illustrate the use of the key elements of the *CAPRAtol* methodology. Figure 3.6 shows the tolerance stack on the solenoid end assembly design as first encountered in Chapter 2. The key requirement was that the plunger displacement, from the sealing face through the solenoid tolerance stack to the plunger end seal, must be within a tolerance of ± 0.2 mm, otherwise fuel flow restriction could occur. The product will be in the warranty return category as it has little effect on user safety if it fails in service, which relates to an FMEA Severity Rating $(S) = 5$.

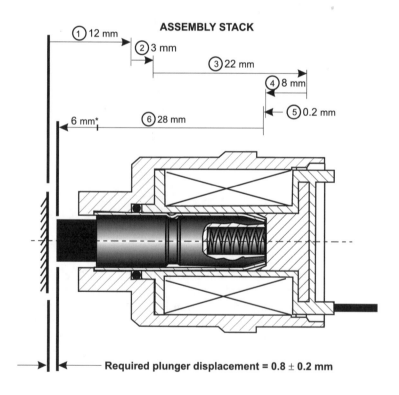

*The 6 mm end seal dimension will be used in the analysis, not 28 mm

Figure 3.6 Solenoid end assembly design

Table 3.1 Results table for a paper-based analysis of the solenoid assembly stack design

Number	m_p	g_p	Design tolerance, t ($\pm$mm)	q_m	σ' (mm)% $(t \cdot q_m^2/12)$	Tolerance contribution (%) $(t^2/t_a^2) \times 100$	Variance contribution (%) $(\sigma'^2/\sigma_a^2) \times 100$
1	1.2	1	0.010	1.7	0.002	0.06	0.04
2	1.2	1.1	0.250	1.7	0.060	37.55	37.70
3	1	1	0.085	1.7	0.020	4.34	4.19
4	1.2	1.7	0.300	1.7	0.072	54.07	54.28
5	1	1	0.004	1.7	0.001	0.01	0.01
6	1	1	0.080	1.7	0.019	3.85	3.78

Assembly tolerance, $t_a = \pm 0.408$ mm
Assembly standard deviation, $\sigma_a = 0.098$ mm
Assembly tolerance, $C_{pk} [\approx t_a/(3\sigma_a)] = 1.39$

The solenoid design will first be analysed in a 'paper-based' approach, followed by a demonstration of the results from the *CAPRAtol* software package[*] which takes advantage of the computer coded algorithms to find an optimized solution. Included as part of the problem this time is the dimensional characteristic on the fuel port block of 12 mm, originally set by a supplier.

3.6.1 Paper-based analysis

This entails assigning design tolerances for each characteristic in the assembly stack based on those given by the process capability maps (Appendix IV), but including the effects of processing the material and geometry of each through the Component Manufacturing Variability Risk Analysis, q_m. The reader is referred to Chapter 2 for a detailed explanation of this part of the analysis. For example, the dimension of 12 mm for characteristic number 1, we can refer to the turning/boring map (Figure 3.2) and locate a tolerance that gives a risk on the limits of feasibility, i.e. $A = 1.7$. This tolerance is ± 0.007 mm from the map. Including the effects of processing mild steel, however ($m_p = 1.2$), this increases the design tolerance to ± 0.01 mm. The final tolerance allocated would be ± 0.01 mm to give $A = 1.7$ and therefore $q_m = 1.7$, because no account of the surface roughness to process risk is made. Similarly, for the other five characteristic dimensions in the assembly stack we can allocate a design tolerance from the respective maps and analysis of its variability risks. This is summarized in Table 3.1.

The final assembly tolerance, t_a, is the statistical sum of the component tolerances, t_i, and can be derived from equation 3.2 due to the fact that the standard deviation multipliers, z_i, for each tolerance are the same, therefore:

$$\sum_{i=1}^{n} t_i^2 \leq t_a^2 \qquad (3.21)$$

[*] The *CAPRAtol* demonstration software is available from the authors on request.

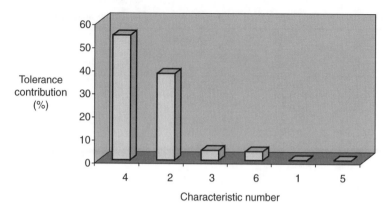

Figure 3.7 Pareto chart showing tolerance contribution of each characteristic to the final assembly tolerance (for the paper-based analysis)

This equation allows us to determine each component tolerance's contribution to the final assembly tolerance by using a similar approach to that presented for sensitivity analysis. It follows from equation 3.21:

$$1 = \left(\frac{t_1^2}{t_a^2}\right) + \left(\frac{t_2^2}{t_a^2}\right) + \cdots + \left(\frac{t_n^2}{t_a^2}\right) \tag{3.22}$$

Although an estimate for the capability of the assembly determined from equation 3.7 is adequate at $C_{pk} = 1.39$ for an FMEA Severity Rating $(S) = 5$, the final assembly tolerance is ± 0.408 mm, greater than the target of ± 0.2 mm. A Pareto chart showing the results of the tolerance sensitivity analysis is shown in Figure 3.7. It indicates that component characteristics 4 and 2 have the greatest contribution. Redesign effort would then focus on these two characteristics to reduce their overall tolerance contribution, as was concluded in the worst case analysis in Chapter 2.

Additionally, we can determine the variance contribution of each component tolerance to the final assembly tolerance using sensitivity analysis, the corresponding chart shown in Figure 3.8. A nominal target value is indicated at $100\%/n$, where 'n' is the number of components in the stack. Figure 3.8 shows this in Pareto chart form again, and reaches the same conclusions above, as expected.

On the generation of a redesign, a similar procedure would be used as that described above to iterate a process capable solution where the target tolerance of ± 0.2 mm and an overall $C_{pk} \geq 1.33$ are achieved. However, the process is speeded up greatly through the use of software incorporating the procedures and knowledge described above.

3.6.2 *CAPRAtol* software analysis

The dimensions of each component in the tolerance stack design (Figure 3.6), their respective materials and manufacturing processes are inputted in sequence into *CAPRAtol*. Also required are the target assembly tolerance and the FMEA Severity

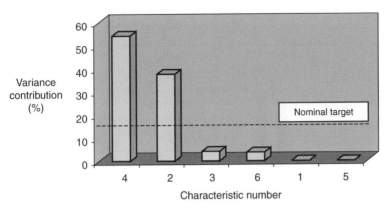

Figure 3.8 Pareto chart showing variance contribution of each characteristic to the final assembly variance (for the paper-based analysis)

Rating (S) to give the target C_{pk}. The program initially calculates a primer tolerance, t_i^p (the statistical sum of the primer tolerances equals the assembly tolerance, t_a) and allocates this to each dimension. The primer tolerances are calculated using equation 3.23, which is derived from equation 3.21, They are used to start the optimization routine and their respective risk values 'A' are calculated from the process capability maps:

$$t_i^p = \frac{t_a}{\sqrt{n}} \qquad (3.23)$$

where:

   t_i^p = bilateral primer tolerance for ith component characteristic

   t_a = bilateral tolerance for assembly

   n = number of component tolerances in the assembly stack.

Figure 3.9 shows the results of the optimization routine for the tolerances. For capability at this first level to be realized, all the component risks must be below the line on completion of the optimization routine, and subsequently the capability required with this design configuration cannot be achieved.

It is evident from Figure 3.9 that two out the six components' characteristics are preventing optimization (as was determined in the paper-based analysis), and therefore redesign effort here is required, specifically the base tolerance of the solenoid body and the pole thickness, which are both impact extruded. A feature of the design related to these two tolerances is the bobbin dimensional tolerance, which sets the position of the pole from the solenoid base. Further redesign data is shown on the final *CAPRAtol* screen in Figure 3.10. Redesign data includes a basic estimate for C_{pk}, and an assembly tolerance calculated from the process capability maps that would be achievable for the given design parameters.

A redesign solution of the tolerance stack is shown in Figure 3.11. It involves a small design alteration (circled) which eliminates the plastic bobbin from the tolerance stack by machining a shoulder on the inside of the solenoid body, up to which the pole piece is precisely located. Also, secondary machining processes are

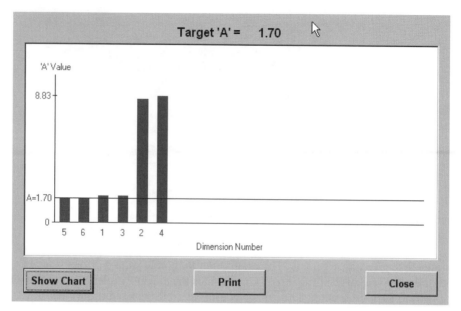

Figure 3.9 Chart showing that capable tolerances cannot be optimized for solenoid tolerance stack design

carried out on the components with the least capability as determined from Figure 3.9. As there were machining operations on these two components anyway, the component cost increase was small. Inputting the design parameters into *CAPRAtol* and proceeding with the optimization routine, we find the largest tolerances for the

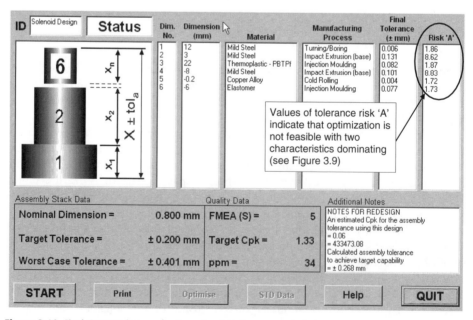

Figure 3.10 Final *CAPRAtol* screen for the solenoid tolerance stack design

ASSEMBLY STACK

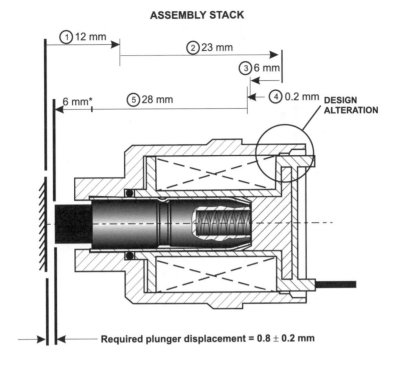

* The 6 mm end seal dimension will be used in the analysis, not 28 mm

Figure 3.11 Solenoid end assembly redesign

least risk are optimized to near equal value as shown in Figure 3.12. The risk values determined at this stage are very low, close to unity in fact, and the situation looks more promising.

Figure 3.13 shows the effects of the material processing and geometry risks for each component from the component manufacturing variability risk, q_m, and these are taken into consideration in the calculation of the final estimates for C_{pk} and C_p for each tolerance.

Part of the design information provided by the software is the standard deviation multiplier, z, for each component tolerance shown in Figure 3.14 in Pareto chart form. Additionally, sensitivity analysis is used to provide a percentage contribution of each tolerance variance to the final assembly tolerance variance as shown in Figure 3.15.

It is evident that characteristic number 5 has the largest contribution (89.8%) and it is likely that if this is shifted, then the final assembly distribution will be shifted from its target value, suggesting a need for SPC in production. However, this redesign solution is very capable (as shown in the final $CAPRAtol$ screen in Figure 3.16) with $C_{pk} = 3.67$, which theoretically relates to no failures. The actual process capability of the assembly tolerance will, in fact, lie somewhere between the two values calculated for C_{pk} and C_p, but using the former we are considering the possibility of shift throughout its life-cycle.

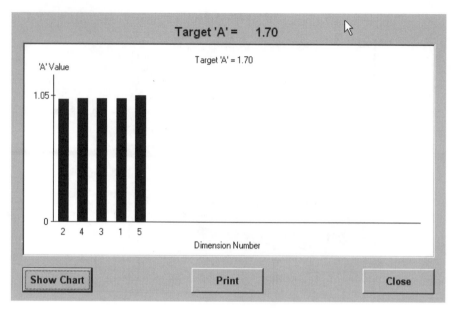

Figure 3.12 Chart showing optimized tolerances for solenoid tolerance stack redesign

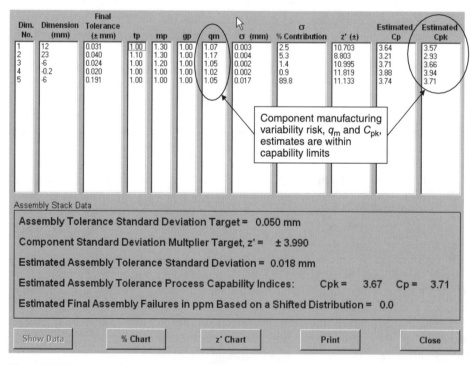

Figure 3.13 Data screen for solenoid tolerance stack redesign

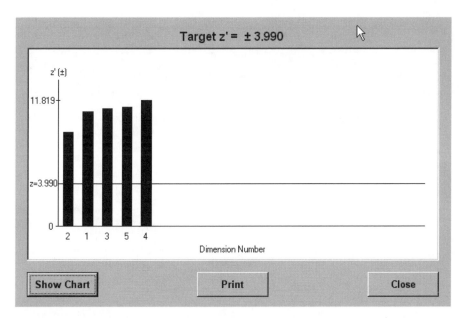

Figure 3.14 Chart of standard deviation multiplier values, z, for each tolerance

The final tolerances the designer would allocate to the component dimensions are also shown in Figure 3.16. The tolerance values are given to three decimal places which, if required for practical use, can be rounded off with minimal effect to the overall assembly tolerance, for example ± 0.191 mm to ± 0.190 mm.

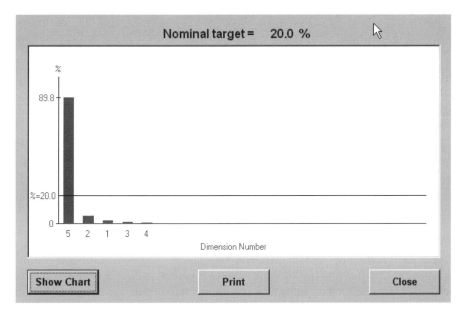

Figure 3.15 Chart showing variance contribution of each tolerance to the final assembly variance

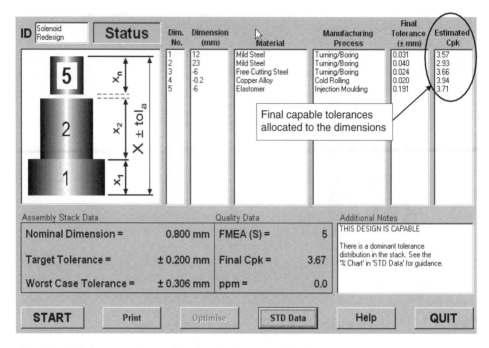

Figure 3.16 Final *CAPRAtol* screen for solenoid tolerance stack redesign

A comparative worst case assembly tolerance based on these capable tolerances optimized using the statistical approach is shown as ±0.306 mm in the lower left-hand corner of Figure 3.16, which is greater than the target of ±0.2 mm. A thorough analysis of the redesign based on the worst case model was presented in Chapter 2. However, the assumption that each component dimension was at its maximum or minimum limit was clearly not the case and the level of variability experienced differed throughout the components in the assembly stack. In practice, the worst case approach gives little indication of the impact of the component tolerance distributions on the final assembly tolerance distribution. Leaving out this data can have as great an impact as leaving out the tolerances in the first place (Hopp, 1993).

The inadequacy of the worst case model is evident and the statistical nature of the tolerance stack is more realistic, especially when including the effects of shifted distributions. This has also been the conclusion of some of the literature discussing tolerance stack models (Chase and Parkinson, 1991; Harry and Stewart, 1988; Wu *et al.*, 1988). Shifting and drifting of component distributions has been said to be the chief reason for the apparent disenchantment with statistical tolerancing in manufacturing (Evans, 1975). Modern equipment is frequently composed of thousands of components, all of which interact within various tolerances. Failures often arise from a combination of drift conditions rather than the failure of a specific component. These are more difficult to predict and are therefore less likely to be foreseen by the designer (Smith, 1993).

3.7 Summary

Recent developments in designing capable tolerance stacks have been reviewed and the application has been demonstrated via an industrial case study. The data used to determine the tolerances and standard deviations has a realistic base and can provide reliable results in the design problem of allocating optimum capable tolerances in assembly stacks. The *CAPRAtol* method uses empirical capability data for a number of processes, including material and geometry effect. Component tolerances with the greatest capability are optimized for the given functional assembly tolerance including the effects of anticipated process shift. The use of the Conformability Map for setting capability targets from FMEA inputs is pivotal in the generation of a capable design solution.

The inadequacy of the worst case approach to tolerance stack design compared to the statistical approach is evident, although it still appears to be popular with designers. The worst case tolerance stack model is inadequate and wasteful when the capability of each dimensional tolerance is high ($C_{pk} \geq 1.33$). Some summarizing comments on the two main approaches are given below.

The 'worst case' tolerance stack approach is characterized by the following:

- Simple to perform
- Assumes tolerance distribution on maximum or minimum limit
- Little information generated for redesign purposes
- Popular as a safeguard, leading to unnecessarily tight tolerances and, therefore, increased costs.

The 'statistical' tolerance stack approach is characterized by:

- More difficult mathematically (computer necessary)
- Assumes tolerances are random variables
- Opportunities for optimization of tolerances in the assembly
- Can perform sensitivity analysis for redesign purposes
- Can include effects of shifting and drifting of component tolerances
- More realistic representation of actual situation.

We must promote the use of statistical methods with integrated manufacturing knowledge, through user-friendly platforms in order to design capable products. Comparing and evaluating assembly stacks as shown in the case study above is an essential way of identifying the capability of designs and indicating areas for redesign.

4

Designing reliable products

4.1 Deterministic versus probabilistic design

For many years, designers have applied so called *factors of safety* in a deterministic design approach. These factors are used to account for uncertainties in the design parameters with the aim of generating designs that will ideally avoid failure in service. Load and stress concentrations were the unknown contributing factors and this led to the term *factor of ignorance* (Gordon, 1991). The factor of safety, or deterministic approach, still predominates in engineering design culture, although the statistical nature of engineering problems has been studied for many years. Many designs are still based on past experience and intuition, rather than on thorough analysis and experimentation (Kalpakjian, 1995).

In general, the deterministic design approach can be shown by equation 4.1. Note that the stress and strength are in the same units:

$$\frac{S}{FS} > L \tag{4.1}$$

where:

$$L = \text{loading stress}$$

$$S = \text{material strength}$$

$$FS = \text{factor of safety.}$$

The factors of safety were initially determined from the sensitivity experienced in practice of a part made from a particular material to a particular loading situation, and in general the greater the uncertainties experienced, the greater the factor of safety used (Faires, 1965). Table 4.1 shows recommended factor of safety values published 60 years apart, first by Unwin (*c*.1905) and by Faires (1965). They are very similar in nature, and in fact the earlier published values are lower in some cases. As engineers learned more about the nature of variability in engineering parameters, but were unable to quantify them satisfactorily, it seems that the factor of safety was increased to accommodate these uncertainties. It is also noticeable that the failure criterion of tensile fracture seems to have been replaced by

Table 4.1 Factors of safety for ductile and brittle materials and various loading conditions (values shown in brackets from 1905, without brackets from 1965) (Su = ultimate tensile strength, Sy = yield strength)

Type of loading	Steel (ductile metals)		Cast iron (brittle metals)
	Based on Su	Based on Sy	Based on Su
Dead load	3 to 4	1.5 to 2	5 to 6
	(3)		(4)
Repeated one direction/mild shock	6	3	7 to 8
	(5)		(6)
Repeated reversed/mild shock	8	4	10 to 12
	(8)		(10)
Shock	10 to 15	5 to 7	15 to 20
	(12)		(15)

ductile yielding (due to the omission of such values $c.1905$), the yield stress being the preferred criterion of failure. This is because in general most machine parts will not function satisfactorily after permanent deformation caused by the onset of yielding.

The factor of safety had little scientific background, but had an underlying empirical and subjective nature. No one can dispute that at the time that stress analysis was in its infancy, this was the best knowledge available, but they are still being applied today! Factors of safety that are recommended in recent literature range from 1.25 to 10 for various material types and loading conditions (Edwards and McKee, 1991; Haugen, 1980).

Figure 4.1 gives an indication that engineers in the 1950s were beginning to think differently about design with the introduction of a 'true' *margin of safety*, and a probabilistic design approach was being advocated. It shows that the design problem was multifactored and variability based. With the increasing use of statistics in engineering around this time, the theories of probabilistic design and reliability were to become established methods in some sectors by the 1960s.

The deterministic approach is not very precise and the tendency is to use it very conservatively resulting in overdesigned components, high costs and sometimes ineffectiveness (Modarres, 1993). Carter (1986) notes that stress rupture was responsible for a sufficient number of failures for us to conclude that deterministic design does not always ensure intrinsic reliability, and that room for improvement still exists. Increasing demands for performance, resulting often in operation near limit conditions, has placed increasing emphasis on precision and realism (Haugen, 1980). There has been a great disenchantment with factors of safety for many years, mainly because they disregard the fact that material properties, the dimensions of the components and the externally applied loads are statistical in nature (Dieter, 1986). The deterministic approach is, therefore, not suitable for today's products where superior functionality and high customer satisfaction should be a design output. The need for more efficient, higher performance products is encouraging more applications of probabilistic methods (Smith, 1995).

Probabilistic design methods have been shown to be important when the design cannot be tested to failure and when it is important to minimize weight and/or cost (Dieter, 1986). In companies where minimizing weight is crucial, for example such as those in the aerospace industry, probabilistic design techniques can be found,

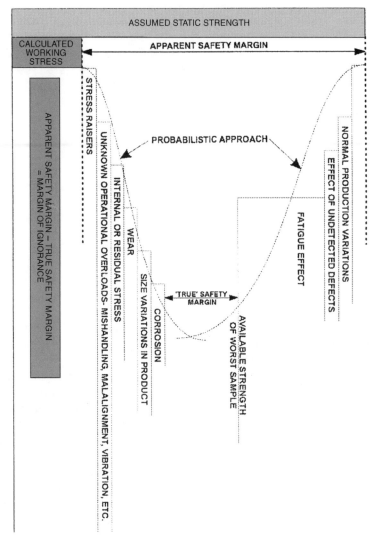

Figure 4.1 The 'true' margin of safety (adapted from Furman, 1981 and Nixon, 1958)

although NASA has found the deterministic approach to be adequate for some structural analysis (NASA, 1995).

Non-complex and/or non-critical applications in mechanical design can also make use of probabilistic design techniques and justify a more in-depth approach if the benefits are related to practitioners and customers alike. Surveys have indicated that many products in the industrial sector have in the past been overdesigned (Kalpakjian, 1995). That is, they were either too bulky, were made of materials too high in quality, or were made with unwarranted precision for the intended use. Overdesign may result from uncertainties in design calculations or the concern of the designer and manufacturer over product safety in order to avoid user injury or death, and can add significantly to the product's cost. Achieving reliability by overdesign, then, is not an economic proposition (Carter, 1997). The use of probabilistic techniques could save money in this way as they provide a basis for a trade-off between design and cost factors (Cruse, 1997a). If they are rejected, however, only the conventional deterministic design approach remains according to which factors of safety are selected based on engineering experience and common sense (Freudenthal *et al.*, 1966).

The random nature of the properties of engineering materials and of applied loads is well known to engineers. Engineers will be familiar with the typical appearance of sets of strength data from tensile tests in which most of the data values congregate around the mid-range with decreasingly fewer values in the upper and lower tails on either side of the mean. For mathematical tractability, the experimental data can be modelled with a Probability Density Function (PDF) or *continuous distribution* that will adequately describe the pattern of the data using just a single equation and its related parameters. In terms of probabilistic design then, the reliability of a component part can based on the interference of its inherent material strength distribution, $f(S)$, and loading stress distribution, $f(L)$, where both are random variables. When we say a function is random, this means that it cannot be precisely predicted in advance. Where stress exceeds strength, failure occurs, the reliability of the part, R, being related to this failure probability, P, by equation 4.2:

$$P(S > L) = R \tag{4.2}$$

Figure 4.2 shows the probabilistic design concept in comparison to the deterministic approach. Not fully understanding the variable nature of the stress and strength, the designer using the deterministic approach would select a suitable factor of safety which would provide adequate separation of the nominal stress and strength values (for argument's sake). Selecting too high a factor of safety results in overdesign; too low and the number of failures could be catastrophically high. In reality, the interference between the actual stress and strength distributions dictates the performance of the product in service and this is the basis of the probabilistic design approach. The degree of interference and hence the failure probability depends on (Mahadevan, 1997):

- The relative position of the two distributions
- The dispersions of the two distributions
- The shapes of the two distributions.

The movement from the deterministic design criteria as described by equation 4.1 to the probability based one described by equation 4.2 has far reaching effects on design (Haugen, 1980). The particular change which marks the development of modern engineering reliability is the insight that probability, a mathematical theory, can be utilized to quantify the qualitative concept of reliability (Ben-Haim, 1994).

The development of the probabilistic design approach, as already touched on, includes elements of probability theory and statistics. The introductory statistical methods discussed in Appendix I provide a useful background for some of the more advanced topics covered next. Wherever possible, the application of the statistical methods is done so through the use of realistic examples, and in some cases with the aid of computer software.

(a) Factor of safety too high leading to overdesign

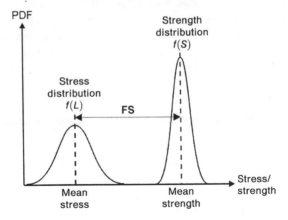

(b) Factor of safety too low leading to a high failure probability

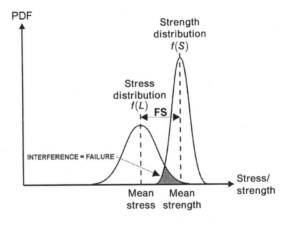

(c) Factor of safety adequate

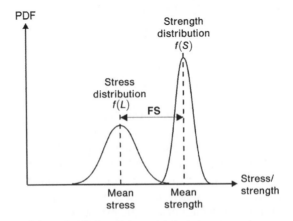

Figure 4.2 Comparison of the probabilistic and deterministic design approaches

4.2 Statistical methods for probabilistic design

4.2.1 Modelling data using statistical distributions

A key problem in probabilistic design is the generation of the statistical distributions from available information about the random variables (Siddal, 1983). The random variable may be a set of real numbers corresponding to the outcome of a series of experiments, tests, data measurements, etc. Usually, information relating to these variables for a particular design is not known beforehand. Even if similar design cases are well documented, there are always particular circumstances affecting the distribution functions (Vinogradov, 1991).

Three statistical distributions that are commonly used in engineering are the Normal (see Figure 4.3(a)), Lognormal (see Figure 4.3(b)) and Weibull, both 2- and 3-parameter (see Figure 4.3(c) for a representation of the 2-parameter type). Each figure shows the characteristic shapes of the distributions with varying parameters for an arbitrary variable. The area under each distribution case is always equal to unity, representing the total probability, hence the varying heights and widths.

Typical applications for the three main distributions have been cited:

- *Normal* – Tolerances, ultimate tensile strength, uniaxial yield strength and shear yield strength of some metallic alloys
- *Lognormal* – Loads in engineering, strength of structural alloy materials, fatigue strength of metals
- *Weibull* – Fatigue endurance strength of metals and strength of ceramic materials.

Other distributions highlighted as being important in reliability engineering are also given below. A summary of all of these distributions in terms of their PDF, notation and variate boundaries is given in Appendix IX. The reader interested in the properties of all the distributions mentioned is referred to Bury (1999).

- Maximum Extreme Value Type I
- Minimum Extreme Value Type I
- Exponential.

Lacking more detailed information regarding the nature of an engineering random variable, it is often assumed that its distribution can be represented by a Normal distribution (Rice, 1997). (The Normal distribution was initially discussed in detail in Appendix I.) The Normal is the most widely used of all distributions and through empirical evidence provides a good representation of many engineering variables and is easily tractable mathematically (Haugen, 1980). If the Normal distribution does not prove to be a good fit to the data, the question should be asked, 'Is the data dependable?' and if it is, 'Are there good reasons for using a different model?' There is no point in fitting a more sophisticated model to untrustworthy data because the end result might prove to be spectacular nonsense.

It has been argued that material properties such as the ultimate tensile strength have only positive values and so the Normal distribution cannot be the true distribution,

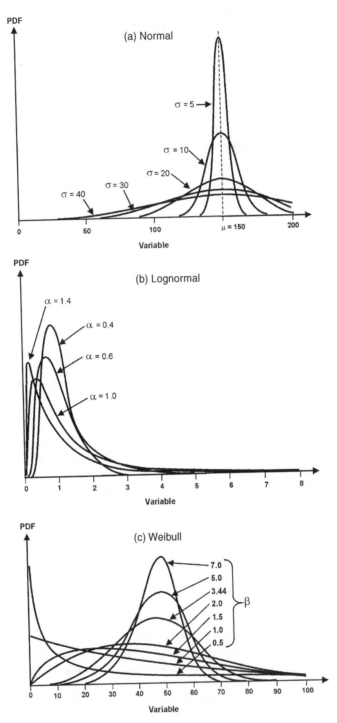

Figure 4.3 Shapes of the probability density function (PDF) for the (a) normal, (b) lognormal and (c) Weibull distributions with varying parameters (adapted from Carter, 1986)

because its range within the statistical model is from $-\infty$ to $+\infty$. However, when the coefficient of variation, C_v, <0.3 using the Normal distribution, the probability of negative values is negligible for strength data (Kapur and Lamberson, 1977). The Lognormal distribution on the other hand, does not admit variate values that are negative. This zero threshold effectively means that the population distribution must start at zero, which makes it useful for modelling some types of material properties as well as several loading conditions. A useful property of the Lognormal distribution is that there is very little difference between it and a Normal distribution when the Lognormal has a coefficient of variation, $C_v < 0.1$.

Data that is not evenly distributed is better represented by a skewed distribution such as the Lognormal or Weibull distribution. The empirically based Weibull distribution is frequently used to model engineering distributions because it is flexible (Rice, 1997). For example, the Weibull distribution can be used to replace the Normal distribution. Like the Lognormal, the 2-parameter Weibull distribution also has a zero threshold. But with increasing numbers of parameters, statistical models are more flexible as to the distributions that they may represent, and so the 3-parameter Weibull, which includes a minimum expected value, is very adaptable in modelling many types of data. A 3-parameter Lognormal is also available as discussed in Bury (1999).

The price of flexibility comes in the difficulty of mathematical manipulation of such distributions. For example, the 3-parameter Weibull distribution is intractable mathematically except by numerical estimation when used in probabilistic calculations. However, it is still regarded as a most valuable distribution (Bompas-Smith, 1973). If an improved estimate for the mean and standard deviation of a set of data is the goal, it has been cited that determining the Weibull parameters and then converting to Normal parameters using suitable transformation equations is recommended (Mischke, 1989). Similar estimates for the mean and standard deviation can be found from any initial distribution type by using the equations given in Appendix IX.

In order to accurately predict the reliability we must carefully establish $f(S)$ and $f(L)$. An important requirement is that the modelling distribution should closely represent the lower tail of the empirical distribution (Bury, 1975). It therefore becomes necessary to collect a great deal of data in order to arrive at a meaningful and adequate distributional model. This is usually not done. The economics require an approximate approach and in practice only sufficient observations are made to determine the mean and standard deviation of the stress and strength leading to the Normal distribution (Mischke, 1970). Therefore, the simplest and most common case has always been when stress and strength are normally distributed (Murty and Naikan, 1997; Vinogradov, 1991). If a complete theory of statistical inference is developed based on the Normal distribution alone, we have a system which may be employed quite generally, because other distributions can be transformed to approximate the Normal form (Mood, 1950). Although certainly not all engineering random variables are normally distributed, a Normal distribution is a good first approximation.

Ullman (1992) argues that the assumption that stress and strength are of the Normal type is a reasonable one because there is not enough data available to warrant anything more sophisticated. A problem with this is that the Normal distribution limits are $-\infty$ to ∞, which produces a conservative estimate for the failure probability when used to model stress and strength compared to other distributions (Haugen, 1982a). Real-life distributions have finite upper and lower limits, though precisely where these are located may be difficult to determine. For practical purposes, the Normal distribution is useful within the range of three standard deviations above and below the mean. Predictions based on extrapolation of the Normal beyond three standard deviations must be regarded as suspect, and this will affect the reliability prediction under some circumstances. The Normal stress–Normal strength distributions give the largest predicted failure probability for the static loading conditions. The Normal distribution, therefore, has an element of conservatism in being unbounded, and distributions such as the 3-parameter Weibull distribution may be better for predicting higher reliabilities (Haugen, 1980). A key problem is that if the use of incorrect distributions was made for stress or strength, the predicted reliability may make nonsense of reliability targets through which a reliable design could be identified.

Some important considerations in the use of statistical distributions have been highlighted, both in terms of the initial data and, more importantly, when modelling the stress and strength for determining the reliability. Stress–Strength Interference (SSI) analysis, which is the main technique used in this connection, will be discussed later.

4.2.2 Fitting distributions to data

The estimation of the mean and standard deviation using the moment equations as described in Appendix I gives little indication of the degree of fit of the distribution to the set of experimental data. We will next develop the concepts from which any continuous distribution can be modelled to a set of data. This ultimately provides the most suitable way of determining the distributional parameters.

The method centres around the *cumulative frequency* of the experimental data. A typical cumulative frequency plot from an arbitrary set of discrete data is shown in Figure 4.4. The horizontal axis is the *independent variable*, being the discrete variable value or the mid-class for the grouped data. The cumulative frequency on the vertical axis is generated by adding subsequent frequencies and is regarded as the *dependent variable*. The original histogram is shown superimposed under the cumulative frequency.

By plotting the cumulative frequency as a relative percentage of the total frequency of the data we get Figure 4.5. (Alternatively, the cumulative frequency can be displayed from 0 to 1.) Overlaid on the tops of the frequency bars is a curve that represents the cumulative function. This curve is drawn by hand in the figure, but the use of polynomial curve fitting software will yield more accurate results. This type of graph and its variants are used to determine the parameters of any distribution. For example, from the points on the x-axis corresponding to the percentile points $\sim$84.1% and $\sim$15.9% on the cumulative frequency (the percentage probabilities at ± 1 of the Standard Normal variate, z, from Table 1 in Appendix I) the standard deviation can be estimated as shown on Figure 4.5. The 50 percentile of the variate determines the median value of the data, which for a symmetrical distribution (such as the Normal) coincides with the mean.

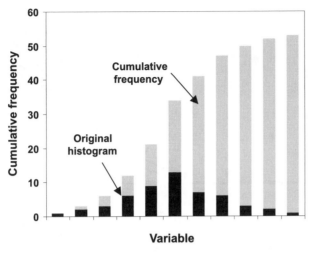

Figure 4.4 Cumulative frequency and histogram for grouped data set

Values for the Cumulative Distribution Function (CDF), given the notation $F(x)$, are generated by integrating the PDF for the distribution in question between the limits 0 and the variate of interest x (see Appendix IX). The value of $F(x)$ then represents the failure probability, P, at that point. Figure 4.6 shows the shape of the Normal CDF with different standard deviations for an arbitrary variable. The CDF equations for all but the Normal and Lognormal distributions are said to be in *closed form*, meaning the equation can be mathematically manipulated as a definite function rather than an integral. Although numerical techniques can be used to integrate the Normal and Lognormal PDFs to obtain the cumulative value of interest, the cumulative SND values as provided in Table 1 in Appendix I are commonly used for convenience.

In reality, it is impossible to know the exact cumulative failure distribution of the random variable, because we are taking only relatively small samples of the

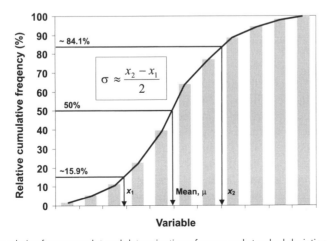

Figure 4.5 Cumulative frequency plot and determination of mean and standard deviation graphically

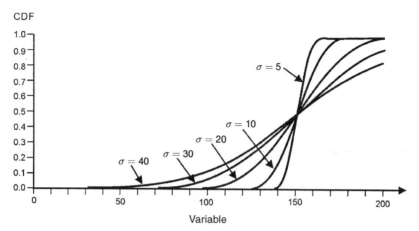

Figure 4.6 Shape of the Cumulative Distribution Function (CDF) for an arbitrary normal distribution with varying standard deviation (adapted from Carter, 1986)

population. We observe that 100% of the sample taken has failed, but this failure distribution does not necessarily match that of the entire population. We can make reasonable judgements as to what the population cumulative failure distribution plotting positions are through the use of empirically based cumulative *ranking equations*. By ranking the cumulative frequency on the vertical axis calculated from one of the many different types of ranking equation, an improved model for the cumulative function can be generated. See Appendix X for a list of the most commonly used ranking equations and the types of distribution they are typically applied to.

An efficient way of using all the information and judgement available in the estimation of the distribution parameters is the use of the 'linearized' cumulative frequency (Siddal, 1983). Essentially, this involves converting the non-linear equation describing the cumulative frequency into a linear one by suitably changing one or more of the axis variables. The mathematical process is called *linear rectification*. A straight line through the plotted data points can then be determined using the *least squares technique* (Burden and Faires, 1997) or 'by eye' to estimate the linear equation parameters, from which the distributional parameters can be determined. For example, converting the curve as shown in Figure 4.5 to a straight line in the form modelled by:

$$y = A0 + A1(x) \tag{4.3}$$

where $A0$ and $A1$ are the linear regression constants.

To determine the best fitting line through a set of data using the least squares technique is automatically performed on most commercial spreadsheet and curve fitting software packages and their use is particularly effective when large amounts of data must be processed. Some would argue the assertion that fitting a straight line by this method is better than fitting by eye. For any set of bi-variate x, y data there are always two regression equations, the regression of y on x and the regression of x on y. The corresponding regression lines both go through the centroid of the data, but at different angles. A true functional relation lies between the two. This is more important when the selection of the dependent and independent variable of the x, y data is difficult to determine.

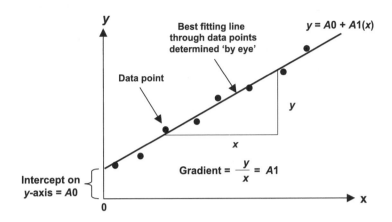

Figure 4.7 Determination of the linear regression equation manually

A reliable method of fitting a straight line by eye is by treating the plotted points in two halves and altering a rule line until it is median in both halves (i.e. equal number of points above and below the line). The determination of the linear regression equation can then be determined by taking the gradient of the line and intercept of the line on the y-axis as shown in Figure 4.7.

For example, to determine whether the Normal distribution is an adequate fit to a set of data requires linearizing the cumulative frequency on the y-axis by converting to the Standard Normal variate, z. To linearize for the Lognormal distribution, the cumulative frequency is converted as for the Normal distribution and the variable on the x-axis is converted to the Natural Logarithmic value (given the shorthand 'ln' as typically shown on calculators). A summary of the linear rectification equations and plotting positions for the common distributions are provided in Appendix X together with the equations for determining the distribution parameters from the linear regression constants $A0$ and $A1$.

The practical utilization of linear rectification is demonstrated later through a worked example. Fitting statistical distributions to sample data using the linear rectification method can be found in Ayyub and McCuen (1997), Edwards and McKee (1991), Kottegoda and Rosso (1997), Leitch (1995), Lewis (1996), Metcalfe (1997), Mischke (1992), Rao (1992), and Shigley and Mischke (1989).

Straight line plots of the cumulative function are commonly used, but there is no foolproof method that will guide the choice of the distribution (Lipson and Sheth, 1973). Additional *goodness-of-fit* tests including the *chi-squared* (χ^2) test and the *Kolmogorov–Smirnov* test are available (Ayyub and McCuen, 1997; Leitch, 1995; Mischke, 1992). The χ^2 test is not applicable when data is sparse ($N < 15$) and relies on grouping the data. This is because of the need to compare the estimated and observed frequencies for the experimental data. The Kolmogorov–Smirnov test statistic is determined from the difference between the observed and estimated cumulative frequencies. It is applicable to small samples and doesn't depend on grouping the data. In both cases, however, their effective use is restricted to the non-linearized domain (Ayyub and McCuen, 1997).

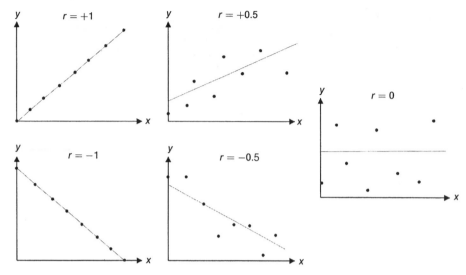

Figure 4.8 Correlation coefficient, r, for several relationships between x and y variables

An alternative method is to fit the 'best' straight line through the linearized set of data associated with distributional models, for example the Normal and 3-parameter Weibull distributions, and then calculate the *correlation coefficient*, r, for each (Lipson and Sheth, 1973). The correlation coefficient is a measure of the degree of (linear) association between two variables, x and y, as given by equation 4.4.

$$r = \frac{\sum x \cdot y - \left(\dfrac{\sum x \cdot \sum y}{N_\mathrm{p}} \right)}{\sqrt{\left(\sum x^2 - \dfrac{(\sum x)^2}{N_\mathrm{p}} \right) \cdot \left(\sum y^2 - \dfrac{(\sum y)^2}{N_\mathrm{p}} \right)}} \tag{4.4}$$

where:

$$N_\mathrm{p} = \text{number of data pairs.}$$

A correlation coefficient of 1 indicates that there is a very strong association between the two variables as shown in Figure 4.8. Lower values of 'r' indicate that the variables have less of an association; until at $r = 0$, no correlation between the variables is evident. A negative value indicates an inverse relationship. Therefore, the maximum value of correlation coefficient for each linear rectification model will give most appropriate distribution that fits the sample data.

The value of the correlation coefficient using the least squares technique and the use of goodness-of-fit tests (in the non-linear domain) together probably provide the means to determine which distribution is the most appropriate (Kececioglu, 1991). However, a more intuitive assessment about the nature of the data must also be made when selecting the correct type of distribution, for example when there is likely to be a zero threshold.

Having introduced the concepts of the correlation coefficient, it becomes straightforward to explain the more involved process of determining the parameters of the 3-parameter Weibull distribution. The procedure for the 3-parameter Weibull distribution is more complex, as you would expect, due to the distribution being modelled by three rather than two parameters. Essentially it requires the determination of the expected minimum value, xo, a location parameter on the x-axis. As shown in Appendix X, the linear rectification equations are a function of $\ln(x_i - xo)$ where $xo < x_{min}$, the minimum variable value on the data. We don't know the value of xo initially, but by searching for a value of xo such that when $\ln(x_i - xo)$ is plotted against $\ln \ln(1/(1 - F_i))$, the correlation coefficient is its highest value, will give a reasonably accurate answer. The process can be easily translated to computer code to speed up the process.

Determining the parameters for the common distributions can also be done by hand using suitably scaled probability plotting paper, a straight line through the data points being determined 'by eye', as described earlier. See Lewis (1996) for examples of probability plotting graph paper for some of the statistical distributions mentioned.

Further improvement in the selection of the best linear rectification model can be performed by comparing the uncertainty in model fit as determined from the *standard error* of the data (Nelson, 1982). Also, the use of *confidence limits* in determining the uncertainty in the estimates from linear regression is useful for assessing the nature of the data, particularly when small samples are taken and/or when outlying data points control the gradient of the regression line. Confidence limits are generally wider than some inexperienced data analysts expect, so they help avoid thinking that estimates are closer to the true value than they really are. A discussion of their application in data analysis can be found in Ayyub and McCuen (1997), Comer and Kjerengtroen (1996), Nelson (1982), and Rice (1997).

Example – fitting a Normal distribution to a set of existing data

We will next demonstrate the use of the linear rectification method described above by fitting a Normal distribution to a set of experimental data. The data to be analysed is in the form of a histogram given in Figure 4.9. It shows the distribution of yield strength for a cold drawn carbon steel (SAE 1018). The data is taken from ASM (1997a), a reference that provides data in the form of histograms for several important mechanical properties of steels. Data collated in this manner has been chosen for analysis because a designer may have to resort to the use of data from existing sources. Also, the analysis of this case raises some interesting questions which may not necessarily be met when analysing data collated in practice and displayed using the methods described in Appendix I.

After a visual inspection, it is evident that the SAE 1018 yield strength data has a distribution approaching the Normal type, although there is an abnormally high frequency value around the mid-range of the data. Further analysis of the data as shown in Table 4.2 using the cumulative frequency modelling approach yields Figure 4.10. Note that the *mean rank* equation is used to determine the plotting positions on the y-axis, F_i, the x-axis plotting positions being the mid-class values for the yield strength in MPa. The class width $w = 13.8 \, \text{MPa}$. The values determined graphically for the mean and standard deviation are also shown.

The estimated cumulative frequency fits the data well, where in fact the curve is modelled with a fifth order polynomial using commercial curve fitting software.

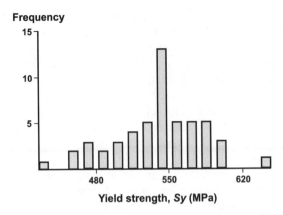

Figure 4.9 Yield strength histogram for SAE 1018 cold drawn carbon steel bar (ASM, 1997a)

(Commercial software such as MS Excel is useful in this connection being widely available.) Omissions in the ranked values of F_i in Table 4.2 reflect the omissions of the data in the original histogram for several classes. As can be judged from Figure 4.10, inclusion of the cumulative probabilities for these classes would not follow the natural pattern of the distribution and are therefore omitted. However, when a very low number of classes exist their inclusion can be justified.

Linear rectification of the cumulative frequency, F_i, is performed by converting to the Standard Normal variate, z. The linear plot together with the straight line equation through the data and the correlation coefficient, r, is shown in Figure 4.11. From Figure 4.11, it is evident that the mean is 530 MPa because the regression line crosses the Standard Normal variate, z, at 0 representing the 50 percentile or median in the non-linearized domain. The mean and standard deviation can also be found from the relationships given in Appendix X. For the Normal distribution

Table 4.2 Analysis of histogram data for SAE 1018 to obtain the Normal distribution plotting positions

Mid-class (MPa)	Frequency (f)	Cumulative freq. (i)	$F_i = \dfrac{i}{N+1}$	$z = \Phi_{\text{SND}}^{-1}(F_i)$
(x-axis)	($N = 52$)		(y-axis)	(y-axis)
431.0	1	1	0.01887	−2.08
444.8	0	1		
458.6	2	3	0.05660	−1.59
472.4	3	6	0.11321	−1.21
486.2	2	8	0.15094	−1.03
500.0	3	11	0.20755	−0.82
513.8	4	15	0.28302	−0.57
527.6	5	20	0.37736	−0.31
541.4	13	33	0.62264	0.31
555.2	5	38	0.71698	0.58
569.0	5	43	0.81132	0.88
582.8	5	48	0.99566	1.32
596.6	3	51	0.96226	1.78
610.4	0	51		
624.2	0	51		
638.0	1	52	0.98113	2.08

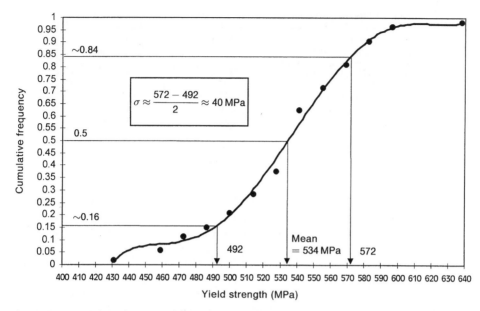

Figure 4.10 Cumulative frequency distribution for SAE 1018 yield strength data

we can calculate the mean and standard deviation from:

$$\mu = -\left(\frac{A0}{A1}\right) = -\left(\frac{-11.663}{0.022}\right) = 530.14 \, \text{MPa}$$

$$\sigma = \left(\frac{1 - A0}{A1}\right) + \left(\frac{A0}{A1}\right) = \left(\frac{1 + 11.663}{0.022}\right) + \left(\frac{-11.663}{0.022}\right) = 45.45 \, \text{MPa}$$

The conclusion is that the Normal distribution is an adequate fit to the SAE 1018 data. A summary of the Normal distribution parameters calculated from Figures 4.10 and 4.11 and other values for the mean and standard deviation from various sources (including commercial software and a package developed at Hull University called *FastFitter*[*]) are given in Table 4.3. The frequency distributions derived from the Normal distribution parameters from source are shown graphically overlaying the original histogram in Figure 4.12 for comparison.

It can be seen from Table 4.3 that there is no positive or foolproof way of determining the distributional parameters useful in probabilistic design, although the linear rectification method is an efficient approach (Siddal, 1983). The choice of ranking equation can also affect the accuracy of the calculated distribution parameters using the methods described. Reference should be made to the guidance notes given in this respect.

The above process above could also be performed for the 3-parameter Weibull distribution to compare the correlation coefficients and determine the better fitting distributional model. Computer-based techniques have been devised as part of the approach to support businesses attempting to determine the characterizing distributions

---

[*] The *FastFitter* software is available from the authors on request.

Table 4.3 Normal distribution parameters for SAE 1018 from various sources

Source of Normal distribution parameters	Mean, μ	Standard deviation, σ
Reference (Mischke, 1992)	541	41
FastFitter	532	44
Moment calculations	537	41
Normal linear rectification	530	45
Cumulative. freq. graph	534	40
Commercial software	545	38
Average	537	42

from sample data. As shown in Figure 4.13, the users screen from the software, called *FastFitter*, is the selection of the best fitting PDF and its parameters representing the sample data, here for the yield strength data for SAE 1018. The software selects the best distribution from the seven common types: Normal, Lognormal, 2-parameter Weibull, 3-parameter Weibull, Maximum Extreme Value Type I, Minimum Extreme Value Type I and the Exponential distribution. Using the *FastFitter* software, it is found that the 3-parameter Weibull distribution gives the highest correlation coefficient of all the models, at $r = 0.995$, compared to $r = 0.991$ for the Normal distribution. The mean and standard deviation in Table 4.3 for *FastFitter* are calculated from the Weibull parameters, the relevant information is provided in Appendix IX.

4.2.3 The algebra of random variables

Typically, if the stress or strength has not been taken directly from the measured distribution, it is likely to be a combination of random variables. For example, a

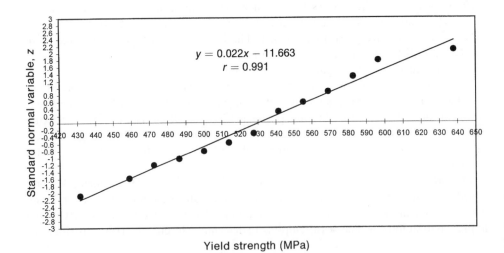

Figure 4.11 Normal distribution linear rectification for SAE 1018 yield strength data

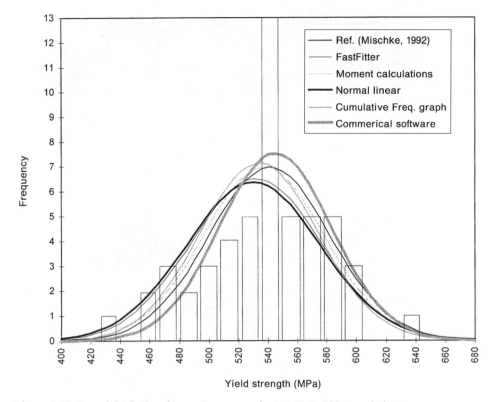

Figure 4.12 Normal distributions from various sources for SAE 1018 yield strength data

failure governing stress is a function of the applied load variation and maybe two- or three-dimensional variables bounding the geometry of the problem. The mathematical manipulation of the failure governing equations and distributional parameters of the random variables used to determine the loading stress in particular are complex, and require that we introduce a new algebra called the *algebra of random variables*.

We need this special algebra to operate on the engineering equations as part of probabilistic design, for example the bending stress equation, because the parameters are random variables of a distributional nature rather than unique values. When these random variables are mathematically manipulated, the result of the operation is another random variable. The algebra has been almost entirely developed with the application of the Normal distribution, because numerous functions of random variables are normally distributed or are approximately normally distributed in engineering (Haugen, 1980).

Engineering variables are found to be either statistically independent or correlated in some way. In engineering problems, the variables are usually found to be unrelated, for instance a dimensional variable is not statistically related to a material strength (Haugen, 1980). Table 4.4 shows some common algebraic functions, typically with one variable, x, or two statistically independent random variables, x and y. The mean and standard deviation of the functions are given in terms of the algebra of

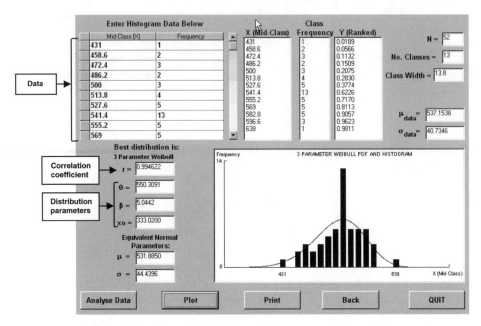

Figure 4.13 *FastFitter* analysis of SAE 1018 yield strength data

random variables. Where the variables x and y are correlated in some way, with correlation coefficient, r, several common functions have also been included.

When a function, ϕ, is a combination of two or more statistically independent variables, x_i, then equation 4.5 can be effectively used to determine their combined variance, V_ϕ (Mischke, 1980).

$$V_\phi \approx \sum_{i=1}^{n} \left(\frac{\partial\phi}{\partial x_i}\right)^2 \cdot \sigma_{x_i}^2 + \frac{1}{2}\sum_{i=1}^{n} \left(\frac{\partial^2\phi}{\partial x_i^2}\right)^2 \cdot \sigma_{x_i}^4 \tag{4.5}$$

To determine the mean value, μ, of the function ϕ:

$$\mu_\phi \approx \phi\left(\mu_{x_1}, \mu_{x_2}, \ldots, \mu_{x_n}\right) + \frac{1}{2}\sum_{i=1}^{n} \frac{\partial^2\phi}{\partial x_i^2} \cdot \sigma_{x_i}^2 \tag{4.6}$$

Equation 4.5 is exact for linear functions, but should only be applied to non-linear functions if the random variables have a coefficient of variation, $C_v < 0.2$ (Furman, 1981; Morrison, 2000). If this is the case, then the approximation using just the first term only differs insignificantly from using higher order terms (Furman, 1981). However, for a function whose first derivative is very small, the higher terms cannot be ignored (Bowker and Lieberman, 1959). Approximate solutions for the mean and standard deviation, σ_ϕ, are provided by omitting the higher order terms, for example equation 4.5 is often written as:

$$\sigma_\phi \approx \left(\sum_{i=1}^{n} \left(\frac{\partial\phi}{\partial x_i}\right)^2 \cdot \sigma_{x_i}^2\right)^{0.5} \tag{4.7}$$

Table 4.4 Mean and standard deviation of statistically independent and correlated random variables x and y for some common functions

Function (ϕ)	Mean (μ_ϕ)	Standard deviation (σ_ϕ)
$\phi = x$	μ_x	σ_x
$\phi = x^2$	$\mu_x^2 + \sigma_x^2$	$2\sigma_x \cdot \mu_x \cdot \left[1 + 0.25\left(\frac{\sigma_x}{\mu_x}\right)^2\right]$
$\phi = x^3$	$\mu_x^3 + 3\sigma_x^2 \cdot \mu_x$	$3\sigma_x \cdot \mu_x^2 \cdot \left[1 + \left(\frac{\sigma_x}{\mu_x}\right)^2\right]$
$\phi = x^4$	$\mu_x^4 + 6\sigma_x^2 \cdot \mu_x^2$	$4\sigma_x \cdot \mu_x^3 \cdot \left[1 + \frac{9}{4}\left(\frac{\sigma_x}{\mu_x}\right)^2\right]$
$\phi = x^n$	$\mu_x^n \cdot \left[1 + 0.5n(n-1)\left(\frac{\sigma_x}{\mu_x}\right)^2\right]$	$n \cdot \sigma_x \cdot \mu_x^{n-1} \cdot \left[1 + 0.25(n-1)^2\left(\frac{\sigma_x}{\mu_x}\right)^2\right]$
$\phi = x^{0.5}$	$\mu_x^{0.5}\left[1 - \frac{1}{8}\left(\frac{\sigma_x}{\mu_x}\right)^2\right]$	$\frac{\sigma_x \cdot \mu_x^{0.5}}{2\mu_x}\left[1 + \frac{1}{16}\left(\frac{\sigma_x}{\mu_x}\right)^2\right]$
$\phi = \frac{1}{x}$	$\frac{1}{\mu_x}\left[1 + \left(\frac{\sigma_x}{\mu_x}\right)^2\right]$	$\frac{\sigma_x}{\mu_x^2}\left[1 + \left(\frac{\sigma_x}{\mu_x}\right)^2\right]$
$\phi = \frac{1}{x^2}$	$\frac{1}{\mu_x^2}\left[1 + 3\left(\frac{\sigma_x}{\mu_x}\right)^2\right]$	$\frac{2\sigma_x}{\mu_x^3}\left[1 + \frac{9}{4}\left(\frac{\sigma_x}{\mu_x}\right)^2\right]$
$\phi = \frac{1}{x^3}$	$\frac{1}{\mu_x^3}\left[1 + 6\left(\frac{\sigma_x}{\mu_x}\right)^2\right]$	$\frac{3\sigma_x}{\mu_x^4}\left[1 + 4\left(\frac{\sigma_x}{\mu_x}\right)^2\right]$
$\phi = x \pm y$	$\mu_x \pm \mu_y$	$(\sigma_x^2 + \sigma_y^2)^{0.5}$
		$(\sigma_x^2 + \sigma_y^2 \pm 2r \cdot \sigma_x \cdot \sigma_y)^{0.5}$
$\phi = x \cdot y$	$\mu_x \cdot \mu_y$	$(\mu_x^2 \cdot \sigma_y^2 + \mu_y^2 \cdot \sigma_x^2 + \sigma_x^2 \cdot \sigma_y^2)^{0.5}$
	$\mu_x \cdot \mu_y + r \cdot \sigma_x \cdot \sigma_y$	$[(\mu_x^2 \cdot \sigma_y^2 + \mu_y^2 \cdot \sigma_x^2 + \sigma_x^2 \cdot \sigma_y^2) + (1 + r^2)]^{0.5}$
$\phi = \frac{x}{y}$	$\frac{\mu_x}{\mu_y} + \frac{\sigma_y^2 \cdot \mu_x}{\mu_y^3}$	$\frac{1}{\mu_y}\left(\frac{\mu_x^2 \cdot \sigma_y^2 + \mu_y^2 \cdot \sigma_x^2}{\mu_y^2 + \sigma_x^2}\right)^{0.5}$
	$\frac{\mu_x}{\mu_y} + \frac{\sigma_y \cdot \mu_x}{\mu_y^2}\left(\frac{\sigma_y}{\mu_y} - r \cdot \frac{\sigma_x}{\mu_x}\right)$	$\frac{\mu_x}{\mu_y} \cdot \left(\frac{\sigma_x^2}{\mu_x^2} + \frac{\sigma_y^2}{\mu_y^2} - 2r \cdot \frac{\sigma_x \cdot \sigma_y}{\mu_x \cdot \mu_y}\right)^{0.5}$

and

$$\mu_\phi \approx \phi(\mu_{x_1}, \mu_{x_2}, \ldots, \mu_{x_n}) \tag{4.8}$$

Equation 4.7 is referred to as the *variance equation* and is commonly used in error analysis (Fraser and Milne, 1990), variational design (Morrison, 1998), reliability analysis (Haugen, 1980) and sensitivity analysis (Parry-Jones, 1999). Most importantly in probabilistic design, through the use of the variance equation we have a means of relating geometric decisions to reliability goals by including the dimensional and load random variables in failure governing stress equations to determine the stress random variable for any given problem.

The variance equation can be solved directly by using the *Calculus of Partial Derivatives*, or for more complex cases, using the *Finite Difference Method*. Another valuable method for 'solving' the variance equation is *Monte Carlo Simulation*. However, rather than solve the variance equation directly, it allows us to simulate the output of the variance for a given function of many random variables. Appendix XI explains in detail each of the methods to solve the variance equation and provides worked examples.

The variance for any set of data can be calculated without reference to the prior distribution as discussed in Appendix I. It follows that the variance equation is also independent of a prior distribution. Here it is assumed that in all the cases the output function is adequately represented by the Normal distribution when the random variables involved are all represented by the Normal distribution. The assumption that the output function is robustly Normal in all cases does not strictly apply, particularly when variables are in certain combination or when the Lognormal distribution is used. See Haugen (1980), Shigley and Mischke (1996) and Siddal (1983) for guidance on using the variance equation.

The variance equation provides a valuable tool with which to draw sensitivity inferences to give the contribution of each variable to the overall variability of the problem. Through its use, probabilistic methods provide a more effective way to determine key design parameters for an optimal solution (Comer and Kjerengtroen, 1996). From this and other information in Pareto Chart form, the designer can quickly focus on the dominant variables. See Appendix XI for a worked example of sensitivity analysis in determining the variance contribution of each of the design variables in a stress analysis problem.

4.3 Variables in probabilistic design

Design models must account for variability in the most important design variables (Cruse, 1997b). If an adequate characterization of these important variables is performed, this will give a cost-effective and a fairly accurate solution for most engineering problems. The main engineering random variables that must be adequately described using the probabilistic approach are shown in Figure 4.14. The variables conveniently divide into two types: *design dependent*, which the designer has the greatest control over, and *service dependent*, which the design has 'limited' control over. Typically, the most important design dependent variables are material strength and dimensional variability. Material strength can be statistically modelled from sample data for the property required, as previously demonstrated; however, difficulties exist in the collation of information about the properties of interest. Dimensional variability and its effects on the stress acting on a component can be great, but information is typically lacking about its statistical nature and its impact on geometric stress concentration values is rarely assessed.

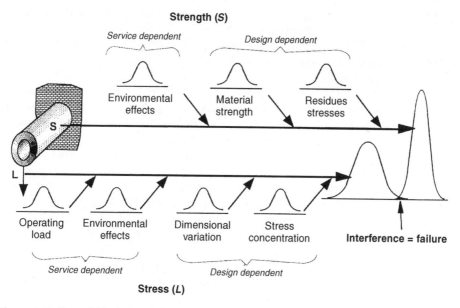

Figure 4.14 Key variables in a probabilistic design approach

Important service dependent variables are related to the loading of the component and stresses resulting from environmental effects. These are generally difficult to determine at the design stage because of the cost of performing experimental data collection, the nature of overloading and abuse in service, and the lack of data about service loads in general. Also, the effect that service conditions have on the material properties is important, the most important considerations arising from extremes in temperature, as there is a tendency towards brittle fracture at low temperatures, and creep rupture at high temperatures. To this end, it has been cited that the quality control of the environment is much more important than quality control of the manufacturing processes in achieving high reliability (Carter, 1986).

Among the most dramatic modifiers encountered in design of those mentioned above are due to thermal effects on strength and stress concentration effects on local stress magnitudes in general (Haugen, 1980). As seen from Figure 4.14, there are several important design dependent variables that terms of an engineering analysis are then:

- Material strength (with temperature and residual processing effects included)
- Dimensional variability
- Geometric stress concentrations
- Service loads.

4.3.1 Material strength

The largest design dependent strength variable is material strength, either ultimate tensile strength (Su), uniaxial yield strength (Sy), shear yield strength (τ_y) or some other failure resisting property. For deflection and instability problems, the Modulus of Elasticity (E) is usually of interest. Shear yield strength, typically used in torsion calculations, is a linear function of the uniaxial yield strength and is likely to have the same distribution type (Haugen, 1980).

With mass produced products, extensive testing can be carried out to characterize the property of interest. When production is small, material testing may be limited to simple tension tests or perhaps none at all (Ayyub and McCuen, 1997). Material properties are often not available with a sufficient number of test repetitions to provide statistical relevance, and remain one of the challenges of greater application of statistical methods, for example in aircraft design (Smith, 1995). Another problem is how close laboratory test results are to that of the material provided to the customer (Welling and Lynch, 1985), because material properties tend to vary from lot to lot and manufacturer to manufacturer (Ireson *et al.*, 1996). However, this can all be regarded as making the case for a probabilistic approach. Ideally, information on material properties should come from test specimens that closely resemble the design configuration and size, and tested under conditions that duplicate the expected service conditions as closely as possible (Bury, 1975). The more information we have about a situation before the trial takes place and the data collected, the more confidence there will be in the final result (Leitch, 1995).

One of the major reasons why design should be based on statistics is that material properties vary so widely, and any general theory of reliability must take this into account (Haugen and Wirsching, 1975). Material properties exhibit variability because of anisotropy and inhomogeneity, imperfection, impurities and defects (Bury, 1975). All materials are, of course, processed in some way so that they are in some useful fabrication condition. The level of variability in material properties associated with the level of processing can also be a major contribution. There are three main kinds of randomness in material properties that are observed (Bolotin, 1994):

- Within specimen – inherent within the microstructure and caused by imperfections, flaws, etc.
- Specimen to specimen – caused by the instabilities and imperfections of the manufacturing processes with the batch.
- Batch to batch – natural variations due to processing, such as material quality, equipment, operator, method, set-up and the environment.

Other uncertainties associated with material properties are due to humidity and ambient chemicals and the effects of time and corrosion (Farag, 1997; Haugen, 1982b). Brittle materials are affected additionally by the presence of imperfections, cracks and internal flaws, which create stress raisers. For example, cast materials such as grey cast iron are brittle due to the graphite flakes in the material causing internal stress raisers. Their low tensile strength is due to these flaws which act as nuclei for crack formation when in tensile loading (Norton, 1996). Subsequently, brittle materials tend to have a large variation in strength, sometimes many times that of ductile materials.

Strain rate also affects tensile properties at test. An increasing strain rate tends to increase tensile properties such as Su and Sy. However, a high loading rate tends to promote brittle fracture (Juvinall, 1967). The average strain rate used in obtaining a stress–strain diagram is approximately 10^{-3} ms/m, and this should be kept in mind when performing experimental testing of materials (Shigley, 1986).

It has been shown that the ultimate tensile strength, Su, for brittle materials depends upon the size of the specimen and will decrease with increasing dimensions, since the probability of having weak spots is increased. This is termed the size effect. This 'size effect' was investigated by Weibull (1951) who suggested a statistical function, the Weibull distribution, describing the number and distribution of these flaws. The relationship below models the size effect for deterministic values of Su (Timoshenko, 1966).

$$\frac{Su_2}{Su_1} = \left(\frac{v_1}{v_2}\right)^{1/\beta} \tag{4.9}$$

where:

Su_1 = ultimate tensile strength of test specimen

Su_2 = ultimate tensile strength of component

v_1 = effective volume of test specimen

v_2 = effective volume of component design

β = shape parameter from Weibull analysis of test specimen data.

As can be seen from the above equation, for brittle materials like glass and ceramics, we can scale the strength for a proposed design from a test specimen analysis. In a more useful form for the 2-parameter Weibull distribution, the probability of failure is a function of the applied stress, L.

$$P = 1 - \exp\left(-\left(\frac{L}{\theta}\right)^\beta\right)^{v_2/v_1} \tag{4.10}$$

where:

P = probability of failure

L = stress applied to component

θ = characteristic value,

and for the 3-parameter Weibull distribution:

$$P = 1 - \exp\left(-\left(\frac{L - xo}{\theta - xo}\right)^\beta\right)^{v_2/v_1} \tag{4.11}$$

where:

xo = expected minimum value.

A high shape factor in the 2-parameter model suggests less strength variability. The Weibull model can also be used to model ductile materials at low temperatures which exhibit brittle failure (Faires, 1965). (See Waterman and Ashby (1991) for a detailed discussion on modelling brittle material strength.)

Several researchers and organizations over the last 50 years have accumulated statistical material property data. However, property data is still not available for many materials or is not made generally available by the companies manufacturing the stock product. This is a problem if you want to design with a specific material in a specific environment. For example, it is not adequate just to say the statistical property of one particular steel is going to be close to that of another similar steel. Approximate values for the mean and standard deviation of the ultimate tensile strength of steel, Su, can be found from a hardness test in Brinnel Hardness (HB). From empirical investigation (Shigley and Mischke, 1989):

$$\mu_{Su} = 3.45\mu_{HB} \tag{4.12}$$

$$\sigma_{Su} = (3.45^2 \cdot \sigma_{HB}^2 + 0.152^2 \cdot \mu_{HB}^2 + 0.152^2 \cdot \sigma_{HB}^2)^{0.5} \tag{4.13}$$

Unfortunately, the statistical data in references such (ASM, 1997a; ASM, 1997b; Haugen, 1980; Mischke, 1992) is the best available to the designer who requires rapid solutions. An example of such data was shown in Figure 4.9. Although the property data strictly applies to US grade ferrous and non-ferrous materials, conversion tables are available which show equivalent material grades for UK, French, German, Swedish and Japanese grades. However, their casual use could make the answers obtained misrepresentative of the problem. They should be treated with caution as a direct comparison is questionable because of small deviations in compositions and processing parameters. Material properties for UK grade materials in statistical form would be advantageous when using probabilistic design techniques. However, there are no immediate plans by the British Standards Institute (BSI) to produce materials property data in a statistical format, and all data currently published is based on values (*pers. comm.*, 1998).

Table 4.5 shows the coefficient of variation, C_v, for various material properties at room temperature compiled from a number of sources (Bury, 1975; Haugen, 1980; Haugen and Wirsching, 1975; Rao, 1992; Shigley and Mischke, 1996; Yokobori, 1965).

Further insight into the statistical strength properties of some commonly used metals is provided by a data sheet in Table 4.6. Again caution should be exercised in their use, but reference will be made to some of these values in the probabilistic design case studies at the end of this section.

Table 4.5 Typical coefficient of variation, C_v, for various materials and mechanical properties (Su = ultimate tensile strength)

Steel, C_v	Other materials, C_v
$Su = 0.05$	Su of cast iron $= 0.09$
Yield Strength (Sy) $= 0.05$ to 0.08	Su of wrought iron $= 0.04$
Endurance limit (Se) $= 0.08$	Su of brittle materials ≤ 0.3
Brinell Hardness (HB) $= 0.05$	Su of glass $= 0.24$
Mod. of Elasticity (E) $= 0.01$ to 0.03	Fracture toughness (K_c) of metallic materials $= 0.07$
Mod. of Rigidity (G) $= 0.02$ to 0.04	Mod. of Elasticity (E) of nodular cast iron $= 0.04$
Fracture toughness (K_c) $= 0.05$ to 0.1	Mod. of Elasticity (E) of titanium $= 0.09$
Poisson's ratio (ν) $= 0.02$ to 0.26	Mod. of Elasticity (E) of aluminium $= 0.03$

Table 4.6 Material strength data sheet

Material	Condition	Ultimate tensile strength (MPa)		Yield strength (MPa)	
		μ_{Su}	σ_{Su}	μ_{Sy}	σ_{Sy}
Free cutting carbon steel BS 220M07	Cold drawn	517	27	447	36
Mild steel BS 070M20	Normalized	506	25	–	–
Low carbon steel SAE 1018 (BS 080A17)	Cold drawn	604	40	540	41
Medium carbon steel SAE 1035 (BS 080A32)	Hot rolled	594	27	342	26
Medium carbon steel SAE 1045 (BS 080M46)	Cold drawn	812	49	658	45
Low alloy steel SAE 4340 (BS 817M40)	Cold drawn annealed $\varnothing \leq 10\,\text{mm}$	803	9	–	–
Structural steel BS Grade 43C	Hot rolled $t \leq 16\,\text{mm}$	–	–	324	16
Stainless steel BS 316S16	Sheet annealed $t \leq 3\,\text{mm}$	579	20	–	–
Aluminium alloy 7075-T6	Sheet aged	555	27.5	484	22
Titanium alloy Ti-6Al-4V	Bar	934	46	900	50

Finally, it is worth investigating how deterministic values of material strength are calculated as commonly found in engineering data books. Equation 4.14 states that the minimum material strength, S_{min}, as used in deterministic calculations, equals the mean value determined from test, minus three standard deviations, calculated for the Normal distribution (Cable and Virene, 1967):

$$S_{min} = \mu - 3\sigma \qquad (4.14)$$

For example, the deterministic value for the yield strength, Sy, for SAE 1018 cold drawn steel for the size range tested is approximately 395 MPa (Green, 1992). Table 4.6 gives the mean and standard deviation as $Sy \sim N(540, 41)$ MPa. The lower bound value as used in deterministic design becomes:

$$Sy_{min} = 540 - 3(41) = 417\,\text{MPa}$$

The values are within 5% of each other. If deterministic values are actually calculated at the negative 3σ limit from the mean, 1350 failures in every million could be expected for an applied stress of the same magnitude as determined from SND theory. It is evident from this that reliability prediction and deterministic design are not compatible, because as the factor of safety is introduced to reduce failures, the probability aspect of the calculation is lost. (Note that the ASTM standard on materials testing suggests setting the minimum material property at -2.33σ from the mean value (Shigley and Mischke, 1989).)

Material properties and temperature

A number of basic material properties useful in static design depend most notably on temperature (Haugen, 1980). For example, Figure 4.15 shows how high temperatures alter the important mechanical properties of a low carbon steel, and the variation that can be experienced. Temperature dependent materials properties are sometimes available in statistical form, as shown in Figure 4.16 where the 3-parameter Weibull distribution is used to model the tensile strength of an alloy steel over a range of temperatures (Lipson *et al.*, 1967). This type of information is strictly for high temperature work where the application of the load lasts approximately 15 to 20 minutes (Timoshenko, 1966).

Experiments at high temperatures also show that tensile tests depend on the duration of the test, because as time increases the load necessary to produce fracture decreases (Timoshenko, 1966). This is the onset of the phenomenon known as *creep*. All materials begin to lose strength at some temperature, and as the temperature increases, the deformations cease to be elastic and become more and more plastic in nature. Given sufficient time, the material may fail by creep usually occurring at a temperature between 30 and 40% of the melting temperature in degrees Kelvin (Ashby and Jones, 1989). In carbon steels, for example, design stresses can be solely based on short-term properties up to an operating temperature of about 400°C, while at temperatures greater than this, creep behaviour is likely to overrule any other design considerations.

Creep stresses used for design purposes are usually determined based on two criteria: the stress for a given acceptable creep deformation after a certain number of hours, which ranges from 0.01 to 1% deformation in 1000 hours; and the nominal

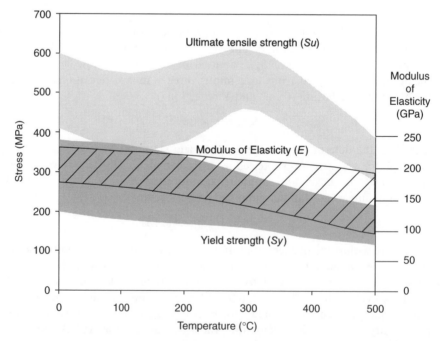

Figure 4.15 Mechanical properties of a low carbon steel as a function of temperature (adapted from Waterman and Ashby, 1991)

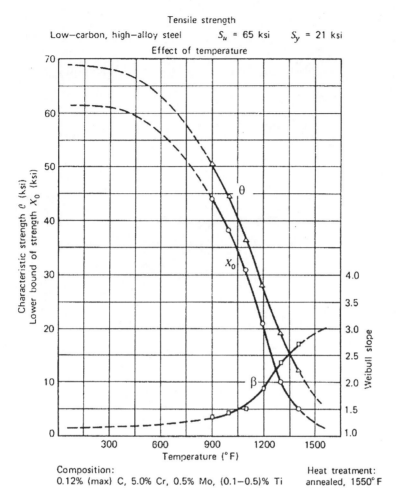

Figure 4.16 Short-term tensile strength Weibull parameters for an alloy steel at various temperatures (Lipson *et al.*, 1967)

stress required to produce rupture after a specified time or at the end of the required life. The creep rupture stress for several steels at 1000 hours is shown in Figure 4.17. It is evident that a large variation exists in the rupture stress values for a given temperature, and in general creep material properties tend to have a coefficient of variation much greater than static properties (Bury, 1974). For example, $C_v = 0.7$ has been cited for the creep time to fracture for copper (Yokobori, 1965). Although little statistical data has been found on the properties highlighted, creep strength data and properties at high temperatures for various materials can be found in ASM (1997a), ASM (1997b), Furman (1980) and Waterman and Ashby (1991).

Many mechanical components also operate at temperatures far lower than room temperature. As the temperature is reduced, both the ultimate tensile strength, *Su*, and tensile yield strength, *Sy*, generally increase for most materials. However, temperatures below freezing have the effect of altering the structure of some ductile

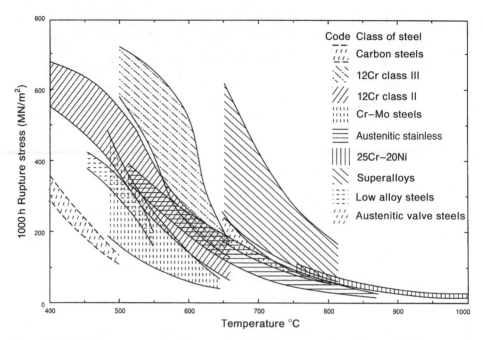

Figure 4.17 1000 hour creep rupture stress as a function of temperature for various steels (Waterman and Ashby, 1991)

metals so they fracture in a brittle manner. The temperature at which this occurs is called the *ductile-to-brittle transition temperature*. Tests to determine this usually involve impact energy tests, for example, *Charpy* or *Izod*, which measures the energy to break the specimen in joules. The plot of the results for a structural steel are shown in Figure 4.18, showing the regions of brittle and ductile behaviour.

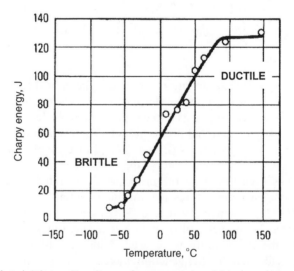

Figure 4.18 Ductile to brittle transition diagram for a structural steel (Mager and Marschall, 1984)

The ductile-to-brittle transition temperature for some steels can be as 'high' as 0°C, depending on the composition of the steel (Ashby and Jones, 1989). However, there is no way of using the data directly from impact tests quantitatively in the design process. Design specifications do usually state a minimum impact strength, but experience suggests that this does not necessarily eliminate brittle failure (Faires, 1965). The *Robertson test* can yield more information than either the Charpy or Izod tests because the transition temperature is statistically correlated with the temperature at which the actual structure has been known to fail in a brittle manner (Benham and Warnock, 1983; Ruiz and Koenigsberger, 1970). The test uses a severely notched specimen tested under static tension, and a plot showing the variation of the nominal stress at fracture with the test specimen temperature drawn. The test gives useful results from which design calculations can be based; however, the test is more expensive and complex compared to other methods. In general, it is dangerous to use a material below its transition temperature because most of its capacity to absorb energy without rupture has been lost and careful design and analysis is required.

Residual stresses and processing

In a component of uniform temperature not acted upon by external loads, any internal stresses that exist are called residual stresses. The material strength, therefore, is dependent not only on the basic material property, but on the residual stresses exhibited by the manufacturing process itself, for example by forging, extrusion or casting. This could be further affected by secondary processing in production such as welding, machining, grinding and surface coating processes, from deliberate or unavoidable heat treatment, and assembly operations such as fastening and shrink fits, all of which promote residual stresses. Many failures result from unsatisfactory welding and joining of parts of engineering components (Heyes, 1989), and this can only be attributable to residual stresses affecting either the static or more commonly the fatigue properties. Additionally manufacturing processes result in variations in surface roughness, sharp corners and other stress raisers (Farag, 1997). It is also evident that the pattern of residual stresses in a component may not be permanent and could change over time due to changing environmental conditions in service (Faires, 1965).

The methods used to measure residual stresses in a component are performed after the manufacturing process, and are broadly classed into two types: mechanical (layer removal, cutting) and physical (X-ray diffraction, acoustic, magnetic). Further reference to the methods used can be found in Chandra (1997), Juvinall (1967), and Timoshenko (1983).

To remove residual stresses usually requires some manner of stress relieving (usually involving heat treatment) or promotion/inducement of opposing stresses by shot peening, or similar methods. Designers need to consider carefully the influence of the residual stresses on component behaviour which may be induced during manufacture, and decide whether stress relief is appropriate (Nicholson *et al.*, 1993). If stress relief is not possible, the problem of how to quantify residual stresses, which will either be detrimental or advantageous to the strength of the material, becomes a difficult one. Significant residual stresses tend to be beneficial if compressive, and detrimental if tensile. For example, residual stresses in brittle materials are problematic if tensile because they have low toughness and this could accelerate catastrophic brittle fracture. The presence of residual stresses are generally detrimental to the product integrity in service and should be eliminated if expected to be harmful (Chandra, 1997).

Theoretically, the effects of the manufacturing process on the material property distribution can be determined, shown here for the case when Normal distribution applies. For an additive case of a residual stress, it follows that from the algebra of random variables (Carter, 1997):

$$\mu_S = \mu_{So} + \mu_V \tag{4.15}$$

$$\sigma_S = (\sigma_{So}^2 + \sigma_V^2)^{0.5} \tag{4.16}$$

where:

μ_S = mean of the final strength

σ_S = standard deviation of the final strength

μ_{So} = mean of the original strength

σ_{So} = standard deviation of the original strength

μ_V = additive quantities of strength from the process

σ_V = standard deviation of the additive strength from the process.

For a proportional improvement in the strength, the product of a function of random variables applies:

$$\mu_S = \mu_{So} \cdot \mu_V \tag{4.17}$$

$$\sigma_S = (\mu_{So}^2 \cdot \sigma_V^2 + \mu_V^2 \cdot \sigma_{So}^2 + \sigma_{So}^2 \cdot \sigma_V^2)^{0.5} \tag{4.18}$$

As can be seen from the above equations, the standard deviation of the strength increases significantly with the number of processes used in manufacture that are adding the residual stresses. This may be the reason for the apparent reluctance of suppliers to give precise statistical data about their product (Carter, 1997).

A practical difficulty using the above approach is that there are too many processes and therefore variables involved. However, in the nuclear industry the additional material factors are often taken into consideration (Carter, 1997). The difficulties associated with the measuring of the variables must also be an inhibitory factor in their use. Residual stresses have been measured and modelled for a number of manufacturing processes, such as welding, cold drawing and heat treatment processes (ASM, 1997a; ASM, 1997b; Osgood, 1982), but their application is so specific that transfer to the general case could be misleading. Statistical descriptions of materials behaviour should take into account the influences of variable metallurgical factors, such as heat treatment and mechanical processing, and this returns to an earlier important statement (Haugen, 1982a). Information on material properties should come from test specimens that closely resemble the design configuration and size, and are tested under conditions that duplicate the expected service conditions as closely as possible.

4.3.2 Dimensional variability

The manufacturing process introduces variations in that absolute dimensional accuracy cannot be attained and variations within specified tolerances are a necessary feature of all manufactured products (Carter, 1986). Proper tolerances are crucial to the proper functioning, reliability and long life of a product and the act of assigning tolerances in fact finalizes reliability (Dixon, 1997; Rao, 1992; Vinogradov, 1991). Large tolerances and/or large variance can result in significant degradation of reliability because the failure probability is a function of the magnitude of dimensional variability and tolerance allocated, and affects load induced stress in a component (Haugen, 1980; Kluger, 1964).

Component reliability will vary as a function of the power of a dimensional variable in a stress function. Powers of dimensional variables greater than unity magnify the effect. For example, the equation for the polar moment of area for a circular shaft varies as the fourth power of the diameter. Other similar cases liable to dimensional variation effects include the radius of gyration, cross-sectional area and moment of inertia properties. Such variations affect stability, deflection, strains and angular twists as well as stresses levels (Haugen, 1980). It can be seen that variations in tolerance may be of importance for critical components which need to be designed to a high reliability (Bury, 1974).

The measures of dimensional variability from Conformability Analysis (CA) (as described in Chapters 2 and 3), specifically the Component Manufacturing Variability Risk, q_m, is useful in the allocation of tolerances and subsequent analysis of their distributions in probabilistic design. The value q_m is determined from process capability maps for the manufacturing process and knowledge of the component's material and geometry compatibility with the process. In the specific case to the ith component bilateral tolerance, t_i, it was shown in Chapter 3 that the standard deviation estimates were:

$$\sigma_i' = \frac{t_i \cdot q_{mi}^2}{12} \tag{4.19}$$

$$\sigma_i = \frac{t_i \cdot q_{mi}^{4/3}}{12} \tag{4.20}$$

The ′ in equation 4.19 relates to the fact that this is not the true standard deviation, but an estimate to measure the process shift (or drift) in the distribution over the expected duration of production. Equation 4.20 is the best estimate for the standard deviation of the distribution as determined by CA with no process shift.

A popular way of determining the standard deviation for use in the probabilistic calculations is to estimate it by equation 4.21 which is based on the bilateral tolerance, t, and various empirical factors as shown in Table 4.7 (Dieter, 1986; Haugen, 1980; Smith, 1995). The factors relate to the fact that the more parts produced, the more confidence there will be in producing capable tolerances:

$$\sigma = \frac{t}{\text{Factor}} \tag{4.21}$$

Historically, in probabilistic calculations, the standard deviation, σ, is expressed as $t/3$ (Dieter, 1986; Haugen, 1980; Smith, 1995; Welling and Lynch, 1985), which

Table 4.7 Empirical factors for determining standard deviation based on tolerance

No. of parts manufactured	Factor
4 to 5	1
10	1.5
25	2
100	2.5
500 to 700	3

relates to a maximum Process Capability Index, $C_p = 1$. This estimate does not take into account process shift, typically $\pm 1.5\sigma$ from the target during a production run due to tooling and production errors (Evans, 1975), and relies heavily on the tolerances being within $\pm 3\sigma$ during inspection and process control. Unless there is 100% inspection, however, there will be some dimensions that will always be out of tolerance (Bury, 1974).

In equations 4.19 and 4.20, improved estimates for the standard deviation are presented based on empirical observations. This is shown in Figure 4.19 for a ± 0.1 mm tolerance on an arbitrary dimensional characteristic, but with an increasing q_m, as would be determined for less capable design schemes. It shows that increasing risk of allocating tolerances that are not capable, increases the estimates for the σ,

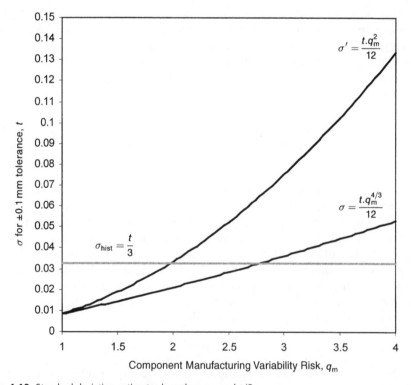

Figure 4.19 Standard deviation estimates based on q_m and $t/3$

more so when process shift is taken into account, as shown by the term σ'. Anticipation of the shift or drift of the tolerances set on a design is therefore an important factor when predicting reliability. It can be argued that the dimensional variability estimates from CA apply to the early part of the product's life-cycle, and may be overconservative when applied to the useful life of the product. However, variability driven failure occurs throughout all life-cycle phases as discussed in Chapter 1.

From the above arguments, it can be seen that anticipation of the process capability levels set on the important design characteristics is an important factor when predicting reliability (Murty and Naikan, 1997). If design tolerances are assigned which have large dimensional variations, the effect on the reliability predicted must be assessed. Sensitivity analysis is useful in this respect.

Stress concentration factors and dimensional variability

Geometric discontinuities increase the stress level beyond the nominal stresses (Shigley and Mischke, 1996). The ratio of this increased stress to the nominal stress in the component is termed the stress concentration factor, Kt. Due to the nature of manufacturing processes, geometric dimensions and therefore stress concentrations vary randomly (Haugen, 1980). The stress concentration factor values, however, are typically based on nominal dimensional values in tables and handbooks. For example, a comprehensive discussion and source of reference for the various stress concentration factors (both theoretical and empirical) for various component configurations and loading conditions can be found in Pilkey (1997).

Underestimating the effects of the component tolerances in conditions of high dimensional variation could be catastrophic. Stress concentration factors based on nominal dimensions are not sufficient, and should, in addition, include estimates based on the dimensional variation. This is demonstrated by Haugen (1980) for a notched round bar in tension in Figure 4.20. The plot shows $\pm 3\sigma$ confidence limits on the stress concentration factor, Kt, as a function of the notch radius, generated using a Monte Carlo simulation. At low notch radii, the stress concentration factor predicted could be as much as 10% in error from the results.

The main cause of mechanical failure is by fatigue with up to 90% failures being attributable. Stress concentrations are primarily responsible for this, as they are among the most dramatic modifiers of local stress magnitudes encountered in design (Haugen, 1980). Stress concentration factors are valid only in dynamic cases, such as fatigue, or when the material is brittle. In ductile materials subject to static loading, the effects of stress concentration are of little or no importance. When the region of stress concentration is small compared to the section resisting the static load, localized yielding in ductile materials limits the peak stress to the approximate level of the yield strength. The load is carried without gross plastic distortion. The stress concentration does no damage (and in fact strain hardening occurs) and so it can be ignored, and no Kt is applied to the stress function. However, stress raisers when combined with factors such as low temperatures, impact and materials with marginal ductility could be very significant with the possibility of brittle fracture (Juvinall, 1967).

For very low ductility or brittle materials, the full Kt is applied unless information to the contrary is available, as governed by the sensitivity index of the material, q_s. For example, cast irons have internal discontinuities as severe as the stress raiser

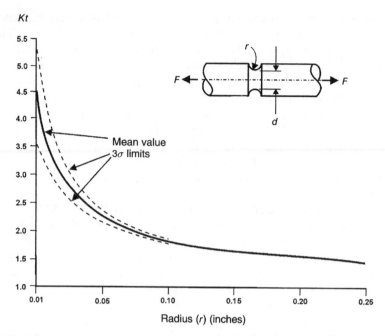

Figure 4.20 Values of stress concentration factor, Kt, as a function of radius, r, with $\pm 3\sigma$ limits for a circumferentially notched round bar in tension [$d \sim N(0.5, 0.00266)$ inches, $\sigma_r = 0.00333$ inches] (adapted from Haugen, 1980)

itself, q_s approaches zero and the full value of Kt is rarely applied under static loading conditions. The following equation is used (Edwards and McKee, 1991; Green, 1992; Juvinall, 1967; Shigley and Mischke, 1996):

$$K' = 1 + q_s(Kt - 1) \tag{4.22}$$

where:

K' = actual stress concentration factor for static loading

q_s = index of sensitivity of the material (for static loading – 0.15 for hardened steels, 0.25 for quenched but untempered steel, 0.2 for cast iron; for impact loading – 0.4 to 0.6 for ductile materials, 0.5 for cast iron, 1 for brittle materials).

In the probabilistic design calculations, the value of Kt would be determined from the empirical models related to the nominal part dimensions, including the dimensional variation estimates from equations 4.19 or 4.20. Norton (1996) models Kt using power laws for many standard cases. Young (1989) uses fourth order polynomials. In either case, it is a relatively straightforward task to include Kt in the probabilistic model by determining the standard deviation through the variance equation.

The distributional parameters for Kt in the form of the Normal distribution can then be used as a random variable product with the loading stress to determine the final stress acting due to the stress concentration. Equations 4.23 and 4.24 show that the mean and standard deviation of the final stress acting (Haugen, 1980):

$$\mu_{L_{Kt}} = \mu_L \cdot \mu_{Kt} \tag{4.23}$$

$$\sigma_{L_{Kt}} = (\mu_L^2 \cdot \sigma_{Kt}^2 + \mu_{Kt}^2 \cdot \sigma_L^2 + \sigma_L^2 \cdot \sigma_{Kt}^2)^{0.5} \tag{4.24}$$

By replacing Kt with K' in the above equations, the stress for notch sensitive materials can be modelled if information is known about the variables involved.

4.3.3 Service loads

One of the topical problems in the field of reliability and fatigue analysis is the prediction of load ranges applied to the structural component during actual operating conditions (Nagode and Fajdiga, 1998). Service loads exhibit statistical variability and uncertainty that is hard to predict and this influences the adequacy of the design (Bury, 1975; Carter, 1997; Mørup, 1993; Rice, 1997). Mechanical loads may not be well characterized out of ignorance or sheer difficulty (Cruse, 1997b). Empirical methods in determining load distribution are currently superior to statistical-based methods (Carter, 1997) and this is a key problem in the development of reliability prediction methods. Probabilistic design then, rather than a deterministic approach, becomes more suitable when there are large variations in the anticipated loads (Welling and Lynch, 1985) and the loads should be considered as being random variables in the same way as the material strength (Bury, 1975).

Loads can be both internal and external. They can be due to weight, mechanical forces (axial tension or compression, shear, bending or torsional), inertial forces, electrical forces, metallurgical forces, chemical or biological effects; due to temperature, environmental effects, dimensional changes or a combination of these (Carter, 1986; Ireson et al., 1996; Shigley and Mischke, 1989; Smith, 1976). In fact some environments may impose greater stresses than those in normal operation, for example shock or vibration (Smith, 1976). These factors may well be as important as any load in conventional operation and can only be formulated with full knowledge of the intended use (Carter, 1986). Additionally, many mechanical systems have a duty cycle which requires effectively many applications of the load (Schatz et al., 1974), and this aspect of the loading in service is seldom reflected in the design calculations (Bury, 1975).

Failures resulting from design deficiencies are relatively common occurrences in industry and sometimes components fail on the first application of the load because of poor design (Nicholson et al., 1993). The underlying assumption of static design is that failure is governed by the occurrence of these occasional large loads, the design failing when a single loading stress exceeds the strength (Bury, 1975). The overload mechanism of mechanical failure (distortion, instability, fracture, etc.) is a common occurrence, accounting for between 11 and 18% of all failures. Design errors leading to overstressing are a major problem and account for over 30% of the cause (Davies, 1985; Larsson et al., 1971). The designer has great responsibility to ensure that they adequately account for the loads anticipated in service, the service life of a product being dependent on the number of times the product is used or operated, the length of operating time and how it is used (Cruse, 1997a).

Some of the important considerations surrounding static loading conditions and static design are discussed next. Initially focusing on static design will aid the development of the more complex dynamic analysis of components in service, for example fatigue design. Fatigue design, although of great practical application, will not be considered here.

The concepts of static design

The most significant factor in mechanical failure analysis is the character of loading, whether static or dynamic. Static loads are applied slowly and remain essentially constant with time, whereas dynamic loads are either suddenly applied (impact loads) or repeatedly varied with time (fatigue loads), or both. The degree of impact is related to the rapidity of loading and the natural frequency of the structure. If the time for loading is three times the fundamental natural frequency, static loading may be assumed (Juvinall, 1967). Impact loading requires the structure to absorb a given amount of energy; static loading requires that it resist given loads (Juvinall, 1967) and this is a fundamental difference when selecting the theory of failure to be used. An analysis guide with respect to the load classification is presented in Table 4.8.

A static load, in terms of a deterministic approach, is a stationary force or moment acting on a member. To be stationary it must be unchanging in magnitude, point or points of application and direction (Shigley and Mischke, 1989). It is a unique value representing what the designer regards as the maximum load in practice that the product will be subjected to in service. Kirkpatrick (1970) defines a limit of strain rate for static loading as less than 10^{-1} but greater than 10^{-5} strain rate/second. This fits in with the strain rate at which tensile properties of materials are tested for static conditions (10^{-3}). With regard to probabilistic design, a load can also be considered static when some variation is expected (Shigley and Mischke, 1989). Static loading also requires that there are less than 1000 repetitions of the load during its designed service (Edwards and McKee, 1991), which introduces the concept of the duty cycle appropriate to reliability engineering. Static loads induce reactions in components and equilibrium usually develops. Where the total static loading on a mechanical system arises from more than one independent source, a statistical model combining the loads may be written for the resultant loading statistics using the algebra of random variables. The correlation of the loading variables is important in this respect. That is the load on the component may be the function of another applied load or associated somehow (Haugen, 1980).

For static design to be valid in practice, we must assume situations where there is no deterioration of the material strength within the time period being considered for the loading history of the product. With a large number of cyclic loads the material will eventually fatigue. With an assumed static analysis, stress rupture is the mechanism of failure to be considered, not fatigue. The number of stress cycles in a problem could

Table 4.8 Loading condition and analysis used (Norton, 1996)

M/c element	Load type	
	Constant	Time-varying
Stationary	Static analysis	Dynamic analysis
Moving	Dynamic analysis	Dynamic analysis

be ignored if the number is small, to create a steady stress or *quasi-static* condition, but one where a variation in loading still exists (Welling and Lynch, 1985). Although, if the loading is treated as a random variable, then this could imply a dynamic analysis (Freudenthal *et al.*, 1966). While static or quasi-static loading is often the basis of engineering design practice, it is often important to address the implications of repeated or fluctuating loads. Such loading conditions can result in fatigue failure, as mentioned above (Collins, 1981).

In a typical load history of a machine element, most of the applied loads are relatively small and their cumulative material damage effects are negligible. When the applied loading stresses are above the material's equivalent endurance limit, the resulting accumulation of damage implies that the component fails by fatigue at some finite number of load repetitions. Relatively large loads occur only occasionally suggesting their cumulative damage effects are negligible (Bury, 1975). Evidently, failure will occur as soon as a single load exceeds the value of the applicable strength criterion.

Some typical load histories for mechanical components and systems are shown in Figure 4.21. The load (dead load, pressure, bending moment, etc.) is subject to variation in all cases, rather than a unique value, the likely shape of the final

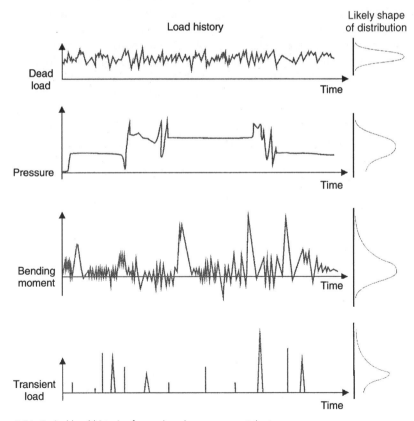

Figure 4.21 Typical load histories for engineering components/systems

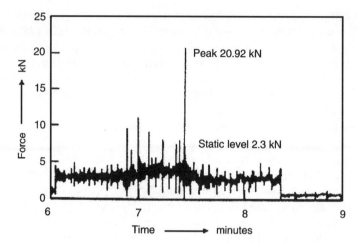

Figure 4.22 Transient loads may be many times the static load in operation (Nicholson *et al.*, 1993)

distribution shown schematically to the right of each load history. Even permanent dead loads that should maintain constant magnitude show a relatively small and slow random variation in practice. At the other end of the loading type spectrum are transient loads which occur infrequently and last for a short period of time – an impulse, for example extreme wind and earthquake loads (Shinozuka and Tan, 1984).

The above would assume that the load distributions for static designs are often highly unsymmetrical, indicating that there is a small proportion of loads that are relatively large (Bury, 1975). During the conditions of use, environmental and service variations give rise to temporary overloads or transients causing failures (Klit *et al.*, 1993). Data collected from mechanical equipment in service has shown that these transient loads developed during operation may be several times the nominal load as shown in Figure 4.22.

There is very little information on the variational nature of loads commonly encountered in mechanical engineering. Several references provide guidance in terms of the coefficient of variation, C_v, for some common loading types as shown in Table 4.9 (Bury, 1975; Ellingwood and Galambos, 1984; Faires, 1965; Lincoln

Table 4.9 Typical coefficient of variation, C_v, for various loading conditions

Loading condition	C_v
Aerodynamic loads in aircraft	0.012 to 0.04
Spring force	0.02
Bolt pre-load using powered screwdrivers	0.03
Powered wrench torque	0.09
Dead load	0.1
Hand wrench torque	0.1
Live load	0.25
Snow load	0.26
Human arm strength	0.3 to 0.4
Wind loads	0.37
Mechanical devices in service	0.5

et al., 1984; Shigley and Mischke, 1996; Smith, 1995; Woodson et al., 1992). These values are representative of the variation experienced during typical duty cycles.

These values can only act as a guide in mechanical design and should be treated with caution and understanding, but they do indicate that loads vary quite considerably, when you observe that the ultimate tensile strength, Su, of steel generally has a $C_v = 0.05$. The smaller the variance in design parameters, the greater will be the reliability of the design to deal with unforeseen events (Suh, 1990), and this would certainly apply to the loads too due to the difficulty in determining the nature of overloading and abuse in service at the design stage.

Example – determining the stress distribution using the coefficient of variation

When dimensional variation is large, its effects must be included in the analysis of the stress distribution for a given situation. However, in some cases the effects of dimensional variation on stress are negligible. A simplified approach to determine the likely stress distribution then becomes available. Given that the mean load applied to the component/assembly is known for a particular situation, the loading stress can be estimated by using the coefficient of variation, C_v, of the load and the mean value for the stress determined from the stress equation for the failure mode of concern.

For example, suppose we are interested in knowing the distribution of stress associated with the tensile static loading on a rectangular bar (see Figure 4.23). It is known from a statistical analysis of the load data that the load, F, has a coefficient of variation $C_v = 0.1$. The stress, L, in the bar is given by:

$$L = \frac{F}{ab}$$

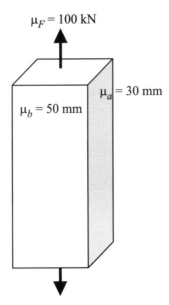

$\mu_F = 100$ kN

$\mu_a = 30$ mm

$\mu_b = 50$ mm

Figure 4.23 Tensile loading on a rectangular bar

where:

$$F = \text{load}$$

$$a \text{ and } b = \text{mean sectional dimensions of the bar.}$$

Assuming the variation encountered in dimensions 'a' and 'b' is negligible, then the mean stress, μ_L, on the bar is:

$$\mu_L = \frac{\mu_F}{\mu_a \times \mu_b} = \frac{100\,000}{0.03 \times 0.05} = 66.67\,\text{MPa}$$

The load coefficient of variation $C_v = 0.1$. Rearranging the equation for C_v to give the standard deviation of the loading stress yields:

$$\sigma_L = C_v \times \mu_L = 0.1 \times 66.67 = 6.67\,\text{MPa}$$

The stress, L, can therefore be approximated by a Normal distribution with parameters:

$$L \sim N(66.67, 6.67)\,\text{MPa}$$

Probabilistic considerations

For many years in mechanical design, load variations have been masked by using factors of safety in a deterministic approach, as shown below (Ullman, 1992):

- 1 to 1.1 – Load well defined as static or fluctuating, if there are no anticipated overloads or shock loads and if an accurate method of analysing the stress has been used.
- 1.2 to 1.3 – Nature of load is defined in an average manner with overloads of 20 to 50% and the stress analysis method may result in errors less than 50%.
- 1.4 to 1.7 – Load not well known and/or stress analysis method of doubtful accuracy.

These factors could be in addition to the factors of safety typically employed as discussed at the beginning of this section. Because of the difficulty in finding the exact distributional nature of the load, this approach to design was considered adequate and economical. It has been argued that it is far too time consuming and costly to measure the load distributions comprehensively (Carter, 1997), but this should not prevent a representation being devised, as the alternative is even more reprehensible (Carter, 1986). Practically speaking, deciding on the service loads is frequently challenging, but some sort of estimates are essential and it is often found that the results are close to the true loads (Faires, 1965).

Since static failures are caused by infrequent large loads, as discussed above, it is important that the distributions that model these extreme events at the right-hand side of the tail are an accurate representation of actual load frequencies. Control of the load tail would appear to be the most effective method of controlling reliability, because the tail of the load distribution dictates the shape of the distribution and hence failure rate (Carter, 1986). The Normal distribution is usually inappropriate, although some applied loads do follow a Normal distribution, for example rocket motor thrust (Haugen, 1965) or the gas pressure in the cylinder heads of reciprocating engines (Lipson *et al.*, 1967). Other loads may be skewed or possess very little scatter

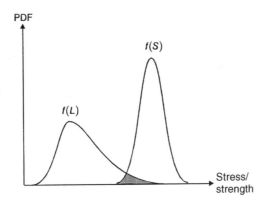

Figure 4.24 A loading stress distribution with extreme events

and in fact load spectrums tend to be highly skewed to the left with high loads occurring only occasionally as shown in Figure 4.24 (Bury, 1974). Civil engineering structures are usually subjected to a Lognormal type of load, but in general the Normal and 3-parameter Weibull distributions are commonly employed for mechanical components, and the Exponential for electrical components (Murty and Naikan, 1997).

It is useful in the first instance to describe the load in terms of a Normal distribution because of the necessity to transform it into the loading stress parameters through the variance equation and dimensional variables. Although the accuracy of the variance equation is dependent on using to near-Normal distributions and variables of low coefficient of variation ($C_v < 0.2$), it is still useful when the anticipated loads have high coefficients of variation. Even if the distribution accounting for the static load is Normal, the stress model is usually Lognormal (Haugen, 1980) due to the complex nature of the variables that make up the stress function.

Experimental load analysis

Information on load distributions was virtually non-existent until quite recently although much of it is very rudimentary, probably because collecting data is very expensive, the measuring transducers being difficult to install on the test product or prototype (Carter, 1997). It has been cited that at least one prototype is required to make a reliability evaluation (Fajdiga *et al.*, 1996), and this must surely be to understand the loads that could be experienced in service as close as possible.

In experimental load studies, the measurable variables are often surface strain, acceleration, weight, pressure or temperature (Haugen, 1980). A discussion of the techniques on how to measure the different types of load parameters can be found in Figliola and Beasley (1995). The measurement of stress directly would be advantageous, you would assume, for use in subsequent calculations to predict reliability. However, no translation of the dimensional variability of the part could then be accounted for in the probabilistic model to give the stress distribution. A better test would be to output the load directly as shown and then use the appropriate probabilistic model to determine the stress distribution.

A key problem in experimental load analysis is the translation of the data yielded from the measurement system (as represented by the load histories in Figure 4.21) into an

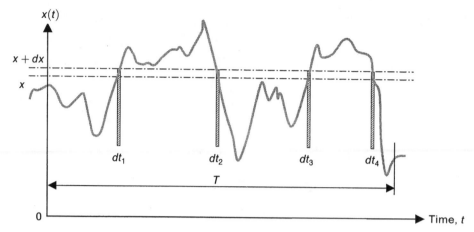

Figure 4.25 Determination of the PDF for a random process (adapted from Newland, 1975)

appropriate distributional form for use in the probabilistic calculations. Measuring the distribution of the load peak amplitudes is a useful model for both static and dynamic loads, as peaks are meaningful values in the load history (Haugen, 1980). A much simpler method, described below, analyses the continuous data from the load history.

Assuming that the statistical characteristics of a load function, $x(t)$, are not changing with time, then we can use the load–time plot, as shown in Figure 4.25, to determine the PDF for $x(t)$. The figure shows a sample history for a random process with the times for which $x \leq x(t) \leq (x + dx)$, identified by the shaded strips. During the time interval, T, $x(t)$ lies in the band of values x to $(x + dx)$ for a total time of $(dt_1 + dt_2 + dt_3 + dt_4)$. We can say that if T is long enough (infinite), the PDF or $f(x)$ is given by:

$f(x)\,dx$ = fraction of total elapsed time for which $x(t)$ lies in the x to $(x + dx)$ band

$$= \frac{(dt_1 + dt_2 + dt_3 + \ldots)}{T} = \frac{\sum_{i=1}^{n} dt_i}{T} \tag{4.25}$$

The fraction of time elapsed for each increment of x can be expressed as a percentage rounded to the nearest whole number for use in the plotting procedure to find the characterizing distributional model. Estimation of the distribution parameters and the correlation coefficient, r, for several distribution types is then performed by using linear rectification and the least squares technique. The distributional model with a correlation coefficient closest to unity would then be chosen as the most appropriate PDF representing the load history. The above can be easily translated into a computer code providing an effective link between the prototype and load model. See the case study later employing this technique for statistically modelling a load history.

For discrete data resulting from many individual load tests, for example spring force for a given deflection as shown in Figure 4.26, a histogram is best constructed. The optimum number of classes can be determined from the rules in Appendix I. Again, the best distribution characterizing the sample data can be selected using the approach in Section 4.2.

Frequency

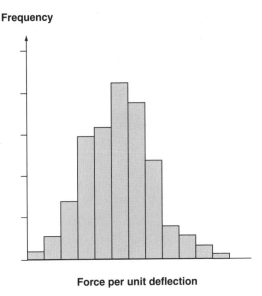

Force per unit deflection

Figure 4.26 Distribution of spring force for a given deflection

Given the lack of available standards and the degree of diversity and types of load application, the approach adopted has focused on supporting the engineer in the early stages of product development with limited experimental load data.

Anthropometric data

Some failures are caused by human related events such as installation, operation or maintenance errors rather than by any intrinsic property of the components (Klit et al., 1993). A number of mechanical systems require operation by a human, who thereby becomes an integral part of the system and has a significant effect on the reliability (Smith, 1976). Consumer products are notorious for the ways in which the consumer can and will misuse the product. It therefore becomes important to make a reasonable determination of the likely loading conditions associated with the different users of the product in service (Cruse, 1997b).

Human force data is typically given as anthropometric strength. An example is given in Table 4.10 for arm strength for males' right arms in the sitting position. This type of data is readily available and presented in terms of the mean and standard deviation for the property of interest (Pheasant, 1987; Woodson et al., 1992). Many studies have been undertaken to collate this data, particularly in the US armed forces and space programmes.

Anthropometric data is also provided for population weights. For example, $C_V = 0.18$ for the weight of males aged between 18 and 70 (Woodson et al., 1992). For English males within this age range, the mean, $\mu = 65\,\text{kg}$, and the standard deviation, $\sigma = 11.7\,\text{kg}$. Therefore, at $+3\sigma$, you would expect 1 in 1350 males in this age range to be over 100 kg in weight, as determined from SND theory. In the US, this probability applies to males of 118 kg in weight. The designer should use this type of data whenever human interaction with the product is anticipated.

Table 4.10 Human arm strength data (Woodson *et al.*, 1992)

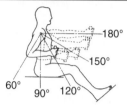

Direction of force	Elbow angle	Mean force μ (N)	Standard deviation σ (N)
Push	60°	410	169
	90°	383	147
	120°	459	192
	150°	548	201
	180°	615	218
Pull	60°	281	103
	90°	392	134
	120°	464	138
	150°	544	161
	180°	535	165

4.4 Stress–Strength Interference (SSI) analysis

Having previously introduced the key methods to determine the important variables with respect to stress and strength distributions, the most acceptable way to predict mechanical component reliability is by applying SSI theory (Dhillon, 1980). SSI analysis is one of the oldest methods to assess structural reliability, and is the most commonly used method because of its simplicity, ease and economy (Murty and Naikan, 1997; Sundararajan and Witt, 1995). It is a practical engineering tool used for quantitatively predicting the reliability of mechanical components subjected to mechanical loading (Sadlon, 1993) and has been described as a simulative model of failure (Dasgupta and Pecht, 1991).

The theory is concerned with the problem of determining the probability of failure of a part which is subjected to a loading stress, L, and which has a strength, S. It is assumed that both L and S are random variables with known PDFs, represented by $f(S)$ and $f(L)$ (Disney *et al.*, 1968). The probability of failure, and hence the reliability, can then be estimated as the area of interference between these stress and strength functions (Murty and Naikan, 1997).

SSI is recommended for situations where a considerable potential for variation exists in any of the design parameters involved, for example where large dimensional variations exist or when the anticipated design loads have a large range. The main reasons for performing it are (Ireson *et al.*, 1996):

- To ensure sufficient strength to operate in its environment and under specified loads
- To ensure no excess of material or overdesign occurs (high costs, weight, mass, volume, etc.).

Some advantages and disadvantages of SSI analysis have also been summarized (Sadlon, 1993):

Advantages:
• Addresses variability of loading stress and material strength
• Gives a quantitative estimate of reliability.

Disadvantages:
• Interference is often at the extremes of the distribution tails
• The stress variable is not always available.

4.4.1 Derivation of reliability equations

A mechanical component is considered safe and reliable when the strength of the component, S, exceeds the value of loading stress, L, on it (Rao, 1992). When the loading stress exceeds the strength, failure occurs, the reliability of the part, R, being related to this failure probability, P, by equation 4.26:

$$P(S > L) = R \qquad (4.26)$$

It is also evident that the reliability can also be derived by finding the probability of the loading stress being less than the strength of the component:

$$P(L < S) = R \qquad (4.27)$$

This is demonstrated graphically in Figure 4.27 which shows the loading stress remains less than some given value of allowable strength (Haugen, 1968; Rao, 1992). The probability of a strength S_1 in an interval dS_1 is:

$$P\left(S_1 - \frac{dS_1}{2} \leq S \leq S_1 + \frac{dS_1}{2}\right) = f(S_1)\,dS_1 = A_1 \qquad (4.28)$$

The probability of occurrence of a loading stress less than S_1 is:

$$P(L < S_1) = \int_{-\infty}^{S} f(L)\,dL = F(L) = A_2 \qquad (4.29)$$

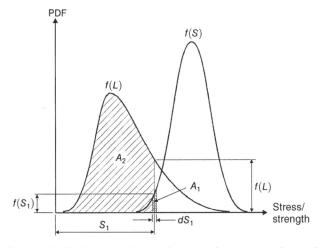

Figure 4.27 SSI theory applied to the case where loading stress does not exceed strength

The probability of these two events occurring simultaneously is the product of equations 4.28 and 4.29 which gives the elemental reliability, dR, as:

$$dR = f(S_1)\, dS_1 \cdot \int_{-\infty}^{S} f(L)\, dL \tag{4.30}$$

For all possible values of strength, the reliability, R, is then given by the integral:

$$R = \int dR = \int_{-\infty}^{\infty} f(S) \left(\int_{-\infty}^{S} f(L)\, dL \right) dS \tag{4.31}$$

This equation can be simplified by omitting negative limits to give the general equation for reliability using the SSI approach. For static design (i.e. no strength deterioration) with a single application of the load, equation 4.31 becomes:

$$R = \int_{0}^{\infty} f(S) \left(\int_{0}^{S} f(L)\, dL \right) dS \tag{4.32}$$

When the loading stress can be determined by a CDF in closed form, this simplifies to:

$$R = \int_{0}^{\infty} F(L) \cdot f(S)\, dS \tag{4.33}$$

In general both the load and strength are functions of time more often than not (Vinogradov, 1991); however, in general mechanical engineering the control of the operation is more relaxed and so it becomes more difficult to define a duty cycle (Carter, 1997). However, in most design applications the number of load applications on a component/system during its contemplated life is a large number (Bury, 1975) and this has a major effect on the predicted reliability. The time dependence of the load becomes a factor that transforms a problem of probabilistic design into one of reliability. The reliability after multiple independent load applications in sequence, R_n, is given by:

$$R_n = \int_{0}^{\infty} f(S) \left(\int_{0}^{S} f(L)\, dL \right)^{n} dS \tag{4.34}$$

where:

n = number of independent load applications in sequence

or from equation 4.33:

$$R_n = \int_{0}^{\infty} F(L)^{n} \cdot f(S)\, dS \tag{4.35}$$

Equation 4.34 represents probably one of the most important theories in reliability (Carter, 1986). The number of load applications defines the useful life of the component and is of appropriate concern to the designer (Bury, 1974). The number of times a load is applied has an effect on the failure rate of the equipment due to the fact that the probability of experiencing higher loads from the distribution population has increased. Each load application in sequence is independent and belongs to the same load distribution and it is assumed that the material suffers no strength deterioration. In fact, the strength distribution would change due to the elimination of the weak items from the population, and would in effect become truncated at the left-hand side. However, no account of strength alteration with time is included in any of the approaches discussed. Monte Carlo simulation can be used to demonstrate this principle, as well as providing another means of solving the general reliability equation. However, the major shortcoming of the Monte Carlo technique is that it requires a very large number of simulations for accurate results and is therefore not suited to low probability levels (Lemaire, 1997).

If we had taken the case where the strength exceeds the applied stress, this would have yielded a slightly different equation as shown below:

$$R = \int_0^\infty f(L) \left(\int_L^\infty f(S) \, dS \right) dL \qquad (4.36)$$

This can also be simplified when the strength CDF is in closed form to give:

$$R = 1 - \int_0^\infty F(S) \cdot f(L) \, dL \qquad (4.37)$$

4.4.2 Reliability determination with a single load application

Analytical solutions to equation 4.32 for a single load application are available for certain combinations of distributions. These *coupling equations* (so called because they couple the distributional terms for both loading stress and material strength) apply to two common cases. First, when both the stress and strength follow the Normal distribution (equation 4.38), and secondly when stress and strength can be characterized by the Lognormal distribution (equation 4.39).

Figure 4.28 shows the derivation of equation 4.38 from the algebra of random variables. (Note, this is exactly the same approach described in Appendix VIII to find the probability of interference of two-dimensional variables.)

$$z = -\frac{\mu_S - \mu_L}{\sqrt{\sigma_S^2 + \sigma_L^2}} \qquad (4.38)$$

where:

$z =$ Standard Normal variate

$\mu_S =$ mean material strength

$\sigma_S =$ material strength standard deviation

$\mu_L =$ mean applied stress

$\sigma_L =$ applied stress standard deviation.

$$z = -\frac{\ln\left(\frac{\mu_S}{\mu_L}\sqrt{\frac{1 + C_L^2}{1 + C_S^2}}\right)}{\sqrt{\ln\left[(1 + C_L^2)(1 + C_S^2)\right]}} \qquad (4.39)$$

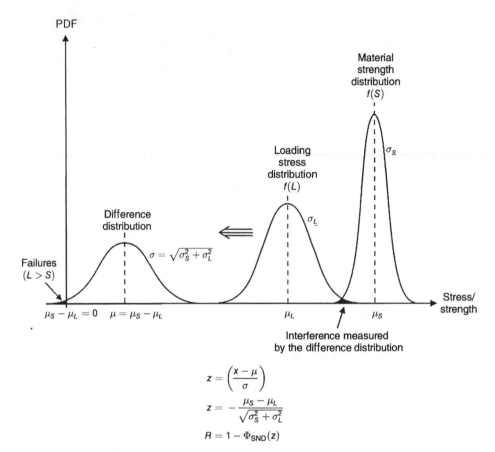

Figure 4.28 Derivation of the coupling equation for the case when both loading stress and material strength are a Normal distribution

where:

$$C_L = \frac{\sigma_L}{\mu_L} \text{ and } C_S = \frac{\sigma_S}{\mu_S} \text{ (the mean and standard deviation determined}$$

from Appendix IX).

Using the SND theory from Appendix I, the probability of failure, P, can be determined from the Standard Normal variate, z, by:

$$P = \Phi_{\text{SND}}(z) \tag{4.40}$$

The reliability, R, is therefore:

$$R = 1 - P = 1 - \Phi_{\text{SND}}(z) \tag{4.41}$$

Using the Normal distribution to model both stress and strength is the most common and the mathematics is easily managed, giving a quick and economical solution. The convenience and flexibility of using a single distribution, such as the Normal, is very attractive and has found favour with many practitioners. However, it has been observed that loading stresses and material strengths do not necessarily follow the

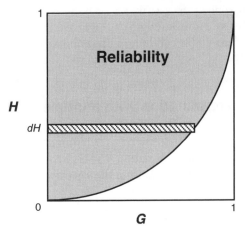

Figure 4.29 Plot of G versus H for the integral transform method

Normal or Lognormal distributions. For the purposes of SSI theory, the 3-parameter Weibull distribution is difficult to work with. The probability of failure cannot be obtained by coupling equations. This means that we need more general approaches to solve the SSI problem for any combination of distribution. One such approach is the *Integral Transform Method* (Haugen, 1968). The integral transform method starts by replacing the terms in equation 4.36 with the following:

$$G = \int_L^\infty f(S)\,dS \tag{4.42}$$

and

$$H = \int_L^\infty f(L)\,dL \tag{4.43}$$

By determining $dH = f(L)\,dL$, this changes the limits of integration to 0 to 1, as shown in Figure 4.29, to give the reliability, R, as:

$$R = \int_0^1 G\,dH \tag{4.44}$$

or

$$R = 1 - \int_0^1 H\,dG \tag{4.45}$$

Equation 4.44 is solved using *Simpson's Rule* to integrate the area of overlap. The method is easily transferred to computer code for high accuracy. See Appendix XII for a discussion of Simpson's Rule used for the numerical integration of a function. Equation 4.44 permits the calculation of reliability for any combinations of distributions for stress and strength provided the partial areas of G and H can be found.

For the general SSI method, its accuracy relies on evaluating the areas under the functions and the numerical methods employed. However, the integral transform method has also been applied to cases where no basis exists for assuming any specific distributions for either stress or strength, but where experimentation has been performed yielding sufficient empirical data (Kapur and Lamberson, 1977; Verma and Murty, 1989).

4.4.3 Reliability determination with multiple load application

The approach taken by Carter (1986, 1997) to determine the reliability when multiple load applications are experienced (equation 4.34) is first to present a Safety Margin, SM, a non-dimensional quantity to indicate the separation of the stress and strength distributions as given by:

$$SM = \frac{\mu_S - \mu_L}{\sqrt{\sigma_S^2 + \sigma_L^2}} \qquad (4.46)$$

This is essentially the coupling equation for the case when both stress and strength are a Normal distribution. A parameter to define the relative shapes of the stress and strength distributions is also presented, called the Loading Roughness, LR, given by:

$$LR = \frac{\sigma_L}{\sqrt{\sigma_S^2 + \sigma_L^2}} \qquad (4.47)$$

The typical loading roughness for mechanical products is high, typically 0.9 (Carter, 1986). A high loading roughness means that the variability of the load is high compared with that of the strength, but does not necessarily mean that the load itself is high (Leitch, 1995) (see Figure 4.30). The general run of mechanical equipment is subject to much rougher loading than, say, electronic equipment, although it has been argued that electronic equipment can also be subjected to rough loading under some conditions (Loll, 1987). It arises largely due to the less precise knowledge and control of the environment, and also from the wide range of different applications of most mechanical components. An implication of this is that the reliability of the system is relatively insensitive to the number of components and the reliability of mechanical systems is determined by its weakest link (Broadbent, 1993; Carter, 1986; Furman, 1981; Roysid, 1992). It also means that the probability of failure per application of load at high SM and LR values where 'n' is much greater than 1 is very close to that for single load applications. For this reason, Carter (1986) notes that it is impossible to design for a specified reliability with lower LR values.

For a given SM, LR and duty cycle, n (as defined by the number of times the load is applied), the failure per application of the load, p, can be determined from Figure 4.31. The figure has been constructed for near-Normal stress and strength interference conditions and where 'n' is much greater than 1. The final reliability is given by (O'Connor, 1995):

$$R_n = (1 - p)^n \qquad (4.48)$$

where:

p = probability of failure per application of load 'n'.

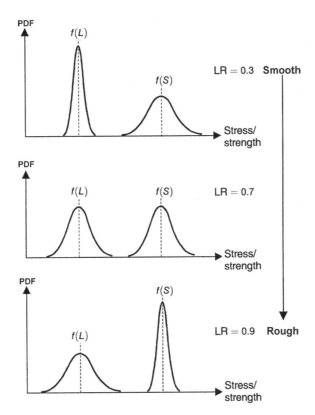

Figure 4.30 Relative shape of loading stress and strength distributions for various loading roughnesses and arbitrary safety margin

Bury (1974; 1975; 1978; 1999) introduces the concept of duty cycles to static design in a different approach. The duty cycle or mission length of a design is equivalent to the number of load applications, n; however, it is only the maximum value which is of design significance. If the load, L, is a random variable then so is its maximum value, $\hat{L}$, and the PDFs of each are related. It can be shown that the CDF, $F(\hat{L})$, of the maximum among 'n' independent observations on L is:

$$G_n(\hat{L}) = [F(\hat{L})]^n \qquad (4.49)$$

$G_n(\hat{L})$ is often difficult to determine for a given load distribution, but when 'n' is large, an approximation is given by the Maximum Extreme Value Type I distribution of the maximum extremes with a scale parameter, Θ, and location parameter, v. When the initial loading stress distribution, $f(L)$, is modelled by a Normal, Lognormal, 2-parameter Weibull or 3-parameter Weibull distribution, the extremal model parameters can be determined by the equations in Table 4.11. These equations include terms for the number of load applications, n. The extremal model for the loading stress can then be used in the SSI analysis to determine the reliability.

For example, to determine the reliability, R_n, for 'n' independent load applications, we can use equation 4.33 when the loading stress is modelled using the Maximum Extreme Value Type I distribution, as for the above approach. The CDF for the

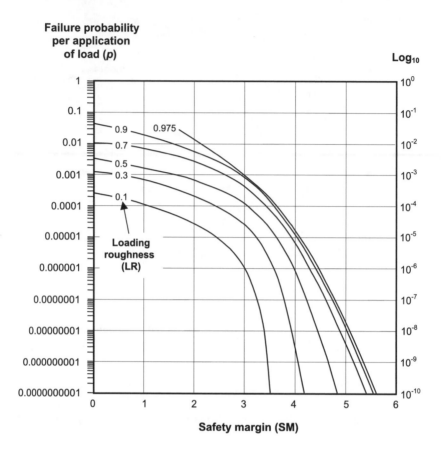

Figure 4.31 Failure probability (per application of load) versus safety margin for various loading roughness values (adapted from Carter, 1997)

Table 4.11 Extremal value parameters from initial loading stress distributions

Initial loading stress distribution	Location parameter (v)	Scale parameter (Θ)
Normal	$\mu + \sigma\left[\dfrac{2\ln(n) - 0.5\ln\ln(n) - 1.2655}{\sqrt{2\ln(n)}}\right]$	$\dfrac{\sigma}{\sqrt{2\ln(n)}}$
Lognormal	$\exp(\lambda + \alpha \cdot z)$ $z = \Theta_{SND}^{-1}\left(1 - \dfrac{1}{n}\right)$	$\alpha \cdot v\left(\dfrac{\left(\dfrac{1}{n}\right)}{\dfrac{1}{\sqrt{2\pi}}\exp\left(-\dfrac{z^2}{2}\right)}\right)$
2-parameter Weibull ($\beta \geq 1$)	$\theta[\ln(n)]^{1/\beta}$	$\left(\dfrac{\theta}{\beta}\right)[\ln(n)]^{(1-\beta/\beta)}$
3-parameter Weibull ($\beta \geq 1$)	$xo + (\theta - xo)[\ln(n)]^{1/\beta}$	$\left(\dfrac{\theta - xo}{\beta}\right)[\ln(n)]^{(1-\beta/\beta)}$

loading stress is given by:

$$F(L) = \exp\left(-\exp\left(-\frac{x - \upsilon_L}{\Theta_L}\right)\right) \qquad (4.50)$$

and the strength, $f(S)$, is given by a Normal distribution, for example by:

$$f(S) = \frac{1}{\sigma_S \sqrt{2\pi}} \exp\left(-\frac{(x - \mu_S)^2}{2\sigma_S^2}\right) \qquad (4.51)$$

the reliability is given by equation 4.33 by substituting in these terms:

$$R_n = \int_0^\infty \left[\exp\left(-\exp\left(-\frac{x - \upsilon_L}{\Theta_L}\right)\right)\right] \cdot \left[\frac{1}{\sigma_S \sqrt{2\pi}} \exp\left(-\frac{(x - \mu_S)^2}{2\sigma_S^2}\right)\right] dx \qquad (4.52)$$

which can be solved numerically using Simpson's Rule as shown in Appendix XII.

The numerical solution of equation 4.35 is sufficient in most cases to provide a reasonable answer for reliability with multiple load applications for any combination of loading stress and strength distribution (Freudenthal *et al.*, 1966).

4.4.4 Reliability determination when the stress is a maximum value and strength is variable

The assumption that a unique maximum loading stress (i.e. variability) is assigned as being representative in the probabilistic model when variability exists in strength sometimes applies, and this is treated as a special case here. The problem is shown in Figure 4.32. We can refer to this maximum stress, L_{max}, from the beginning of application until it is removed. Several time dependent loading patterns may be treated as maximum loading cases, for example the torque applied to a bolt or pressure applied to a rivet. If the applied load is short enough in duration not to cause weakening of the strength due to fatigue, then it may be represented by a maximum load. The resulting reliability does not depend on time and is simply the

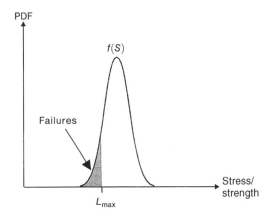

Figure 4.32 Maximum static loading stress and variable strength

probability that the system survives the application of the load. However, in reality loads are subject to variability and if the component does not have the strength to sustain any one of these, it will fail (Lewis, 1996).

Therefore, the reliability $R_{L_{max}}$ is given by:

$$R_{L_{max}} = 1 - P = 1 - \int_0^{L_{max}} f(S) \, dS \qquad (4.53)$$

where:

$$f(S) = \text{strength distribution.}$$

Equation 4.53 can be solved by integrating the $f(S)$ using Simpson's Rule or by using the CDF for the strength directly when in closed form, i.e. $R = 1 - F(L_{max})$. In the case of the Normal and Lognormal distributions, the use of SND theory makes the calculation straightforward. The above formulation suggests that all strength values less than the maximum loading stress will fail irrespective of any actual variation on the loading conditions which may occur in practice.

4.4.5 Example – calculation of reliability using different loading cases

Consider the situation where the loading stress on a component is given as $L \sim N(350, 40)$ MPa relating to a Normal distribution with a mean of $\mu_L = 350$ MPa and standard deviation $\sigma_L = 40$ MPa. The strength distribution of the component is $S \sim N(500, 50)$ MPa. It is required to find the reliability for these conditions using each approach above, given that the load will be applied 1000 times during a defined duty cycle.

Maximum static loading stress, L_max, with variable strength

If we assume that the maximum stress applied is $+3\sigma$ from the mean stress, where this loading stress value covers 99.87% those applied in service:

$$L_{max} = 350 + 3(40) = 470 \, \text{MPa}$$

Because the strength distribution is Normal, we can determine the Standard Normal variate, z, as:

$$z = \left(\frac{x - \mu}{\sigma}\right) = \left(\frac{470 - 500}{50}\right) = -0.6$$

From Table 1 in Appendix I, the probability of failure $P = 0.274253$. The reliability, R, is given by:

$$R = 1 - P = 1 - 0.274253$$

$$R_{L_{max}} = 0.725747$$

The reliability, $R_{L_{max}}$, as a function of the maximum stress value used is shown in Figure 4.33. The reliability rapidly falls off at higher values of stress chosen, such

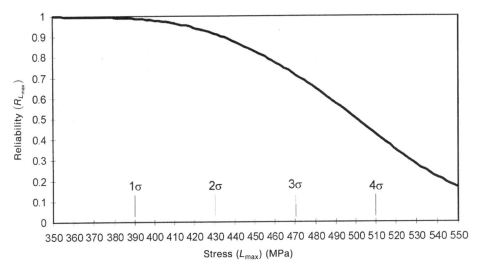

Figure 4.33 Reliability as a function of maximum stress, L_{max}

as 4σ. A particular difficulty in this approach is, then, the choice of maximum loading stress that reflects the true stress of the problem.

Single application of a variable static loading stress with variable strength

Substituting the given parameters for stress and strength in the coupling equation (equation 4.38) gives:

$$z = -\frac{500 - 350}{\sqrt{50^2 + 40^2}} = -2.34$$

From Table 1 in Appendix I, the probability of failure $P = 0.009642$. The reliability is then:

$$R = 1 - P = 1 - 0.009642$$

$$R_1 = 0.990358$$

This value is much more optimistic, as you would assume, because the reliability is the probability of both stress and strength being interfering, not just the strength being equal to a maximum value.

Variable static loading stress with a defined duty cycle of 'n' load applications with variable strength using approaches by Bury (1974), Carter (1997) and Freudenthal et al. (1966)

Using Carter's approach first, from equation 4.47 we can calculate LR to be:

$$\text{LR} = \frac{40}{\sqrt{50^2 + 40^2}} = 0.62$$

We already know SM $= 2.34$ because it is the positive value of the Standard Normal variate, z, calculated above. The probability of failure per application of load

$p \approx 0.0009$ from Figure 4.31. Using equation 4.48, and given that $n = 1000$, gives the reliability as:

$$R_n = (1 - p)^n = (1 - 0.0009)^{1000}$$

$$R_{1000} = 0.406405$$

Next using Bury's approach, from Table 4.11 the extremal parameters, v and Θ, from an initial Normal loading stress distribution are determined from:

$$v = \mu + \sigma \left[\frac{2\ln(n) - 0.5\ln\ln(n) - 1.2655}{\sqrt{2\ln(n)}} \right]$$

$$= 350 + 40 \left[\frac{2\ln(1000) - 0.5\ln\ln(1000) - 1.2655}{\sqrt{2\ln(1000)}} \right] = 474.66\,\text{MPa}$$

and

$$\Theta = \frac{\sigma}{\sqrt{2\ln(n)}} = \frac{40}{\sqrt{2\ln(1000)}} = 10.76\,\text{MPa}$$

Substituting in the parameters for both stress and strength into equation 4.52 and solving using Simpson's Rule (integrating between the limits of 1 and 1000, for example) gives that the reliability is:

$$R_{1000} = 0.645$$

This value is more optimistic than that determined by Carter's approach.

Next, solving equation 4.35 directly using Simpson's Rule for R_n as described by Freudenthal *et al.* (1966):

$$R_n = \int_0^\infty F(L)^n \cdot f(S)\, dS$$

A 3-parameter Weibull approximates to a Normal distribution when $\beta = 3.44$, and so we can convert the Normal stress to Weibull parameters by using:

$$xo_L \approx \mu_L - 3.1394473\sigma_L$$

$$\theta_L \approx \mu_L + 0.3530184\sigma_L$$

$$\beta_L = 3.44$$

Therefore, the loading stress CDF can be represented by a 3-parameter Weibull distribution:

$$F(L) = 1 - \exp\left(-\left(\frac{x - xo_L}{\theta_L - xo_L} \right)^{\beta_L} \right)$$

and the strength is represented by a Normal distribution, the PDF being as follows:

$$f(S) = \frac{1}{\sigma_S\sqrt{2\pi}} \exp\left(-\frac{(x - \mu_S)^2}{2\sigma_S^2} \right)$$

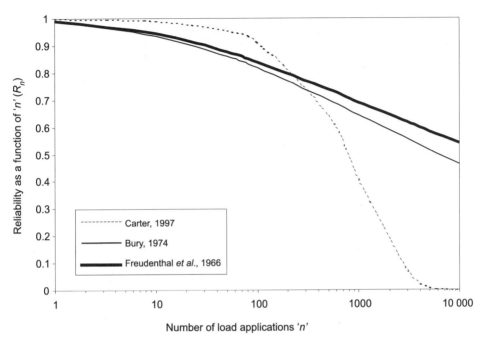

Figure 4.34 Reliability as a function of number of load applications using different approaches for LR = 0.62 (medium loading roughness) and SM = 2.34

Therefore, substituting these into equation 4.35 gives the reliability, R_n, as:

$$R_n = \int_{S=xo_L}^{\infty} \left[1 - \exp\left(-\left(\frac{x - xo_L}{\theta_L - xo_L} \right)^{\beta_L} \right) \right]^n \cdot \left[\frac{1}{\sigma_S \sqrt{2\pi}} \exp\left(-\frac{(x - \mu_S)^2}{2\sigma_S^2} \right) \right] dx$$

From the solution of this equation numerically for $n = 1000$, the reliability is found to be:

$$R_{1000} = 0.690$$

Figure 4.34 shows the reliability as a function of the number of load applications, R_n, using the three approaches described to determine R_n. There is a large discrepancy between the reliability values calculated for $n = 1000$. Repeating the exercise for the same loading stress, $L \sim N(350, 40)$ MPa, but with a strength distribution of $S \sim N(500, 20)$ MPa increases the LR value to 0.89 and SM = 3.35. Figure 4.35 shows that at higher LR values, the results are in better agreement, up to approximately 1000 load applications, which is the limit for static design.

The above exercise suggests that if we had used equation 4.32 to determine the reliability of a component when it is known that the load may be applied many times during its life, an overoptimistic value would have been obtained. This means that the component could experience more failures than that anticipated at the design stage. This is common practice and a fundamentally incorrect approach (Bury, 1975). A high confidence in the reliability estimates is accepted for the situation where a single application of the load is experienced, R_1. However, the confidence is

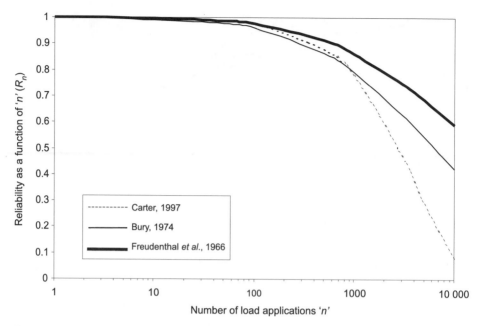

Figure 4.35 Reliability as a function of number of load applications using different approaches for LR = 0.89 (rough loading) and SM = 3.35

lower when determining the reliability as a function of the number of load applications, R_n, when $n \gg 1$, using the various approaches outlined at low LR values. At higher LR values, the three approaches to determine the reliability for $n \gg 1$ do give similar results up to $n = 1000$, this being the limit of the number of load applications valid for static design. It can also be seen that a very high initial reliability is required from the design at R_1 to be able to survive many load applications and still maintain a high reliability at R_n.

4.4.6 Extensions to SSI theory

The question arises from the above as to the amount of error in reliability calculations due to the assumption of normally distributed strength and particularly the loading stress, when in fact one or both could be Lognormal or Weibull. Distributions with small coefficients of variations ($C_v \leq 0.1$ for the Lognormal distribution) tend to be symmetrical and have a general shape similar to that of the Normal type with approximately the same mean and standard deviation. Differences do occur at the tail probabilities (upper tail for stress, lower tail for strength) and significant errors could occur from substituting the symmetrical form of the Normal distribution for a skewed distribution.

If the form of the distributional model is only approximately correct, then the tails may differ substantially from the tails of the actual distribution. This is because the model parameters, related to low order moments, are determined from typical rather than rare events. In this case, design decisions will be satisfactory for bulk occurrences, but may be less than optimal for rare events. It is the rare events, catastrophes, for example, which are often of greatest concern to the designer. The suboptimality may be manifested as either an overconservative or an unsafe design (Ben-Haim, 1994). Many authors have noted that the details of PDFs are often difficult to verify or justify with concrete data at the tails of the distribution. If one fails to model the tail behaviour of the basic variables involved correctly, then the resulting reliability level is questionable as noted (Maes and Breitung, 1994). It may be advantageous, therefore, to use statistics with the aim of determining certain tail characteristics of the random variables. Tail approximation techniques have developed specifically to achieve this (Kjerengtroen and Comer, 1996).

Confidence levels on the reliability estimates from the SSI model can be determined and are useful when the PDFs for stress and strength are based on only small amounts of data or where critical reliability projects are undertaken. However, approaches to determine these confidence levels only strictly apply when stress and strength are characterized by the Normal distribution. Detailed examples can be found in Kececioglu (1972) and Sundararajan and Witt (1995).

Another consideration when using the approach is the assumption that stress and strength are statistically independent; however, in practical applications it is to be expected that this is usually the case (Disney *et al.*, 1968). The random variables in the design are assumed to be independent, linear and near-Normal to be used effectively in the variance equation. A high correlation of the random variables in some way, or the use of non-Normal distributions in the stress governing function are often sources of non-linearity and transformations methods should be considered. These are generally called *Second Order Reliability Methods*, where the use of independent, near-Normal variables in reliability prediction generally come under the title *First Order Reliability Methods* (Kjerengtroen and Comer, 1996). For economy and speed in the calculation, however, the use of *First Order Reliability Methods* still dominates presently.

4.5 Elements of stress analysis and failure theory

The calculated loading stress, L, on a component is not only a function of applied load, but also the stress analysis technique used to find the stress, the geometry, and the failure theory used (Ullman, 1992). Using the variance equation, the parameters for the dimensional variation estimates and the applied load distribution, a statistical failure theory can then be formulated to determine the stress distribution, $f(L)$. This is then used in the SSI analysis to determine the probability of failure together with material strength distribution $f(S)$.

Use of the classical stress analysis theories to predict failure involves firstly identifying the maximum or effective stress, L, at the critical location in the part and then comparing that stress condition with the strength, S, of the part at that location (Shigley and Mischke, 1996). Among such maximum stress determining factors are: stress concentration factors; load factors (static, dynamic, impact) applied to axial, bending and torsional loads; temperature stress factors; forming or manufacturing stress factors (residual stresses, surface and heat treatment factors); and assembly stress factors (shrink fits and press fits) (Haugen, 1968). The most significant factor in failure theory is the character of loading, whether static or dynamic. In the discussion that follows, we constrain the argument to failure by static loading only.

Often in stress analysis we may be required to make simplified assumptions, and as a result, uncertainties or loss of accuracy are introduced (Bury, 1975). The accuracy of calculation decreases as the complexity increases from the simple case, but ultimately the component part will still break at its weakest section. Theoretical failure formulae are devised under assumptions of ideal material homogeneity and isotropic behaviour. Homogeneous means that the materials properties are uniform throughout; isotropic means that the material properties are independent of orientation or direction. Only in the simplest of cases can they furnish us with the complete solution of the stress distribution problem. In the majority of cases, engineers have to use approximate solutions and any of the real situations that arise are so complicated that they cannot be fully represented by a single mathematical model (Gordon, 1991).

The failure determining stresses are also often located in local regions of the component and are not easily represented by standard stress analysis methods (Schatz et al., 1974). Loads in two or more axes generally provide the greatest stresses, and should be resolved into principal stresses (Ireson et al., 1996). In static failure theory, the error can be represented by a coefficient of variation, and has been proposed as $C_v = 0.02$. This margin of error increases with dynamic models and for static finite element analysis, the coefficient of variation is cited as $C_v = 0.05$ (Smith, 1995; Ullman, 1992).

Understanding the potential failure mechanisms of a product is also necessary to develop reliable products. Failure mechanisms can be broadly grouped into overstress (for example, brittle fracture, ductile fracture, yield, buckling) and wear-out (wear, corrosion, creep) mechanisms (Dasgupta and Pecht, 1991). Gordon (1991) argues that the number of failure modes observed probably increases with complexity of the system, therefore effective failure analysis is an essential part of reliability work (Burns, 1994). The failure governing stress must be determined for the failure mode in question and the use of FMEA in determining possible failure modes is crucial in this respect.

4.5.1 Simple stress systems

In postulating a statistical model for a static stress variable, it is important to distinguish between brittle and ductile materials (Bury, 1975). For simple stress systems, i.e. uniaxial or pure torsion, where only one type of stress acts on the component, the following equations determine the failure criterion for ductile and brittle types to predict the reliability (Haugen, 1980):

For ductile materials in uniaxial tension, the reliability is the probabilistic requirement to avoid yield:

$$R = P(Sy > L) \tag{4.54}$$

For brittle materials in tension, the reliability is given by the probabilistic requirement to avoid tensile fracture:

$$R = P(Su > L) \tag{4.55}$$

For ductile materials subjected to pure shear, the reliability is the probabilistic requirement to avoid shear yielding:

$$R = P(\tau_y = 0.577Sy > L) \tag{4.56}$$

where:

$$Sy = \text{yield strength}$$

$$Su = \text{ultimate tensile strength}$$

$$L = \text{loading stress}$$

$$\tau_y = \text{shear yield strength.}$$

The formulations for the failure governing stress for most stress systems can be found in Young (1989). Using the variance equation and the parameters for the dimensional variation estimates and applied load, a statistical failure theory can be formulated for a probabilistic analysis of stress rupture.

4.5.2 Complex stress systems

Predicting failure and establishing geometry that will avert failure is a relatively simple matter if the machine is subjected to uniaxial stress or pure torsion. It is far more difficult if biaxial or triaxial states of stress are encountered. It is therefore desirable to predict failure utilizing a theory that relates failure in the multiaxial state of stress by the same mode in a simple tension test through a chosen modulus, for example stress, strain or energy. In order to determine suitable allowable stresses for the complicated stress conditions that occur in practical design, various theories have been developed. Their purpose is to predict failure (yield or rupture) under combined stresses assuming that the behaviour in a tension or compression test is known. In general, ductile materials in static tensile loading are limited by their shear strengths while brittle materials are limited by their tensile strengths (though there are exceptions to these rules when ductile materials behave as if they were brittle) (Norton, 1996). This observation required the development of different failure theories for the two main static failure modes, ductile and brittle fracture.

Ductile fracture
Of all the theories dealing with the prediction of yielding in complex stress systems, the *Distortion Energy Theory* (also called the *von Mises Failure Theory*) agrees best with experimental results for ductile materials, for example mild steel and aluminium (Collins, 1993; Edwards and McKee, 1991; Norton, 1996; Shigley and Mischke, 1996). Its formulation is given in equation 4.57. The right-hand side of the equation is the effective stress, L, for the stress system.

$$2Sy^2 = (s_1 - s_2)^2 + (s_2 - s_3)^2 + (s_3 - s_1)^2 \tag{4.57}$$

where s_1, s_2, s_3 are the principal stresses.

Therefore, for ductile materials under complex stresses, the reliability is the probabilistic requirement to avoid yield as given by:

$$R = P\left(Sy > \sqrt{\frac{(s_1 - s_2)^2 + (s_2 - s_3)^2 + (s_3 - s_1)^2}{2}} \right) \qquad (4.58)$$

Again, using the variance equation, the parameters for the dimensional variation estimates and applied load to determine the principal stress variables, a statistical failure theory can be determined. The same applies for brittle material failure theories described next. In summary, the Distortion Energy Theory is an acceptable failure theory for ductile, isotropic and homogeneous materials in static loading, where the tensile and compressive strengths are of the same magnitude. Most wrought engineering metals and some polymers are in this category (Norton, 1996).

Brittle fracture

The arbitrary division between brittle and ductile behaviour is when the elongation at fracture is less than 5% (Shigley and Mischke, 1989) or when Sy is greater than $E/1034.2$ (in Pa) (Haugen, 1980). Most ductile materials have elongation at fracture greater than 10% (Norton, 1996). However, brittle failure may be experienced in ductile materials operating below their transition temperature (as described in Section 4.3). Brittle failure can also occur in ductile materials at sharp notches in the component's geometry, termed triaxiality of stress (Edwards and McKee, 1991). Strain rate as well as defects and notches in the materials also induce ductile-to-brittle behaviour in a material. Such defects can considerably reduce the strength under static loading (Ruiz and Koenigsberger, 1970).

Brittle fracture is the expected mode of failure for materials like cast iron, glass, concrete and ceramics and often occurs suddenly and without warning, and is associated with a release of a substantial amount of energy. Brittle materials, therefore, are less suited for impulsive loading than ductile materials (Faires, 1965). In summary, the primary factors promoting brittle fracture are then (Juvinall, 1967):

- Low temperature – increases the resistance of the material to slip, but not cleavage
- Rapid loading – shear stresses set up in impact may be accompanied by high normal stresses which exceed the cleavage strength of the material
- Triaxial stress states – high tensile stresses in comparison to shear stresses
- Size effect on thick sections – may have lower ductility than sample tests.

By definition, a brittle material does not fail in shear; failure occurs when the largest principal stress reaches the ultimate tensile strength, Su. Where the ultimate compressive strength, Su_c, and Su of brittle material are approximately the same, the *Maximum Normal Stress Theory* applies (Edwards and McKee, 1991; Norton, 1996). The probabilistic failure criterion is essentially the same as equation 4.55.

Materials such as cast-brittle metals and composites do not exhibit these uniform properties and require more complex failure theories. Where the properties Su_c and Su of a brittle material vary greatly (approximately 4 : 1 ratio), the *Modified Mohr Theory* is preferred and is good predictor of failure under static loading conditions (Norton, 1996; Shigley and Mischke, 1989).

The stress, L, determined using the Modified Mohr method effectively accounts for all the applied stresses and allows a direct comparison to a materials strength property to be made (Norton, 1996), as was established for the Distortion Energy Theory for ductile materials. The set of expressions to determine the effective or maximum stress are shown below and involve all three principal stresses (Dowling, 1993):

$$C_1 = \frac{1}{2}\left[|s_1 - s_2| + \frac{Su_c - 2Su}{Su_c}(s_1 + s_2)\right] \tag{4.59}$$

$$C_2 = \frac{1}{2}\left[|s_2 - s_3| + \frac{Su_c - 2Su}{Su_c}(s_2 + s_3)\right] \tag{4.60}$$

$$C_3 = \frac{1}{2}\left[|s_3 - s_1| + \frac{Su_c - 2Su}{Su_c}(s_3 + s_1)\right] \tag{4.61}$$

where:

$$L = \max(C_1, C_2, C_3, s_1, s_2, s_3) \tag{4.62}$$

The effective stress is then compared to the materials ultimate tensile strength, Su; the reliability is given by the probabilistic requirement to avoid tensile fracture (Norton, 1996):

$$R = P(Su > L) \tag{4.63}$$

In a probabilistic sense, this is the same as equation 4.55, but for brittle materials under complex stresses.

Stress raisers, whether caused by geometrical discontinuities such as notches or by localized loads should be avoided when designing with brittle materials. The geometry and loading situation should be such as to minimize tensile stresses (Ruiz and Koenigsberger, 1970). The use of brittle materials is therefore dangerous, because they may fail suddenly without noticeable deformation (Timoshenko, 1966). They are not recommended for practical load bearing designs where tensile loads may be present.

4.5.3 Fracture mechanics

The static failure theories discussed above all assume that the material is perfectly homogeneous and isotropic, and thus free from defects, such as cracks that could serve as stress raisers. This is seldom true for real materials, which could contain cracks due to processing, welding, heat treatment, machining or scratches through mishandling. Localized stresses at the crack tips can be high enough for even ductile materials to fracture suddenly in a brittle manner under static loading. If the zone of yielding around the crack is small compared to the dimensions of the part (which is commonly the case), then *Linear Elastic Fracture Mechanics* (LEFM) theory is applicable (Norton, 1996). In an analysis, the largest crack would be examined which is perpendicular to the line of maximum stress on the part.

In general, a stress intensity factor, K, can be determined for the stress condition at the crack tip from:

$$K = \beta\sigma_{\text{nom}}\sqrt{\pi a} \tag{4.64}$$

where:

K = stress intensity factor

β = factor depending on the part geometry and type of loading

σ_{nom} = nominal stress in absence of the crack

a = crack length.

The stress intensity factor can then be compared to the fracture toughness for the material, K_c, which is a property of the material which measures its resistance against crack formation, where K_c can be determined directly from tests or by the equation below (Ashby and Jones, 1989):

$$K_c = \sqrt{EG_c} \tag{4.65}$$

where:

K_c = fracture toughness

E = Modulus of Elasticity

G_c = toughness.

As long as the stress intensity factor is below the fracture toughness for the material, the crack can be considered to be in a stable mode (Norton, 1996), i.e. fast fracture occurs when $K = K_c$. The development of a probabilistic model which satisfies the above can be developed and reference should be made to specialized texts in this field such as Bloom (1983), but in general, the reliability is determined from the probabilistic requirement:

$$R = P(K_c > K) \tag{4.66}$$

Also see Furman (1981) and Haugen (1980) for some elementary examples. For a comprehensive reference for the determination of stress intensity factors for a variety of geometries and loading conditions, see Murakami (1987).

4.6 Setting reliability targets

4.6.1 Reliability target map

The setting of quality targets for product designs has already been explored in the Conformability Analysis (CA) methodology in Chapter 2. During the development of CA, research into the effects of non-conformance and associated costs of failure found that an area of acceptable design can be defined for a component characteristic on a graph of Occurrence (or ppm) versus Severity as shown in Figure 2.22. Here then we have the two elements of risk – Occurrence, or How many times do we expect the event to occur? – and Severity, What are the consequences on the user or environment? Furthermore, it was possible to plot points on this graph and construct lines of equal quality cost (% isocosts) which represent a percentage of the total product cost. See Figure 2.20 for a typical FMEA Severity Ratings table.

The acceptable design area was defined by a minimum acceptable quality cost line of 0.01%. The 0.01% line implies that even in a well-designed product there is a quality cost; 100 dimensional characteristics on the limit of acceptable design are likely to incur 1% of the product cost in failures. Isocosts in the non-safety critical region (FMEA Severity Rating ≤ 5) come from a sample of businesses and assume levels of cost at internal failure, returns from customer inspection or test (80%) and warranty returns (20%). The costs in the safety critical region (FMEA Severity Rating > 5) are based on allowances for failure investigations, legal actions and product recall. In essence, as failures get more severe, they cost more, so the only approach available to a business is to reduce the probability of occurrence. Therefore, the quality–cost model or Conformability Map enables appropriate capability levels to be selected based on the FMEA Severity Rating (S) and levels of design acceptability, that is, acceptable or special control.

Reliability as well as safety are important quality dimensions (Bergman, 1992) and design target reliabilities should be set to achieve minimum cost (Carter, 1997). The situation for quality–cost described above is related to reliability. The above assumed a failure cost of 0.01% of the total product cost, where typically 100 dimensional characteristics are associated with the design, giving a total failure cost of 1%. In mechanical design, it is a good assumption that the product fails from its weakest link, this assumption being discussed in detail below, and so an acceptable failure cost for reliability can be based on the 1% isocost line. Also assuming that 100% of the failures are found in the field (which is the nature of stress rupture), it can be shown that this changes the location of the acceptable design limits as redrawn on the proposed reliability target map given in Figure 4.36.

The figure also includes areas associated with overdesign. The overdesign area is probably not as important as the limiting failure probability for a particular Severity Rating, but does identify possible wasteful and costly designs. Failure targets are a central measure and are bounded by some range which spans a space of credibility, never a point value because of the confidence underlying the distributions used for prediction (Fragola, 1996).

Reliability targets are typically set based on previous product failures or existing design practice (Ditlevsen, 1997); however, from the above arguments, an approach based on FMEA results would be useful in setting reliability targets early in the design process. Large databases and risk analyses would become redundant for use at the design stage and the designer could quickly assess the design in terms of unreliability, reliability success or overdesign when performing an analysis. Various workers in this area have presented target failure probabilities ranging from 10^{-3} for unstressed applications to 10^{-9} for intrinsic reliability (Carter, 1997; Dieter, 1986; Smith, 1993), but with limited consideration of safety and/or cost. These values fit in well with the model proposed.

4.6.2 Example – assessing the acceptability of a reliability estimate

From the example in Section 4.4.5, we found that for a single load application when stress and strength are variable gave a reliability $R_1 = 0.990358$. We assume that this loading condition reflects that in service. We can now consult the reliability target

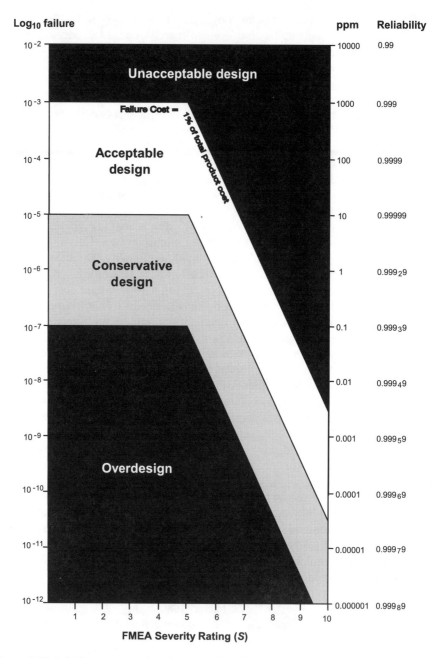

Figure 4.36 Reliability target map based on FMEA Severity

map in Figure 4.36 to assess the level of acceptability for the product design. Given that the FMEA Severity Rating $(S) = 6$ for the design (i.e. safety critical), a target value of $R = 0.99993$ is an acceptable reliability level. The design as it stands is not reliable, the reliability estimate being in the 'Unacceptable Design' region on the map.

It is left to the designer to seek a way of meeting the target reliability values required in a way that does not compromise the safety and/or cost of the product by the methods given.

4.6.3 System reliability

Most products have a number of components, subassemblies and assemblies which all must function in order that the product system functions. Each component contributes to the overall system performance and reliability. A common configuration is the series system, where the multiplication of the individual component reliabilities in the system, R_i, gives the overall system reliability, R_{sys}, as shown by equation 4.67. It applies to system reliability when the individual reliabilities are statistically independent (Leitch, 1995):

$$R_{sys} = R_1 \cdot R_2 \ldots R_m \qquad (4.67)$$

where:

$$m = \text{number of components in series.}$$

In reliability, the objective is to design all the components to have equal life so the system will fail as a whole (Dieter, 1986). It follows that for a given system reliability, the reliability of each component for equal life should be:

$$R_i = \sqrt[m]{R_{sys}} \qquad (4.68)$$

This is shown graphically in Figure 4.37. As can be seen, small changes in component reliability cause large changes in the overall system reliability using this approach

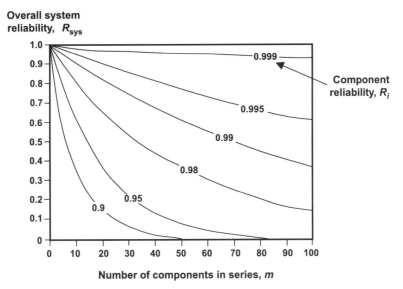

Figure 4.37 Component reliability as a function of overall system reliability and number of components in series (adapted from Michaels and Woods, 1989)

(Amster and Hooper, 1986). Other formulations exist for components in parallel with equal reliability values, as shown in equation 4.69, and for combinations of series, parallel and redundant components in a system (Smith, 1997). The complexity of the equations to find the system reliability further increases with redundancy of components in the system and the number of parallel paths (Burns, 1994):

$$R_i = 1 - (1 - R_{\text{sys}})^{1/m} \tag{4.69}$$

where:

$$m = \text{number of components in parallel.}$$

In very complex systems, grave consequences can result from the failure of a single component (Kapur and Lamberson, 1977), therefore if the weakest item can endure the most severe duty without failing, it will be completely reliable (Bompas-Smith, 1973). It follows that relationships like those developed above must be treated with caution and understanding (Furman, 1981). The simple models of 'in series' and 'in parallel' configurations have seldom been confirmed in practice (Carter, 1986). The loading roughness of most mechanical systems is high, as discussed earlier. The implication of this is that the reliability of the system is relatively insensitive to the number of components, and therefore their arrangement, and the reliability of mechanical systems is determined by their weakest link (Broadbent, 1993; Carter, 1986; Furman, 1981; Roysid, 1992).

Carter (1986) illustrates this rule using Figure 4.38 relating the loading roughness and the number of components in the system. Failure to understand it can lead to errors of judgement and wrong decisions which could prove expensive and/or

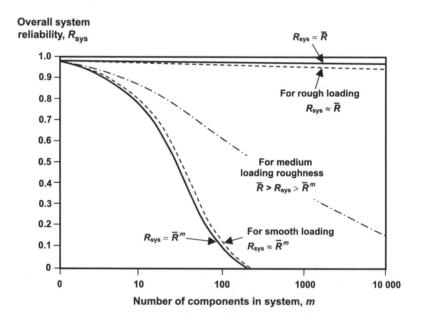

Figure 4.38 Overall system reliability as a function of the mean component reliability, $\bar{R}$, for various loading roughnesses (adapted from Carter, 1986)

catastrophic during development or when the equipment comes into service (Leitch, 1990). In conclusion, the overall reliability of a system with a number of components in series lies somewhere between that of the product of the component reliabilities and that of the least reliable component. System reliability could also be underestimated if loading roughness is not taken into account at the higher values (Leitch, 1990).

In the case where high loading roughness is expected, as in mechanical design, simply referring to the reliability target map is sufficient to determine a reliability level which is acceptable for the given failure severity for the component/system early in the design process.

4.7 Application issues

The reliability analysis approach described in this text is called *CAPRAstress* and forms part of the *CAPRA* methodology (**CA**pabilty and **PR**obabilistic Design Analysis). Activities within the approach should ideally be performed as capability knowledge and knowledge of the service conditions accumulate through the early stages of product development, together with qualitative data available from an FMEA. The objectives of the approach are to:

- Model the most important design dependent variables (material strength, dimensions, loads)
- Determine reliability targets and failure modes taken from design FMEA inputs
- Provide reliability estimates
- Provide redesign information using sensitivity analysis
- Solve a wide range of mechanical engineering problems.

The procedure for performing an analysis using the probabilistic design technique is shown in Figure 4.39 and has the following main elements:

- Determination of the material strength from statistical methods and/or database (**Stage 1**)
- Determination of the applied stress from the operating loads (**Stage 2**)
- Reliability estimates – determined from the appropriate failure mode and failure theory using Stress–Strength Interference (SSI) analysis (**Stage 3**)
- Comparison made to the target reliability (**Stage 4**)
- Redesign if unable to meet target reliability.

In the event that the reliability target is not met, there are four ways the designer can increase reliability (Ireson *et al.*, 1996):

- Increase mean strength (increasing size, using stronger materials)
- Decrease average stress (controlling loads, increasing dimensions)
- Decrease strength variations (controlling the process, inspection)
- Decrease stress variations (limitations on use conditions).

Through the use of techniques like sensitivity analysis, the approach will guide the designer to the key parameters in the design.

A key problem in probabilistic design is the generation of the PDFs from available information of the random nature of the variable (Siddal, 1983). The methods

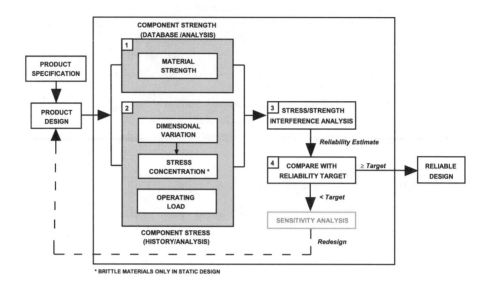

Figure 4.39 The *CAPRAstress* methodology (static design)

described allow the most suitable distribution (Weibull, Normal, Extreme Value Type I, etc.) to be used to model the data. If during the design phase there isn't sufficient information to determine the distributions for all of the input variables, probabilistic methods allow the user to assume distributions and then perform sensitivity studies to determine the critical values which affect reliability (Comer and Kjerengtroen, 1996). However, without the basic information on all aspects of component behaviour, reliability prediction can be little more than conjecture (Carter, 1986). It is largely the appropriateness and validity of the input information that determines the degree of realism of the design process, the ability to accurately predict the behaviour and therefore the success of the design (Bury, 1975). A key objective of the methodology is to provide the designer with a deeper understanding of these critical design parameters and how they influence the adequacy of the design in its operating environment. The design intent must be to produce detailed designs that reflect a high reliability when in service.

The use of computers is essential in probabilistic design (Siddal, 1983). However, research has shown that even the most complete computer supported analytical methods do not enable the designer to predict reliability with sufficiently low statistical risk (Fajdiga *et al.*, 1996). Far more than try to decrease the statistical risk, which is probably impossible, it is hoped that the approach will make it possible to model a particular situation more completely, and from this provide the necessary redesign information which will generate a reliable design solution.

It will be apparent from the discussions in the previous sections that an absolute value of reliability is at best an educated guess. However, the risk of failure determined is a quantitative measure in terms of safety and reliability by which various parts can be defined and compared (Freudenthal *et al.*, 1966). In developing a reliable product, a number of design schemes should be generated to explore each for their ability to meet the target requirements. Evaluating and comparing alternative designs and choosing the one with the greatest predicted reliability will provide the most effective design solution, and this is the approach advocated here for most applications, and by many others working in this area (Bieda and Holbrook, 1991; Burns, 1994; Klit *et al.*, 1993).

An alternative approach to the designer selecting the design with the highest reliability from a number of design schemes is to make small redesign improvements in the original design, especially if product development time is crucial. The objective could be to maximize the improvement in reliability, this being achieved by many systematic changes to the design configuration (Clausing, 1994). Although high reliability cannot be measured effectively, the design parameters that determine reliability can be, and the control and verification of these parameters (along with an effective product development strategy) will lead to the attainment of a reliable design (Ireson *et al.*, 1996). The designer should keep this in mind when designing products, and gather as much information about the critical parameters throughout the product development process before proceeding with any analysis. The achievement of high reliability at the design stage is mainly the application of engineering common sense coupled with a meticulous attention to trivial details (Carter, 1986).

The range of problems that probabilistic techniques can be applied to is vast, basically anywhere where variability dominates that problem domain. If the component is critical and if the parameters are not well known, then their uncertainty must be included in the analysis. Under these sorts of requirements, it is essential to quantify the reliability and safety of engineering components, and probabilistic analysis must be performed (Weber and Penny, 1991). In terms of SSI analysis, the main application modes are:

- Stress rupture – ductile and brittle fracture for simple and complex stresses
- Assembly features – torqued connections, shrink fits, snap fits, shear pins and other weak link mechanisms.

The latter is an area of special interest. Stress distributions in joints due to the mating of parts on assembly are to be investigated. Stresses are induced by the assembly operation and have effects similar to residual stresses (Faires, 1965). This is an important issue since many industrial problems result from a failure to anticipate production effects in mating components. Also, the probabilistic analysis of problems involving deflection, buckling or vibration is made possible using the methods described.

We will now go on to illustrate the application of the methodology to a number of problems in engineering design.

4.8 Case studies

4.8.1 Solenoid torque setting

The assembly operation of a proposed solenoid design (as shown in Figure 4.40) has two failure modes as determined from an FMEA. The first failure mode is that it

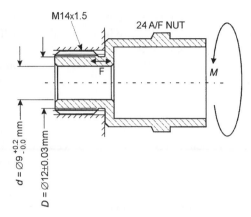

Figure 4.40 Solenoid arrangement on assembly

could fail at the weakest section by stress rupture due to the assembly torque, and secondly that the pre-load, F, on the solenoid thread is insufficient and could cause loosening in service. The FMEA Severity Rating (S) for the solenoid is 5 relating to a warranty return if it fails in service. The objective is to determine the mean torque, M, to satisfy these two competing failure modes using a probabilistic design approach.

The material used for the solenoid body is 220M07 free cutting steel. It has a minimum yield strength $Sy_{min} = 340\,\text{MPa}$ and a minimum proof stress $Sp_{min} = 300\,\text{MPa}$ for the size of bar stock (BS 970, 1991). The outside diameter, D, at the relief section of the M14 × 1.5 thread is turned to the tolerance specified and the inside diameter, d, is drilled to tolerance. Both the solenoid body and housing are cadmium plated. The solenoid is assembled using an air tool with a clutch mechanism giving a 30% scatter in the pre-load typically (Shigley and Mischke, 1996). The thread length engagement is considered to be adequate to avoid failure by pullout.

Probabilistic design approach

Stress on first assembly
Figure 4.41 shows the Stress–Strength Interference (SSI) diagrams for the two assembly operation failure modes. The instantaneous stress on the relief section on first assembly is composed of two parts: first the applied tensile stress, s, due to the pre-load, F, and secondly, the torsional stress, τ, due to the torque on assembly, M, and this is shown in Figure 4.41(a) (Edwards and McKee, 1991). This stress is at a maximum during the assembly operation. If the component survives this stress, it will not fail by stress rupture later in life.

Therefore:

$$s = \frac{F}{A} = \frac{4F}{\pi(D^2 - d^2)} \tag{4.70}$$

$$\tau = \frac{Mr}{J} \tag{4.71}$$

(a) Situation on first asembly

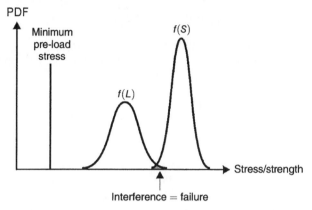

(b) Situation after relaxation of shear stress

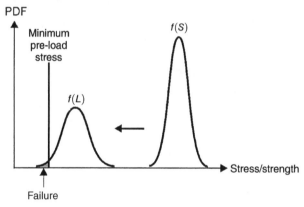

Figure 4.41 SSI models for the solenoid assembly failure modes

where:

$$A = \text{cross-sectional area}$$

$$J = \text{polar second moment of area.}$$

An approximate relationship is commonly used to determine the torque for assembly, M, for a given pre-load, F (Shigley and Mischke, 1996). It is a standard formulation for bolts and fasteners determined from experiment and is related to the friction found in the contacting surfaces of the parts on assembly.

$$M = KFD \tag{4.72}$$

where:

$$K = \text{torque coefficient (or nut factor).}$$

Therefore, combining the above equations in terms of the shear stress gives:

$$\tau = \frac{32KFDr}{\pi(D^4 - d^4)} \tag{4.73}$$

The principal stresses at the relief section, s_1 and s_2, are found from:

$$s_{1,2} = 0.5s \pm \sqrt{0.25s^2 + \tau^2} \qquad (4.74)$$

Using *von Mises* Theory from equation 4.58, the probabilistic requirement, P, to avoid yield in a ductile material, but under a biaxial stress system, is used to determine the reliability, R, as:

$$R = P\left(Sy > \sqrt{s_1^2 + s_2^2 - s_1 s_2} \right) \qquad (4.75)$$

Stress in service
The shear stress, τ, due to the assembly torque diminishes to zero with time, the pre-load, F, remaining constant, and so the stress on the solenoid section is only the direct stress, s, as given in equation 4.75 (see Figure 4.41(b)) (Edwards and McKee, 1991). A second reliability can then be determined by considering the requirement that the pre-load stress remains above a minimum level to avoid loosening in service ($0.5Sp_{min}$ from experiment) (Marbacher, 1999). The reliability, R, can then be determined from the probabilistic requirement, P, to avoid loosening:

$$R = P(L > 0.5Sp_{min}) \qquad (4.76)$$

Determining the design variables

Before a probabilistic model can be developed, the variables involved must be determined. It is assumed that the variables all follow the Normal distribution and that they are statistically independent, i.e. not correlated in anyway. The scatter of the pre-load, F, using an air tool with a clutch is approximately $\pm 30\%$ of the mean, which gives the coefficient of variation, $C_v = 0.1$, assuming $\pm 3\sigma$ covers this range, therefore:

$$\sigma_F = 0.1\mu_F \qquad (4.77)$$

For the torque coefficient, K, reported values range from 0.153 to 0.328 for cadmium plated parts, with a mean of 0.24 (Shigley and Mischke, 1996). Therefore, applying the same reasoning as above:

$$\mu_K = 0.24 \quad \text{and} \quad \sigma_K = 0.0292$$

For the solenoid dimensions, D, r and d, we can use Conformability Analysis (CA) to predict the standard deviations based on a shifted distribution, σ', which provides the largest estimate (or worst case) anticipated during a production run. Given that the dimension $D = \emptyset 12 \pm 0.03$ mm is turned, the material to process risk can be shown to be $m_p = 1.2$, and the geometry to process risk, $g_p = 1$. An adjusted tolerance is then given by:

$$\text{Adjusted tolerance} = \frac{\text{Design tolerance } (t)}{m_p \times g_p} = \frac{0.03}{1.2 \times 1} = 0.025$$

Looking at the process capability map for turning/boring in Figure 4.42 gives a risk value, $A = 1.05$, for a dimension of $\emptyset 12$ mm. This value defaults to the component manufacturing variability risk, q_m, when there is no consideration of surface finish capability in an analysis.

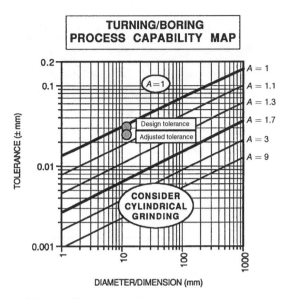

Figure 4.42 Process capability map for turning/boring

The shifted standard deviation, σ', for the dimensional tolerance on the diameter, D, can then be predicted from equation 4.28:

$$\sigma'_D = \frac{t \cdot q_m^2}{12} = \frac{0.03 \times 1.05^2}{12} = 0.0028 \, \text{mm}$$

Therefore:

$$\mu_D = 12 \, \text{mm} \quad \text{and} \quad \sigma_D = 0.0028 \, \text{mm}$$

It follows that for the radius, r, the variability is half of that of the diameter, D:

$$\mu_r = 6 \, \text{mm} \quad \text{and} \quad \sigma_r = 0.0014 \, \text{mm}$$

Similarly, for the internal diameter, d, the process capability map for drilling is provided in Figure 4.43. Note that the tolerance in the case of a drilled dimension is given as a '+' only.

$$\text{Adjusted tolerance} = \frac{\text{Design tolerance } (T)}{m_p \cdot g_p} = \frac{+0.2}{(1 \times 1) \times (1.2)} = +0.167 \, \text{mm}$$

A risk value, $A = 1.02$, is interpolated for a dimension of $\emptyset 9 \, \text{mm}$ and the adjusted tolerance. This value again defaults to the component manufacturing variability risk, to give $q_m = 1.02$.

The standard deviation for one half of the tolerance can be estimated by:

$$\sigma'_d \approx \frac{(T/2) \cdot q_m^2}{12} = \frac{0.1 \times 1.02^2}{12} = 0.0087 \, \text{mm}$$

Therefore,

$$\mu_d \approx 9.1 \, \text{mm} \quad \text{and} \quad \sigma_d = 0.0087 \, \text{mm}$$

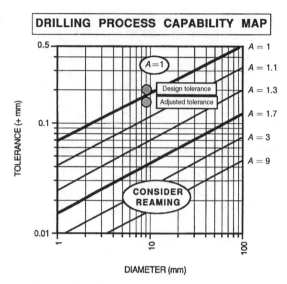

Figure 4.43 Process capability map for drilling

Determining the stress distribution using the variance equation

Assuming that all the variables follow a Normal distribution, a probabilistic model can be created to determine the stress distribution for the first failure mode using the variance equation and solving using the Finite Difference Method (see Appendix XI). The function for the *von Mises* stress, L, on first assembly at the solenoid section is taken from equation 4.75 and is given by:

$$L = \sqrt{[(0.5s + \sqrt{0.25s^2 + \tau^2})^2 + (0.5s - \sqrt{0.25s^2 + \tau^2})^2}$$
$$- (0.5s + \sqrt{0.25s^2 + \tau^2})(0.5s - \sqrt{0.25s^2 + \tau^2})] \qquad (4.78)$$

where:

$$s = \frac{4F}{\pi(D^2 - d^2)} \quad \text{and} \quad \tau = \frac{32KFDr}{\pi(D^4 - d^4)}$$

As there are five variables involved, F, K, D, r and d, the variance equation becomes:

$$\sigma_L = \left[\left(\frac{\partial L}{\partial F}\right)^2 \cdot \sigma_F^2 + \left(\frac{\partial L}{\partial K}\right)^2 \cdot \sigma_K^2 + \left(\frac{\partial L}{\partial D}\right)^2 \cdot \sigma_D^2 + \left(\frac{\partial L}{\partial r}\right)^2 \cdot \sigma_r^2 + \left(\frac{\partial L}{\partial d}\right)^2 \cdot \sigma_d^2 \right]^{0.5}$$
$$(4.79)$$

The Finite Difference Method can be used to approximate each term in this equation by using the difference equation for the first partial derivative. The values of the function at two points either side of the point of interest, k, are determined, y_{k+1} and y_{k-1}. These are equally spaced by an increment Δx. The finite difference equation approximates the value of the partial derivative by taking the difference of these values and dividing by the increment range. The terms subscripted by i indicate that only x_i is incremented by Δx_i for calculating y_{k+1} and y_{k-1}, holding the other independent variables constant at their k value points.

$$\left(\frac{\partial y}{\partial x_i}\right)_k \approx \frac{(y_{k+1} - y_{k-1})_i}{2\Delta x_i}$$

The variables F, K, D, r and d are all assumed to be random in nature following the Normal distribution, with the parameters shown in common notational form:

$$F \sim N(\mu_F, 0.1\mu_F)\,\text{N}$$

$$K \sim N(0.24, 0.0292)$$

$$D \sim N(12, 0.0028)\,\text{mm}$$

$$R \sim N(6, 0.0014)\,\text{mm}$$

$$D \sim N(9.1, 0.0087)\,\text{mm}$$

Consider the point x_k to be the mean value of a variable and x_{k+1} and x_{k-1} the extremes of the variable. The extremes can be determined for each variable by assuming they exist $\pm 4\sigma$ away from the mean, which covers approximately 99.99% of situations. For example, for the pre-load force, F, the extremes become (for $\mu_F = 10\,000\,\text{N}$):

$$F_{k+1} = 14\,000\,\text{N}$$

$$F_{k-1} = 6000\,\text{N}$$

and

$$\Delta F = 4(1000) = 4000\,\text{N}$$

For the first variable, F, in Equation 4.84.

$$\left(\frac{\partial L}{\partial F}\right) \approx \frac{(L_{k+1} - L_{k-1})}{2\Delta F}$$

Letting the variable F be its maximum value and the other variables kept at their mean values, gives L_{k+1} when applied to the stress function, L:

$$L_{k+1} = 423.6 \times 10^6 \quad \text{and} \quad L_{k-1} = 181.5 \times 10^6$$

Therefore,

$$\left(\frac{\partial L}{\partial F}\right) = \frac{423.6 \times 10^6 - 181.5 \times 10^6}{2 \times 4000} = 30\,257.25$$

Repeating the above for each variable and substituting in equation 4.79 gives:

$$\sigma_L = \begin{bmatrix} (30\,257.3^2 \times 1000^2) + ((6.445 \times 10^8)^2 \times 0.0292^2) \\ +((1.223 \times 10^8)^2 \times 0.0000028^2) \\ +((2.6518 \times 10^7)^2 \times 0.0000014^2) \\ +((7.6178 \times 10^7)^2 \times 0.0000087^2) \end{bmatrix}^{0.5}$$

$$\sigma_L = 35.6\,\text{MPa}$$

The mean value of the *von Mises* stress can be approximated by substituting in the mean values of each variable in equation 4.78 to give:

$$\mu_L = 302.6 \, \text{MPa}$$

Therefore, the loading stress can be approximated by:

$$L \sim N(302.6, 35.6) \, \text{MPa}$$

The *von Mises* stress, L, is then determined for various values of pre-load, F, using the above method. Equally, we could have used Monte Carlo Simulation to determine an answer for the stress standard deviation. The answer using this approach is in fact $\sigma_L \approx 36 \, \text{MPa}$ over a number of trial runs.

Stress–Strength Interference (SSI) models

A statistical representation of the yield strength for BS 220M07 is not available; however, the coefficient of variation, C_v, for the yield strength of steels is commonly given as 0.08 (Furman, 1981). For convenience, the parameters of the Normal distribution will be calculated by assuming that the minimum value is -3 standard deviations from the expected mean value (Cable and Virene, 1967):

$$Sy_{\min} = \mu_{Sy} - 3\sigma_{Sy}$$

Therefore,

$$340 = \mu_{Sy} - 3(0.08\mu_{Sy})$$

$$\mu_{Sy} = \frac{340}{0.76} = 447 \, \text{MPa}$$

and

$$\sigma_{Sy} = 0.08 \times 447 = 36 \, \text{MPa}$$

The yield strength for 220M07 can be approximated by:

$$Sy \sim N(447, 36) \, \text{MPa}$$

In the stress rupture case, the interference of the stress, L, and strength, Sy, both following a Normal distribution can be determined from the coupling equation:

$$z = -\frac{\mu_{Sy} - \mu_L}{\sqrt{\sigma_{Sy}^2 + \sigma_L^2}}$$

and the reliability, R, can be determined as:

$$R = 1 - \Theta_{\text{SND}}(z)$$

For the loosening case, the probability that the loading stress is less than a unique value of stress $(0.5Sp_{\min})$ is used, given by the following equation:

$$z = \left(\frac{x - \mu}{\sigma}\right) = \left(\frac{0.5Sp_{\min} - \mu_L}{\sigma_L}\right)$$

the reliability, R, again being determined by:

$$R = 1 - \Theta_{\text{SND}}(z)$$

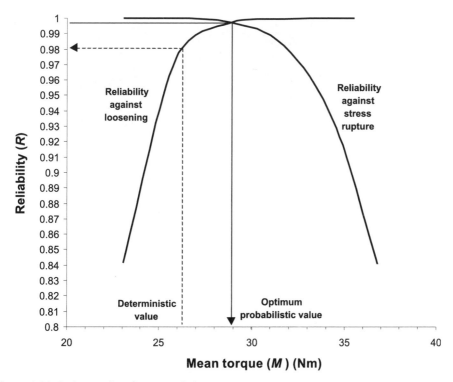

Figure 4.44 Optimum value of mean applied torque

The reliabilities due to the competing failure modes of stress rupture and loosening can be superimposed on a graph against mean torque, M, as shown in Figure 4.44. An optimum value can then be selected based on these conditions. The optimum assembly torque $M = 28.9\,\text{Nm}$ giving a reliability of $R = 0.997$, close to the target of 0.999 for FMEA $(S) = 5$ from Figure 4.36. Without changing the design scheme, an obvious improvement would be to tighten the tolerances on the drilled internal diameter by reaming to finished size.

Deterministic design approach
The maximum pre-load, F_{max}, can be taken as a proportion of the proof load, F_{p}, typically 0.9 for permanent assemblies (Shigley, 1986):

$$F_{\text{max}} = 0.9F_{\text{p}} = 0.9Sp_{\text{min}}\frac{\pi}{4}\left(D^2 - d^2\right) = 0.9 \times 300 \times 10^6 \times \frac{\pi}{4}\left(0.012^2 - 0.0091^2\right)$$

$$= 12976\,\text{N}$$

The mean pre-load, μ_F, applied can be found from:

$$\mu_F = F_{\text{max}} - 3\sigma_F$$
$$\mu_F = 12\,976 - 3(0.1 \times 12\,976) = 9083\,\text{N}$$

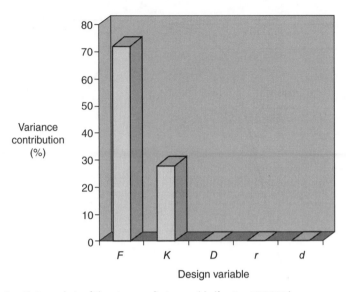

Figure 4.45 Sensitivity analysis of the stress on first assembly (for $F = 10\,000\,\text{N}$)

The mean or target assembly torque is calculated from equation 4.72:

$$\mu_M = \mu_K \cdot \mu_F \cdot \mu_D$$

$$\mu_M = 0.24 \times 9083 \times 0.012 = 26.2\,\text{Nm}$$

The assembly torque, M, is 10% lower using the deterministic approach than that using probabilistic analysis. This would result in more of the solenoids becoming loose and failing in service, approximately 2% from Figure 4.44.

Sensitivity analysis

The contribution of each variable to the final stress distribution in the case of stress rupture can be examined using sensitivity analysis. From the variance equation:

$$\sigma_L^2 = 1.2674 \times 10^{15} = (9.155 \times 10^{14}) + (3.5417 \times 10^{14}) + (117\,265.2)$$
$$+ (1378.3) + (439\,235.7)$$

Therefore, the contribution of the pre-load, F, to the overall stress is given by:

$$\frac{9.155 \times 10^{14}}{1.2674 \times 10^{15}} \times 100 = 72.1\%$$

The same process can be performed with the other variables to provide Figure 4.45. It shows that the largest variance contribution is provided by the torque on assembly which affects the pre-load, F.

4.8.2 Foot pedal optimum design

The objective of this case study is to determine the optimum depth, d, of the foot pedal section as shown in Figure 4.46 using static probabilistic design methods. We are

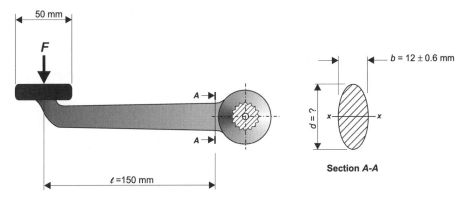

Figure 4.46 Foot pedal arrangement

given that the breadth, b, of the elliptical section is set at 12 ± 0.6 mm and that a load, F, described by a random function is applied about a couple length, ℓ, of 150 mm from the section A–A. We assume that failure will occur due to ductile fracture at this section due to a pure bending stress being greater than the yield strength of the material (ignoring any stresses due to shear or direct stresses due to the direction of the load deviating from the perpendicular). Failure will lead to a violation of a statutory requirement during operation, which relates to an FMEA Severity Rating $(S) = 6$. Although not safety critical to the operator, it is required that a degree of safety back-up is provided by the allocation of a high severity rating. It is anticipated that 10 000 will be produced *per annum* and based on this information, the part will be forged from a medium carbon steel, SAE 1035, with a yield strength $Sy \sim W(272.4, 350.3, 2.88)$ MPa (Mischke, 1992).

Figure 4.47 shows a typical load history (in Newtons force) measured experimentally during the operation of a similar product in service. It is also anticipated that the

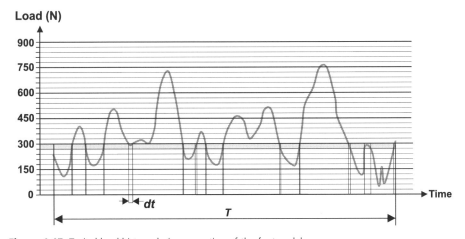

Figure 4.47 Typical load history during operation of the foot pedal

pedal will be subjected to no more than 1000 independent load applications during its designed service life.

Determining the load distribution

There is sufficient information to generate a PDF for the load assuming that the load history over the time interval, T, is representative of the actual load history in service. The approach used was discussed in Section 4.3.3. First we can divide the load on the y-axis into classes with a class width, w, of 30 N for convenience. By summing the amount of time, dt, that the load signal falls within each class, we can obtain a relative measure of the load frequency with respect to the time interval, T. For example, the shaded strip in Figure 4.47 represents the load class from 270 to 300 N with a mid-class value of 285 N. The approximate amount of time that the signal is within this load class is approximately 7.9% of the total time, T. Repeating this process for each load class builds up a frequency distribution for the load in percentage of the total, as shown in Figure 4.48.

From a visual inspection of the histogram, it is evident that the load frequency approaches zero at zero load and is slightly skewed to the left. The 2-parameter Weibull distribution can be effectively used to model this shape of frequency distribution with a zero threshold. For conciseness, it is the only distribution type considered, although comparison of the load data with the Lognormal distribution may also be performed. The frequency values in Figure 4.48 are in percentages to one decimal place, but we can simplify for the process of fitting the distributional model to the data by multiplying the frequency by 10 to convert to whole numbers, therefore $N = 1000$.

The analysis of the frequency data is shown in Table 4.12. Note the use of the Median Rank equation, commonly used for both Weibull distributions. Linear rectification equations provided in Appendix X for the 2-parameter Weibull model are used to

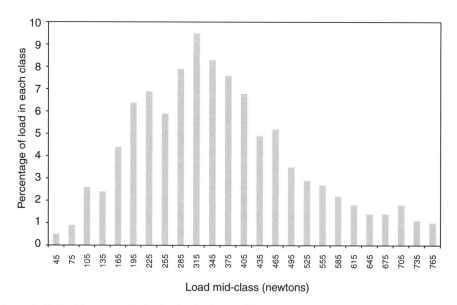

Figure 4.48 Load frequency distribution for the foot pedal

Table 4.12 Analysis of load frequency data and plotting positions for the 2-parameter Weibull distribution

Mid-class (newtons)	Frequency (%)	Frequency (×10) $N = 1000$	Cumulative frequency (i)	$F_i = \dfrac{i - 0.3}{N + 0.4}$	ln(mid-class) (x-axis)	ln ln(1/(1 − F_i)) (y-axis)
45	0.5	5	5	0.0047	3.8067	−5.3578
75	0.9	9	14	0.0137	4.3175	−4.2835
105	2.6	26	40	0.0397	4.6540	−3.2062
135	2.4	24	64	0.0637	4.9053	−2.7208
165	4.4	44	108	0.1077	5.1059	−2.1720
195	6.4	64	172	0.1716	5.2730	−1.6700
225	6.9	69	241	0.2406	5.4161	−1.2902
255	5.9	59	300	0.2996	5.5413	−1.0325
285	7.9	79	379	0.3785	5.6525	−0.7431
315	9.5	95	474	0.4735	5.7526	−0.4439
345	8.3	83	557	0.5565	5.8435	−0.2070
375	7.6	76	633	0.6324	5.9270	0.0008
405	6.8	68	701	0.7004	6.0039	0.1867
435	4.9	49	750	0.7494	6.0753	0.3249
465	5.2	52	802	0.8014	6.1420	0.4802
495	3.5	35	837	0.8364	6.2046	0.5935
525	2.9	29	866	0.8654	6.2634	0.6959
555	2.7	27	893	0.8923	6.3190	0.8013
585	2.2	22	915	0.9143	6.3716	0.8989
615	1.8	18	933	0.9323	6.4216	0.9905
645	1.4	14	947	0.9463	6.4693	1.0731
675	1.4	14	961	0.9603	6.5147	1.1714
705	1.8	18	979	0.9783	6.5582	1.3420
735	1.1	11	990	0.9903	6.5999	1.5338
765	1	10	1000	0.9993	6.6399	1.9830

linearize the plotting positions and provide values for the x- and y-axes in the last two columns. The plotted results are shown in Figure 4.49 where the best straight line through the data has been determined using MS Excel. The correlation coefficient, r, is calculated to be 0.998 and indicates that there is a strong relationship between the 2-parameter Weibull distribution and the load frequency distribution.

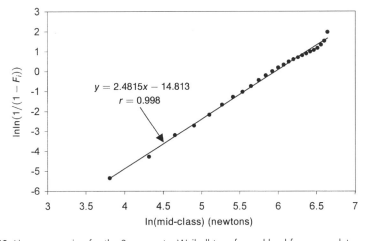

Figure 4.49 Linear regression for the 2-parameter Weibull transformed load frequency data

The characteristic value, θ, and shape parameter, β, for the 2-parameter Weibull distribution can be determined from the equation for the line in the form $y = A1x + A0$ and from the equations given in Appendix X, where:

$$\theta = \exp\left(-\frac{A0}{A1}\right) = \exp\left(-\left(\frac{-14.813}{2.4815}\right)\right) = 391.3 \, \text{N}$$

and

$$\beta = A1 = 2.48$$

Therefore, the load, F, can be characterized by a 2-parameter Weibull distribution with notation:

$$F \sim W(391.3, 2.48) \, \text{N}$$

The equivalent mean $\mu_F = 347.1 \, \text{N}$ and $\sigma_F = 149.6 \, \text{N}$.

We can compare the calculated 2-parameter Weibull distribution with the original frequency distribution by multiplying the $f(x)$ or PDF by the scaling factor, Nw, as determined from equation 4.86. The variate, x, is the mid-class value over the load range. Also note that the population, N, is divided by 10 to change the frequency back to a percentage value. The results of this exercise are shown in Figure 4.50.

$$f(x) = Nw\left(\frac{\beta}{\theta}\right)\left(\frac{x}{\theta}\right)^{\beta-1} \exp\left(-\left(\frac{x}{\theta}\right)^{\beta}\right)$$

$$= \frac{1000}{10} \times 10\left(\frac{2.48}{391.3}\right)\left(\frac{x}{391.3}\right)^{1.48} \exp\left(-\left(\frac{x}{391.3}\right)^{2.48}\right) \qquad (4.80)$$

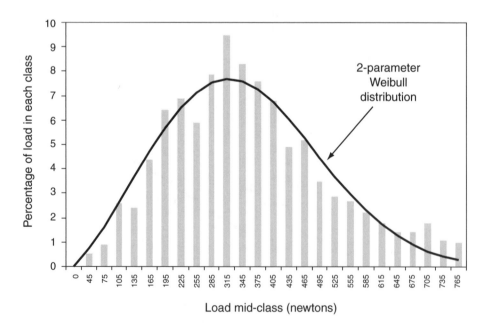

Figure 4.50 Comparison of the calculated 2-parameter Weibull distribution with the original load frequency distribution

Determining the stress variable

The stress, L, due to pure bending at the section A–A on the pedal is given by equation 4.81:

$$L = \frac{My}{I_{xx}} = \frac{F\ell\left(\frac{d}{2}\right)}{I_{xx}} \qquad (4.81)$$

where:

M = bending moment

y = distance from x–x axis to extreme fibre

I_{xx} = second moment of area about the axis x–x

F = load

ℓ = couple length

d = depth of section.

For the elliptical cross-section specified, the second moment of area, I, about x–x axis is given by:

$$I_{xx} \approx 0.04909d^3b \qquad (4.82)$$

where:

b = breadth of the section.

Therefore, substituting equation 4.82 into equation 4.81 gives:

$$L \approx \frac{10.18537F\ell}{d^2b} \qquad (4.83)$$

Equation 4.83 states that there are four variables involved. We have already determined the load variable, F, earlier. The load is applied at a mean distance, μ_ℓ, of 150 mm representing the couple length, and is normally distributed about the width of the foot pad. The standard deviation of the couple length, σ_ℓ, can be approximated by assuming that 6σ covers the pad, therefore:

$$\sigma_\ell = \frac{50}{6} = 8.333 \text{ mm}$$

$$\ell \sim N(150, 8.333) \text{ mm}$$

The width of the elliptical cross-section, b, has a mean $\mu_b = 12$ mm. The standard deviation can be determined from equation 4.28 and reference to the closed die forging process capability map for low to medium carbon and low alloy steels for the weight range given, shown in Figure 4.51.

Because the width of the section is over a parting line, the only process risk is a geometry to process risk $g_p = 1.7$ which gives the adjusted tolerance as:

$$\text{Adjusted tolerance} = \frac{\text{Design tolerance } (T)}{m_p \cdot g_p} = \frac{1.2}{1 \times 1.7} = 0.706 \text{ mm}$$

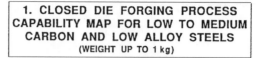

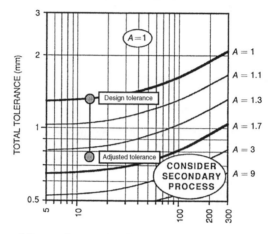

Figure 4.51 Process capability map for closed die forging of steel

From Figure 4.51, the risk 'A' and hence $q_m = 1.5$ because there are no other risk factors to take into account. The standard deviation can be approximated from:

$$\sigma_b' \approx \frac{(T/2) \cdot q_m^2}{12} = \frac{0.6 \times 1.5^2}{12} = 0.113 \, \text{mm}$$

Therefore,

$$b \sim N(12, 0.113) \, \text{mm}$$

The original objective of the exercise was to find the optimum depth of the section with regard to its failure severity. Values ranging from $d = 14$ to $25 \, \text{mm}$ in steps of $1 \, \text{mm}$ will be used in the calculation of the reliability. The corresponding standard deviations for each depth are again calculated from the process capability map for closed die forging, the total tolerance for each value taken from the $A = 1.7$ line giving $q_m = 1.7$ as there are no other process risks to take into account. For example at $d = 14 \, \text{mm}$ the total tolerance at $A = 1.7$ is $T = 0.66 \, \text{mm}$. Therefore, the standard deviation is given by:

$$\sigma_d' \approx \frac{(T/2) \cdot q_m^2}{12} = \frac{0.33 \times 1.7^2}{12} = 0.080 \, \text{mm}$$

$$d \sim N(14, 0.080) \, \text{mm}$$

This can be repeated for the remaining values of the section depth chosen.

We can use a Monte Carlo simulation of the random variables in equation 4.83 to determine the likely mean and standard deviation of the loading stress, assuming that this will be a Normal distribution too. Except for the load, F, which is modelled by a 2-parameter Weibull distribution, the remaining variables are characterized by the Normal distribution. The 3-parameter Weibull distribution can be used to model

Table 4.13 Loading stress Normal distribution parameters for a range of section depth values

d (mm)	μ_L (MPa)	σ_L (MPa)
14	225.0	98.2
15	196.0	85.7
16	172.2	75.1
17	152.6	66.6
18	136.1	59.4
19	122.1	53.3
20	110.2	48.1
21	100.0	43.6
22	91.1	39.7
23	83.4	36.6
24	76.6	33.4
25	70.5	30.8

the Normal when $\beta = 3.44$. The procedure is discussed in detail in Appendix XI, including the sample computer code that can be used for a 10 000 trial simulation.

Table 4.13 gives the simulated mean, μ_L, and standard deviation, σ_L, of the loading stress for the range of section depths, d, from 14 to 25 mm. The values are average results from five simulations at each value of d. Because we are using a non-symmetrical distribution to model the load, the mean loading stress should be determined using simulation rather than determination from equivalent mean values substituted in equation 4.89. However, the load distribution is near-Normal from a visual inspection of Figure 4.50 and there is very little difference. The load, F, has a very dominant variance contribution in the determination of the final loading stress, with $C_v = 0.43$. This is greater than the $C_v = 0.2$ as generally recommended earlier in Section 4.2.3, but an approximation using Monte Carlo simulation is still valid.

Calculating the reliability

We can now use equation 4.35 to determine the reliability for a given number of independent load applications, n:

$$R_n = \int_0^\infty F(L)^n \cdot f(S)\, dS \qquad (4.35)$$

where:

$$R_n = \text{reliability at } n\text{th application of load}$$

$$n = \text{number of independent load applications in sequence}$$

$$F(L) = \text{loading stress CDF}$$

$$f(S) = \text{strength PDF}.$$

In the problem here, the loading stress is a Normal distribution and the strength is a 3-parameter Weibull distribution. Because the Normal distribution's CDF is not in closed form, the 3-parameter Weibull distribution can be used as an approximating distribution when $\beta = 3.44$. The parameters for the 3-parameter Weibull distribution,

xo and θ, can be estimated given the mean, μ, and standard deviation, σ, for a Normal distribution (assuming $\beta = 3.44$) by:

$$xo \approx \mu - 3.1394473\sigma$$

$$\theta \approx \mu + 0.3530184\sigma$$

For example, the loading stress at $d = 14\,\text{mm}$ in terms of Weibull parameters becomes:

$$xo_L = \mu_L - 3.1394473\sigma_L = 225 - 3.1394473(98.2) = -83.3\,\text{MPa}$$

$$\theta_L = \mu_L + 0.3530184\sigma_L = 225 + 0.3530184(98.2) = 259.7\,\text{MPa}$$

$$\beta_L = 3.44$$

Although xo_L is negative, it will not affect the determination of the reliability because we are only interested in the right-hand side of the distribution for stress–strength interference analysis.

The final reliability formulation for the interference of two 3-parameter Weibull distributions subjected to multiple load applications is given in equation 4.84:

$$R_n = \int_{S=xo_S}^{\infty} \left[1 - \exp\left(-\left(\frac{x - xo_L}{\theta_L - xo_L} \right)^{\beta_L} \right) \right]^n \cdot \left[\left(\frac{\beta_S}{\theta_S - xo_S} \right) \left(\frac{x - xo_S}{\theta_S - xo_S} \right)^{\beta_S - 1} \right.$$

$$\left. \cdot \exp\left(-\left(\frac{x - xo_S}{\theta_S - xo_S} \right)^{\beta_S} \right) \right] dx$$

The limits of integration are from the expected minimum value of yield strength, $xo_S = 272.4\,\text{MPa}$ to $1000\,\text{MPa}$, representing ∞. The solution of this equation numerically using Simpson's Rule is described in Appendix XII. For the case when $d = 20\,\text{mm}$ and the number of load applications $n = 1000$, the reliability, R_n, is found to be:

$$R_{1000} = 0.997856$$

Finding an optimum design

The boldest line on Figure 4.52 shows R_{1000} calculated for each value of the section depth from 14 to 25 mm. The approach described above is based on Freudenthal

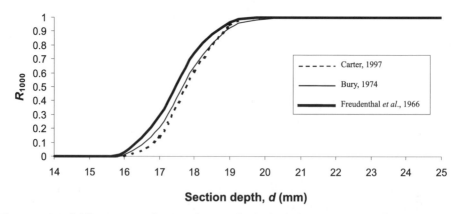

Figure 4.52 Reliability, R_{1000}, as a function of section depth, d, calculated using three different approaches

et al. (1966). Two further approaches to determine R_n by Bury (1974) and Carter (1997) were also identified in Section 4.4.3, the results from these approaches are also shown on Figure 4.52. Note that the approach by Carter requires that an equivalent mean and standard deviation is calculated for the material's yield strength, which are found to be $\mu_S = 342\,\text{MPa}$ and $\sigma_S = 26\,\text{MPa}$.

The results from the three different methods are in partial agreement and it is evident that a suitable design exists with a section depth, d, greater than 20 mm, but probably less than 24 mm to avoid overdesigning the part. Consulting the reliability target map in Figure 4.53, for an FMEA Severity Rating $(S) = 6$, a target value of $R = 0.99993$ is required for an acceptable design. The reliabilities calculated for $d = 22\,\text{mm}$ show quite a large spread, reflecting the differences of the methods. However, the final selection of a pedal section depth $d = 22\,\text{mm}$ would be justified.

Deterministic approach

When designing using a deterministic approach, it is a fair assumption that a generous factor of safety (FS) would be allocated to determine the allowable working stress from the minimum material strength. For the variables in the problem such as dimensions, mean values would be chosen, except for the load, which would be the maximum load expected throughout the service life. The minimum yield strength, S_{min} in this case can be approximated from the mean value minus three standard deviations, as discussed in Section 4.3.1. The working stress and the section depth of the pedal can be calculated for a range of factor of safety values, where the working stress is substituted for L in equation 4.83. Therefore:

$$S_{min} = \mu_S - 3\sigma_S$$
$$S_{min} = 342 - 3(26) = 264\,\text{MPa}$$

Rearranging equation 4.83 for the section depth, d, we obtain:

$$d = \sqrt{\frac{10.18537 F_{max}\mu_\ell}{\mu_b\left(\dfrac{S_{min}}{FS}\right)}} \tag{4.84}$$

For example, for FS $= 3$ and anticipating that the maximum load is 765 N from Figure 4.47:

$$d = \sqrt{\frac{10.18537 \times 765 \times 0.15}{0.012\left(\dfrac{264 \times 10^6}{3}\right)}} = 0.03327\,\text{m} = 33.3\,\text{mm}$$

Repeating this exercise for a range of FS values gives Figure 4.54. To have determined the optimum section depth of 22 mm, FS $= 1.3$ would need to have been specified. From Table 4.1, a typical factor of safety applied might have been 2 or 3 (or even greater!) for this type of problem and subsequently the part would have been over-designed, increasing the volume and therefore the material cost by approximately 50%.

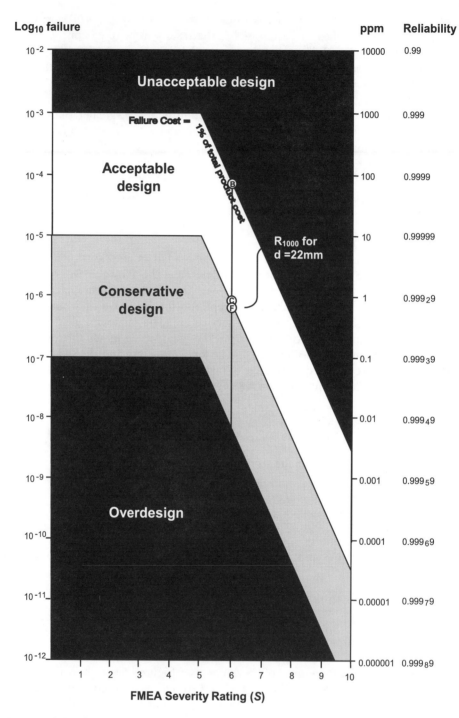

Figure 4.53 Reliability target map showing the range of the reliability values, R_{1000}, calculated using three different approaches for a section depth $d = 22$ mm (B = Bury (1974), C = Carter (1997), F = Freudenthal *et al.* (1966))

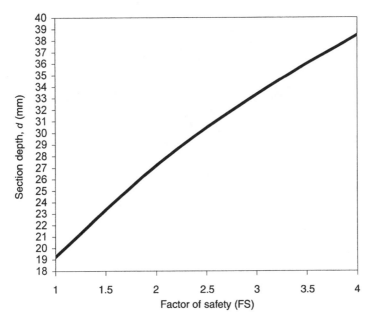

Figure 4.54 Section depth, d, based on factor of safety

4.8.3 Torque transmitted by a shrink fit

An effective way of assembling a machine element such as a gear or pulley to a shaft is by shrink fitting. This involves setting up a radial pressure between a shaft with a slightly larger diameter than the inside diameter of the machine element, termed an interference fit. The part is heated to allow assembly and then on cooling the pressure on the inside diameter is established which maintains its ability to transmit torque from the shaft.

It is required to find the torque without slippage that can be transmitted by a hub that is assembled by an interference fit to a powered shaft. The hub outside diameter $D = \varnothing 70$ mm, and the shaft diameter $d = \varnothing 50$ mm, as shown in Figure 4.55. The length of the hub is 100 mm. Both hub and shaft are machined from hot rolled steel SAE 1035 with a yield strength $Sy \sim N(342, 26)$ MPa (see Table 4.6). Given that the hub is stopped suddenly in service due to a malfunction, and considering only the torsional stresses, what is the probability that the shaft will yield?

Governing equations
The torque that can be transmitted over the length of the hub without slipping, called the holding torque, is a function of the friction and the radial pressure between the hub and shaft, the contact surface area and the radius of the shaft itself. The holding torque, M_H, is given by:

$$M_H = \frac{f \cdot p \cdot \pi \cdot d_S^2 \cdot l_H}{2} \qquad (4.85)$$

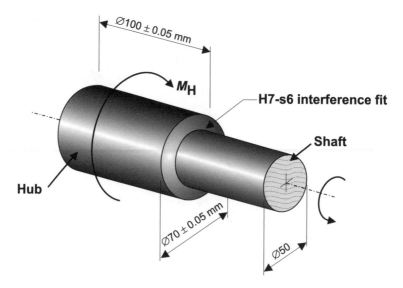

Figure 4.55 Arrangement of the hub and shaft

where:

$$M_H = \text{holding torque (hub)}$$

$$f = \text{coefficient of friction between hub and shaft}$$

$$p = \text{radial pressure}$$

$$d_S = \text{shaft diameter}$$

$$l_H = \text{length of hub.}$$

The radial pressure, p, set up on assembly is given by equation 4.86 (Timoshenko, 1966)[*]:

$$p = \frac{E \cdot \gamma \left(D_H^2 - d_S^2\right)}{2d_S \cdot D_H^2} \tag{4.86}$$

where:

$$E = \text{Modulus of Elasticity}$$

$$D_H = \text{hub outside diameter}$$

$$\gamma = \text{interference between the shaft and hub.}$$

Substituting equation 4.86 into equation 4.85 gives:

$$M_H = 0.7854f \cdot l \cdot E \cdot \gamma \cdot d_S \left(1 - \left(\frac{d_S}{D_H}\right)^2\right) \tag{4.87}$$

---

[*] The radial pressure is not constant over the length of the hub, but in fact peaks at the projecting portions of the shaft which resist compression resulting in an increased pressure at the ends of the hub, or stress concentration. For this reason, fretting fatigue failure may be anticipated when the applied torque is alternating.

Determining the variables

For an adequate interference fit selected on the basis of the hole, the tolerances are taken from BS 4500A (1970) as H7-s6. This translates to a tolerance on the hub bore H7 = +0.030, +0.000 mm and for the shaft diameter, s6 = +0.072, +0.053 mm. Given the notation for hub bore diameter is d_H, and converting to mean values and bilateral tolerances, these dimensions become:

$$d_H = \varnothing 50.015 \pm 0.015 \, \text{mm}$$

$$d_S \approx \varnothing 50.063 \pm 0.010 \, \text{mm}$$

and from Figure 4.55,

$$D_H = \varnothing 70 \pm 0.05 \, \text{mm}$$

$$l_H = 100 \pm 0.05 \, \text{mm}$$

The hub inside diameter, outside diameter and length are machined using a lathe and so we employ the turning/boring map as shown previously in Figure 4.42. The material specified for the hub is mild steel giving a material to process risk $m_p = 1.3$. The geometry to process risk $g_p = 1.02$ due to a 2 : 1 length to diameter ratio. An adjusted tolerance for the hub bore, d_H, is then given by:

$$\text{Adjusted tolerance} = \frac{\text{Design tolerance } (t)}{m_p \times g_p} = \frac{0.015}{1.3 \times 1.02} = 0.011$$

With reference to the process capability map for turning/boring, $A = 1.7$ for a dimension of $\varnothing 50$ mm. This value defaults to the component manufacturing variability risk, q_m, when there is no consideration of surface finish capability in an analysis[*]. The shifted standard deviation, σ', for the dimensional tolerance on the hub bore can then be predicted from equation 4.28:

$$\sigma'_{d_H} = \frac{t \cdot q_m^2}{12} = \frac{0.015 \times 1.7^2}{12} = 0.004 \, \text{mm}$$

Therefore:

$$d_H \sim N(50.015, 0.004) \, \text{mm}$$

Similarly for an analysis on the hub outside diameter and length, which are turned, and the shaft diameter, which is finished using cylindrical grinding, we get:

$$D_H \sim N(70, 0.005) \, \text{mm}$$

$$l_H \sim N(100, 0.006) \, \text{mm}$$

$$d_S \sim N(50.063, 0.0012) \, \text{mm}$$

---

[*] Note that the surface roughness to process risk, s_p, is not included in the formulation to determine the standard deviation of the tolerance. However, a 0.4 μm Ra surface finish is required to provide adequate frictional adhesion between the shaft and hub bore (Bolz, 1981). This capability requirement is met by the manufacturing processes selected when analysed using CA.

The mean of the interference between the shaft and hub bore is given by:

$$\mu_\gamma = \mu_{d_S} - \mu_{d_H} = 50.063 - 50.015 = 0.048 \, \text{mm}$$

The standard deviation of the interference is given by:

$$\sigma_\gamma = \left(\sigma'^2_{d_S} + \sigma'^2_{d_H}\right)^{0.5} = \left(0.0012^2 + 0.004^2\right)^{0.5} = 0.0042 \, \text{mm}$$

Therefore,

$$\gamma \sim N(0.048, 0.0042) \, \text{mm}$$

The maximum coefficient of variation for the Modulus of Elasticity, E, for carbon steel was given in Table 4.5 as $C_v = 0.03$. Typically, $E = 208$ GPa and therefore we can infer that E is represented by a Normal distribution with parameters:

$$E \sim N(208, 6.24) \, \text{GPa}$$

Values typically range from 0.077 to 0.33 for the static coefficient of friction for steel on steel under an interference fit with no lubrication (Kutz, 1986). The interference and coefficient of friction are correlated in practice but for the example here, we assume statistical independence. Also, assuming that 6 standard deviations cover the range given, we can derive that:

$$f \sim N(0.2, 0.04)$$

Determining the probability of interference

The mean value for the holding torque capacity, μ_{M_H}, is found by inserting the mean values of all the variables into equation 4.87. To find the standard deviation of the holding torque, σ_{M_H}, we can apply the Finite Difference Method. Finally, we arrive at:

$$M_H \sim N(3.84, 0.85) \, \text{kNm}$$

At this stage it is worth highlighting the relative contribution of each variable to the holding torque variance. The results from the Finite Difference Method are used to construct Figure 4.56 which shows the sensitivity analysis of all the variables to the variance of the holding torque. It is clear that accurate and representative data for the coefficient of friction, f, for a particular situation and the control of the interference fit dimensions are crucial in the determination of the holding torque distribution.

The torque that can be transmitted by the shaft without yielding, M_S, is given by:

$$M_S = \frac{\tau_y \cdot J}{r}$$

where:

$$\tau_y = \text{shear yield strength}$$

$$J = \text{polar second moment of area}$$

$$r = \text{radius of shaft.}$$

In terms of the shaft diameter, this simplifies to:

$$M_S = 0.1963495 \tau_y \cdot d_S^3 \tag{4.88}$$

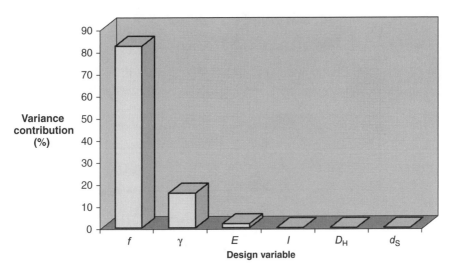

Figure 4.56 Sensitivity analysis for the holding torque variables

The shear yield strength for ductile metals is a linear function of the uniaxial yield strength. Therefore, for pure torsion from equation 4.56:

$$\tau_y = 0.577 Sy$$

Applying this conversion to the Normal distribution parameters for SAE 1035 steel gives:

$$\tau_y \sim N(197.3, 15.2)\,\text{MPa}$$

Using Monte Carlo simulation applied to equation 4.88, the shaft torque capacity is found to be:

$$M_S \sim N(4.86, 0.38)\,\text{kNm}$$

Both of the torque capacities calculated, the holding torque of the hub and the shaft torque at yield, are represented by the Normal distribution, therefore we can use the coupling equation to determine the probability of interference, where:

$$z = -\frac{\mu_{M_S} - \mu_{M_H}}{\sqrt{\sigma_{M_S}^2 + \sigma_{M_H}^2}} = -\frac{4.86 - 3.84}{\sqrt{0.38^2 + 0.85^2}} = -1.10$$

From Table 1 in Appendix I, the probability of failure $P = 0.135666$. This suggests that when the holding torque level is reached in service, due to a malfunction stopping the hub from rotating, there is about a 1 in 7 chance that the shaft will yield before the hub slips. Clearly this is not adequate if we want to protect the shaft and its transmission from damage.

Rather than change the specifications of the design as a whole, a small increase in the yield strength of the shaft is favoured. This is because there are only two variables in the shaft strength and it is obvious that the major contribution to the torque is from the shear yield strength. Repeating the calculation to determine the shaft torque at yield using cold drawn SAE 1018 with $Sy \sim N(540, 41)\,\text{MPa}$ gives $P = 0.000104$ or

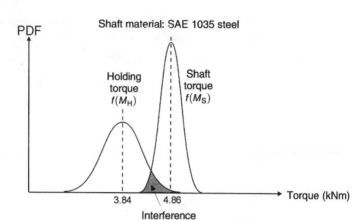

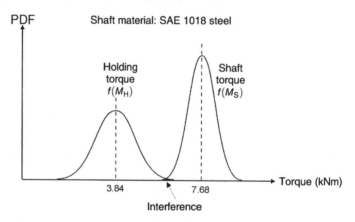

Figure 4.57 Distributions and relative interference of the holding torque and shaft torque capacities for two different shaft steels

a reliability $R = 0.999896$. Although not safety critical, the degree of protection from the shaft yielding is now adequate after reference to the reliability target map. The relative shape of the torque distributions and degree of interference are shown in Figure 4.57 for the two situations where different shaft steels are used.

4.8.4 Weak link design

The major assumption in weak link design is that the cost of failure of the machine that is to be protected from an overload situation in service is much greater than the cost of failure of a weak link placed in the system which is designed to fail first. The situation is primarily driven by various costs which must be balanced to avoid at one extreme the cost of failure of the system, and the other overdesign of the elements in the system. The cost factors involved are typically:

- Cost of introducing the weak link
- Cost of replacement of the weak link

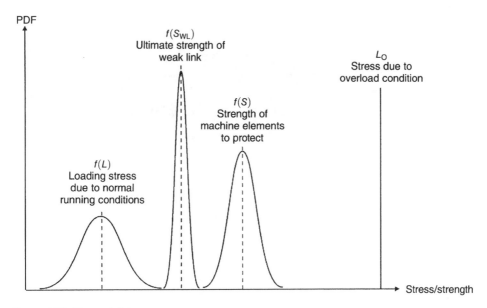

Figure 4.58 The weak link concept

- Cost of failure with the weak link relative to the cost of failure without the weak link
- Cost due to machine downtime if the weak link fails prematurely
- Cost of increasing strength of machine elements to accommodate the weak link.

Figure 4.58 shows the concept of weak link design. The loading stress distribution is determined from the normal operating conditions found in the system and this stress is used to determine the dimensions of the weak link. The failure mode for the weak link is stress rupture and so the ultimate tensile or ultimate shear strength is specified. The appropriate level of interference between the loading stress and weak link strength is determined from the consequences of machine downtime if the weak link fails prematurely. The use of the target reliability map with FMEA Severity Ratings (S) for production processes is useful in this respect (see the Process FMEA Severity Ratings provided by Chrysler Corporation *et al.*, 1995). For example, in Section 2.6.4, the characteristic dimension on the cover support leg was critical to the success of the automated assembly process, the potential failure mode being a major disruption to the production line on failure to meet the required capability, therefore $S = 8$ was allocated. If the weak link was designed to a very low level of failure probability, meaning the separation between the two distributions was greater, the machine elements could be overdesigned.

The variability of the weak link strength should be well known because it is a critical component in the system and experimental testing to determine the ultimate tensile strength, Su, is recommended where data is lacking. The interference between the weak link strength and machine strength is usually always smaller comparatively than that between the loading stress and weak link strength. This is due to the fact that we must see the failure of the weak link before the machine in all situations.

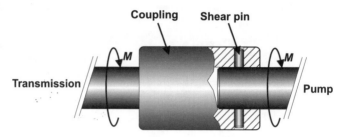

Figure 4.59 Weak link arrangement

However, too great a separation, and overdesign may occur. The overload condition is represented by a unique stress, which is very much greater than the working stress, applied suddenly which causes only the weak link to failure due to stress rupture.

In the following case study we will examine the concepts of weak link design. Figure 4.59 shows the arrangement of a coupling with a shear pin acting as a weak link between a transmission system and a pump, the assumption being that the cost of failure of the pump is much greater than the cost of failure of the weak link. In the event that the pump suddenly stops due to a blockage, the shear pin must protect the system from damage. The applied torque through the transmission shaft under normal running conditions is $M = 3.2 \text{ kNm}$ with a coefficient of variation $C_v = 0.1$, and the diameter variables of the transmission and pump shafts are $D \sim N(60, 0.004) \text{ mm}$. The FMEA Severity Rating $(S) = 5$, relating to a minor disruption if the weak link fails prematurely causing the pump to experience downtime. Steel is to be used as the material for all the machine elements.

Experimental determination of ultimate tensile strength of the weak link material

Because the design of the shear pin is critical, the ultimate tensile strength of the steel selected for the weak link material was in this case measured statistically by performing a simple experimental hardness test. The grade of steel selected is 220M07 cold drawn free cutting steel. The size tested is $\varnothing 16 \text{ mm}$, estimated as the approximate diameter of the pin, and 30 samples are selected from the stock material. The Brinnel Hardness (HB) value of each sample is measured, the results of which are shown in Table 4.14. Rather than develop a histogram for the data, we can determine the Normal distribution plotting positions using the mean rank equation for the individual values. The results are plotted in Figure 4.60, the equation of the straight line and the correlation coefficient, r, determined.

The mean and standard deviation of the hardness for the steel can be determined from the regression constants $A0$ and $A1$ as:

$$\mu_{HB} = -\left(\frac{A0}{A1}\right) = -\left(\frac{-34.801}{0.2319}\right) = 150.07$$

$$\sigma_{HB} = \left(\frac{1 - A0}{A1}\right) + \left(\frac{A0}{A1}\right) = \left(\frac{1 + 34.801}{0.2319}\right) + \left(\frac{-34.801}{0.2319}\right) = 4.31$$

Table 4.14 Analysis of hardness data and plotting positions for the Normal distribution

HB (x-axis)	Cum. freq. (i) ($N = 30$)	$F_i = \dfrac{i}{N + 1}$	$z = \Phi_{SND}^{-1}(F_i)$ (y-axis)
141.8	1	0.0323	−1.849
144.5	2	0.0645	−1.518
145.4	3	0.0968	−1.300
145.4	4	0.1290	−1.131
145.4	5	0.1613	−0.989
145.9	6	0.1935	−0.865
147.8	7	0.2258	−0.753
147.8	8	0.2581	−0.649
147.8	9	0.2903	−0.552
147.8	10	0.3226	−0.460
147.8	11	0.3548	−0.372
149.3	12	0.3871	−0.287
150.3	13	0.4192	−0.204
150.3	14	0.4516	−0.121
150.3	15	0.4839	−0.041
150.3	16	0.5161	0.041
150.3	17	0.5484	0.121
150.3	18	0.5806	0.204
150.3	19	0.6129	0.287
150.3	20	0.6452	0.372
151.3	21	0.6774	0.460
152.8	22	0.7097	0.552
152.8	23	0.7419	0.649
152.8	24	0.7742	0.753
152.8	25	0.8065	0.865
154.4	26	0.8387	0.989
155.4	27	0.8710	1.131
156.5	28	0.9032	1.300
157.0	29	0.9355	1.518
158.1	30	0.9677	1.849

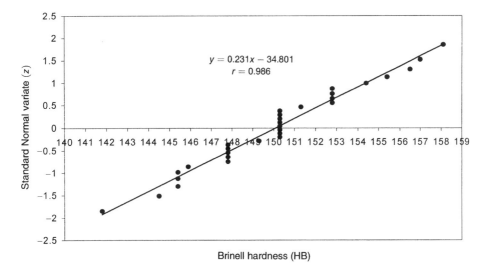

Figure 4.60 Linear regression for the Normal distribution transformed hardness data

From equations 4.12 and 4.13, the mean and standard deviation for the ultimate tensile strength, Su, for steel can be derived:

$$\mu_{Su} = 3.45\mu_{HB} = 3.45 \times 150.07 = 517.7\,\text{MPa}$$

$$\sigma_{Su} = (3.45^2 \cdot \sigma_{HB}^2 + 0.152^2 \cdot \mu_{HB}^2 + 0.152^2 \cdot \sigma_{HB}^2)^{0.5}$$

$$= (3.45^2 \times 4.31^2 + 0.152^2 \times 150.07^2 + 0.152^2 \times 4.31^2)^{0.5}$$

$$\sigma_{Su} = 27.2\,\text{MPa}$$

Therefore:

$$Su \sim N(517.7, 27.2)\,\text{MPa}$$

Typically for ductile steels, the ultimate shear strength, τ_u is 0.75 of Su (Green, 1992), therefore:

$$\tau_u \sim N(388.3, 20.4)\,\text{MPa}$$

Determining the diameter of the shear pin

Assuming that an adequate transition fit is specified for both the pin hole and coupling bore, the pin is in double pure shear with the bore of the coupling, and so the shear stress, L, is given by:

$$L = \frac{F}{2A}$$

where:

$$F = \text{tangential force}$$

$$A = \text{area of pin.}$$

The tangential force acting on the pin at a radius, r, due to the applied torque, M, is given by:

$$F = \frac{M}{r}$$

Therefore, combining the above equations and substituting the radius for the diameter variable gives:

$$L = \frac{1.27324M}{D \cdot d^2} \tag{4.89}$$

Rearranging equation 4.89 in terms of the pin diameter, d, gives:

$$d = \sqrt{\frac{1.27324M}{D \cdot L}} \tag{4.90}$$

where:

$$M = \text{applied torque}$$

$$D = \text{shaft diameter}$$

$$L = \text{loading stress.}$$

For FMEA $(S) = 5$, the reliability of the weak link in service is required to be $R = 0.999$ with reference to the reliability target map given in Figure 4.36. This relates to a failure probability $P = 0.001$ or when working with the SND, $z = -3.09$ from Table 1 in Appendix I. For a given value of the Standard Normal variate, z, the coupling equation for the interference of two Normal distributions can be used to determine the loading stress, and hence the diameter of the shear pin. From the coupling equation we get:

$$3.09 = \frac{\mu_{\tau u_{WL}} - \mu_L}{\sqrt{\sigma_{\tau u_{WL}}^2 + \sigma_L^2}} \tag{4.91}$$

We know that the coefficient of variation, C_v, of the applied torque is approximately 0.1, and that the final loading stress variable will have a similar level of variation because the dimensional variables have a very small variance contribution in comparison. We also know the ultimate shear strength parameters of the weak link material, therefore substituting in equation 4.91 and rearranging to set the right-hand side to zero gives:

$$3.09 \sqrt{20.4^2 + (0.1\mu_L)^2} + \mu_L - 388.3 = 0$$

which yields $\mu_L \approx 281$ MPa when solved by iteration.

From equation 4.90, the mean pin diameter, μ_d, is found to be:

$$\mu_d = \sqrt{\frac{1.27324 \cdot \mu_M}{\mu_D \cdot \mu_L}} = \sqrt{\frac{1.27324 \times 3200}{0.06 \times 281 \times 10^6}} = 0.015545 \, \text{m} = 15.545 \, \text{mm}$$

The pin is machined and cylindrically ground to size. It can be shown that the Normal distribution parameters of the diameter $d \sim N(15.545, 0.0005)$ mm for a tolerance of ± 0.002 mm chosen from the relevant process capability map.

Solving equation 4.89 using Monte Carlo simulation for the variables involved, the shear stress in the pin is found to have a Normal distribution of $L \sim N(281, 28.3)$ MPa. Calculating the reliability using the coupling equation for the given stress and strength parameters gives $R = 0.998950$ which is almost exactly that specified initially.

Selecting the pump shaft material

The torque capacity of the pump shaft must be greater than the torque capacity of the shear pin in all cases. We assume that failure of the pump shaft occurs at the interference of these two torque distributions. From equation 4.89, the torque capacity of the shear pin can be determined by substituting the ultimate shear strength of the weak link material, $\tau_{u_{WL}}$ for L, giving:

$$M_{WL} = 0.785398 D \cdot d^2 \cdot \tau_{u_{WL}} \tag{4.92}$$

Solving equation 4.92 using Monte Carlo simulation for the variables involved, the torque capacity of the shear pin is found to have a Normal distribution of $M_{WL} \sim N(4421.7, 234.1)$ Nm.

The coefficient of variation of the yield strength, and hence the shear yield strength for steels, is typically $C_v = 0.08$. This means that the coefficient of variation of the torque capacity of the pump shaft will approximately be the same, because the dimensional variable of the shaft diameter is very small in comparison to the shear yield strength. We also need to speculate about the likely failure probability acceptable between the weak link and pump shaft torque capacities. Assuming that the overload situation is only likely to occur once in 1000 operating cycles, the coupling equation can be written as:

$$3.09 = \frac{\mu_{M_\mathrm{P}} - \mu_{M_\mathrm{WL}}}{\sqrt{\sigma^2_{M_\mathrm{P}} + \sigma^2_{M_\mathrm{WL}}}} \tag{4.93}$$

Rearranging equation 4.93 to set the right-hand side to zero and substituting in the known parameters gives:

$$3.09\sqrt{(0.08\mu_{M_\mathrm{P}})^2 + 234.1^2} + 4421.7 - \mu_{M_\mathrm{P}} = 0$$

which yields $\mu_{M_\mathrm{P}} \approx 6092.5$ Nm by iteration.

Therefore, the Normal distribution parameters for the pump shaft torque capacity are:

$$M_\mathrm{P} \sim N(6092.5, 487.4)\,\mathrm{Nm}$$

The torque that can be transmitted by a shaft without yielding was given in equation 4.88. Rearranging for the shear yield strength and the variables in this example gives:

$$\tau_\mathrm{y} = \frac{5.09296 M_\mathrm{P}}{D^3} \tag{4.94}$$

Solving equation 4.94 using Monte Carlo simulation for the variables involved, the shear yield strength required for the pump shaft material is found to have a Normal distribution with parameters:

$$\tau_\mathrm{y} \sim N(143.7, 11.6)\,\mathrm{MPa}$$

Therefore, the Normal distribution parameters for the material's tensile yield strength are $1/0.577$ times greater, giving:

$$Sy \sim N(249, 20.1)\,\mathrm{MPa}$$

A suitable material would be hot rolled mild steel 070M20, which has a minimum yield strength, $Sy_\mathrm{min} = 215$ MPa (BS 970, 1991). By considering that the minimum yield strength is -3 standard deviations from the mean and that the typical coefficient of variation $C_\mathrm{v} = 0.08$ for the yield strength of steel, the Normal distribution parameters for 070M20 can be approximated by:

$$Sy \sim N(282.9, 22.6)\,\mathrm{MPa}$$

From equation 4.94 and using Monte Carlo simulation again, the actual torque capacity of the pump shaft using the 070M20 steel grade is found to be:

$$M_\mathrm{P} \sim N(6922.9, 560.1)\,\mathrm{Nm}$$

Finally, the interference between the applied torque and the pump shaft torque capacity can be analysed to determine if separation between them is too great leading

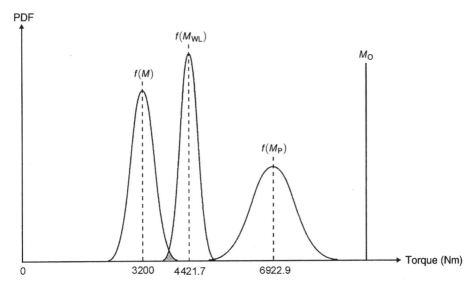

Figure 4.61 Weak link torque capacity shown relative to the applied torque and the torque capacity of the pump shaft

to overdesign. The Safety Margin (SM) was given in equation 4.46 (Carter, 1986):

$$SM = \frac{\mu_{M_P} - \mu_M}{\sqrt{\sigma_{M_P}^2 + \sigma_M^2}} = \frac{6922.9 - 3200}{\sqrt{560.1^2 + 320^2}} = 5.77$$

As a guide, SM should be less than 10 for all cases of failure severity to avoid over-design. The distributions of the applied torque, weak link torque capacity and pump torque capacity are plotted to scale in Figure 4.61 for comparison.

4.8.5 Design of a structural member

The use of probabilistic concepts in structural steelwork design could potentially reduce material costs by delivering optimized designs with standard section sizes, such sections being typically used in large volumes and repetitive applications. Here we will demonstrate this point by selecting the optimum section size for a given situation where a standard structural member must be utilized.

Figure 4.62 shows the arrangement of one of a pair of hangers, which suspend a self-driven belt conveyor unit 3 m in length above a factory floor. The conveyor unit is part of 50 that comprise the materials handling system in the factory. The hanger is essentially a cantilever beam made from an unequal angle section and is securely attached to a column. It is nominally 1250 mm in length from the column face to the hole for the vertical tie rod, and has a fabrication tolerance of ±5 mm. At its free end, the hanger carries a load, F, which hangs vertically, but has an acceptable misalignment tolerance of ±1.5°, based on installation experience.

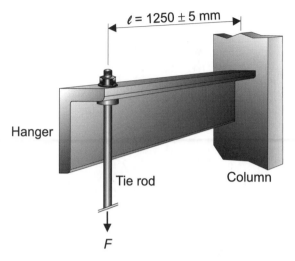

Figure 4.62 Hanger arrangement

The load carried comprises approximately half the mass of the conveyor unit $(50 \pm 9\,\text{kg})$, and half the mass of the items being conveyed at any one time. The mass of the items being conveyed on half of each conveyor unit fluctuates from 0 to 72 kg, approximately following a Normal distribution. The material specified for the hanger is hot rolled Grade 43C structural steel, which has a minimum yield strength $Sy_{\text{min}} = 275\,\text{MPa}$ for a thickness $t \leq 16\,\text{mm}$ (BS 4360, 1990).

From an FMEA of the system design, a Severity Rating $(S) = 7$ was allocated, relating to a safety critical failure in service. It is required to find the optimum unequal angle section size from the standard sizes available. It is assumed that the load is carried at the section's centre of gravity, G, and only stresses due to bending of the section are considered, that is, the torsional effects are minimal. The combined weight of the beam and tie rod are not to be taken into account.

The general dimensional properties of the unequal angle section used for the hanger are shown in Figure 4.63. Note that $a < b$ and $t < a$ and that the leg radii are square for mathematical simplification of the problem.

Determining the stress variables

In terms of the dimensions, a, b and t for the section, several area properties can be found about the x–x and y–y axes, such as the second moment of area, I_{xx}, and the product moment of area, I_{xy}. However, because the section has no axes of symmetry, unsymmetrical bending theory must be applied and it is required to find the principal axes, u–u and v–v, about which the second moments of area are a maximum and minimum respectively (Urry and Turner, 1986). The principal axes are again perpendicular and pass through the centre of gravity, but are a displaced angle, α, from x–x as shown in Figure 4.63. The objective is to find the plane in which the principal axes lie and calculate the second moments of area about these axes. The following formulae will be used in the development of the problem.

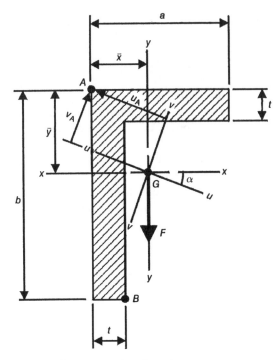

Figure 4.63 General dimensions for an unequal angle section

The position of the centre of gravity, G, is located by:

$$\bar{x} = \frac{a^2 + (b-t)t}{2(a+(b-t))} \tag{4.95}$$

$$\bar{y} = \frac{b^2 + (a-t)t}{2(a+(b-t))} \tag{4.96}$$

The second moments of area about x–x and y–y respectively are:

$$I_{xx} = \tfrac{1}{3}\left[t(b-\bar{y})^3 + a\bar{y}^3 - (a-t)(\bar{y}-t)^3\right] \tag{4.97}$$

$$I_{yy} = \tfrac{1}{3}\left[t(a-\bar{x})^3 + b\bar{y}^3 - (b-t)(\bar{x}-t)^3\right] \tag{4.98}$$

The product moment of area is:

$$I_{xy} = \left[\left((a-\bar{x})-\frac{a}{2}\right)\times\left(\bar{y}-\frac{t}{2}\right)\times at\right] + \left[\left(\bar{x}-\frac{t}{2}\right)\times\left(b-\bar{y}-\frac{b-t}{2}\right)\times(b-t)t\right] \tag{4.99}$$

The principal second moments of area about the principal axes, u–u and v–v, respectively are given by:

$$I_{uu} = \tfrac{1}{2}\left(I_{xx}+I_{yy}\right) + \tfrac{1}{2}\sqrt{\left(I_{xx}-I_{yy}\right)^2 + 4I_{xy}^2} \tag{4.100}$$

$$I_{vv} = \tfrac{1}{2}\left(I_{xx}+I_{yy}\right) - \tfrac{1}{2}\sqrt{\left(I_{xx}-I_{yy}\right)^2 + 4I_{xy}^2} \tag{4.101}$$

where the principal plane is at an angle:

$$\alpha = \frac{\tan^{-1}\left(\dfrac{2I_{xy}}{I_{yy} - I_{xx}}\right)}{2} \tag{4.102}$$

The above equations can all be written in terms of the nominal dimensions, a, b and t, for the section. Solutions for the mean and standard deviation of each property, for any section, can be found using Monte Carlo simulation with knowledge of the likely dimensional variation for hot rolling of structural steel sections. The coefficient of variation for this process/material combination is $C_v = 0.0083$ (Haugen, 1980).

To determine the stress at any point on the section requires that the load be resolved into components parallel to the principal axes. Each component will cause bending in the plane of a principal axis and the total stress at a given point is the sum of the stress due to the load components considered separately. However, first we must consider the nature of the loading distribution and how it is resolved about the principal axes.

The mean of the mass of half the conveyor unit was 50 kg. Assuming that 6 standard deviations adequately characterizes the tolerance range of ± 9 kg, the standard deviation approximates to 3 kg. The mass of the items conveyed was stated to range from 0 to 72 kg and it was assumed this mass varies randomly according to a Normal distribution. Again we assume that this mass range approximates to 6 standard deviations, which gives a mean of 36 kg and a standard deviation of 12 kg. The total mass can be represented by a Normal distribution with a mean $\mu_m = 50 + 36 = 86$ kg, and the standard deviation of the total mass, σ_m, is the statistical sum of the independent variables given by:

$$\sigma_m = \sqrt{3^2 + 12^2} = 12.37 \text{ kg}$$

Converting to newtons force by multiplying by 9.807 gives the load in terms of the Normal distribution as:

$$F \sim N(843, 121) \text{ newtons}$$

The allowable misalignment tolerance for the vertical tie rod, $\psi = \pm 1.5°$, is also considered to be normally distributed in practice. With the assumption that approximately 6 standard deviations are covering this range, the standard deviation becomes $\sigma_\psi = 0.5°$. The mean of the angle on which the principal plane lies is μ_α, and the loads must be resolved for this angle, but its standard deviation is the statistical sum of σ_α and σ_ψ, as given by equation 4.103:

$$\sigma_{\alpha'} = \sqrt{\sigma_\alpha^2 + \sigma_\psi^2} \tag{4.103}$$

and

$$\mu_{\alpha'} = \mu_\alpha$$

The Normal distribution parameters of the length, ℓ, can be developed in the same manner as above to give:

$$\ell \sim N(1250, 1.667) \text{ mm}$$

Therefore, the bending moments resolved about the principal axes are:

$$M_{uu} = F \cdot \cos \alpha' \cdot \ell \qquad (4.104)$$

$$M_{vv} = F \cdot \sin \alpha' \cdot \ell \qquad (4.105)$$

Finally, it is evident from Figure 4.63 that the maximum tensile stress on the section due to the load components about the principal axes will be at point A. The maximum compressive stress will be at point B. From trigonometry, the distances from the centre of gravity to point A on the section in the directions of the principal axes are:

$$u_A = \bar{x} \cos \alpha + \bar{y} \sin \alpha \qquad (4.106)$$

$$v_A = \bar{y} \cos \alpha - \bar{x} \sin \alpha \qquad (4.107)$$

The tensile stress at point A on the section L_A can then be determined by applying simple bending theory:

$$L_A = \frac{M_{uu} \cdot v_A}{I_{uu}} + \frac{M_{vv} \cdot u_A}{I_{vv}} \qquad (4.108)$$

A similar approach can be used to determine the compressive stress at point B on the section L_B where:

$$u_B = \cos \alpha [\tan \alpha (b - \bar{y}) - (\bar{x} - t)] \qquad (4.109)$$

$$v_B = \left(\frac{b - \bar{y}}{\cos \alpha} \right) - \sin \alpha [\tan \alpha (b - \bar{y}) - (\bar{x} - t)] \qquad (4.110)$$

$$L_B = -\frac{M_{uu} \cdot v_B}{I_{uu}} - \frac{M_{vv} \cdot u_B}{I_{vv}} \qquad (4.111)$$

Stress–strength interference analysis

Several standard section sizes for unequal angles are listed in Table 4.15 (BS 4360, 1990). For each standard section, first the statistical variation of the area properties, distances and angles can be estimated using Monte Carlo simulation, which are then used to determine stresses at points A and B on the section from solution of equations 4.108 and 4.111. The stresses found at points A and B for the sections listed are also shown in Table 4.11.

A specific statistical representation of the yield strength for Grade 43C hot rolled steel is not available; however, the coefficient of variation, C_v, for the yield strength of British structural steels is given as 0.05 for a thickness $t \leq 12.7$ mm (Rao, 1992). For convenience, the parameters of the Normal distribution will be calculated by assuming that the minimum value is -3 standard deviations from the expected

Table 4.15 Section size ($t \leq 16$), loading stress parameters and reliability for a range of standard unequal angle sections

a, b, t (mm)	μ_{L_A} (MPa)	σ_{L_A} (MPa)	μ_{L_B} (MPa)	σ_{L_B} (MPa)	R_A	R_B
30 × 60 × 5	305.2	45.1	−346.9	50.8	0.648724	0.330391
30 × 60 × 6	264.2	39.2	−297.0	43.6	0.918951	0.715319
50 × 65 × 5	211.1	31.2	−264.8	38.8	0.999299	0.918648
50 × 65 × 6	183.0	27.0	−225.9	32.9	0.999995	0.996112
50 × 65 × 8	148.2	22.0	−176.2	26.0	1.000000	0.999999
50 × 75 × 6	142.6	21.0	−172.2	25.1	1.000000	1.000000
50 × 75 × 6	114.0	16.8	−134.5	19.6	1.000000	1.000000

mean value as described by equation 4.23:

$$Sy_{min} = \mu_{Sy} - 3\sigma_{Sy}$$

Therefore,

$$275 = \mu_{Sy} - 3(0.05\mu_{Sy})$$

$$\mu_{Sy} = \frac{275}{0.85} = 323.5 \, \text{MPa} \quad \text{and} \quad \sigma_{Sy} = 0.05 \times 323.5 = 16.2 \, \text{MPa}$$

The yield strength of Grade 43C structural steel can be approximated by:

$$Sy \sim N(323.5, 16.2) \, \text{MPa}$$

The coupling equation for the stress–strength interference analysis for the problem is given by:

$$z = -\frac{\mu_{Sy} - \mu_{L_{A,B}}}{\sqrt{\sigma_{Sy}^2 + \sigma_{L_{A,B}}^2}}$$

where the Standard Normal variate, z, can be related to the reliability, R, as shown previously. The results are provided in Table 4.15. The FMEA Severity Rating $(S) = 7$ for the system indicates a target reliability $R \approx 0.999995$ from Figure 4.36. Based on this specification, the $50 \times 65 \times 8$ section has a reliability $R_B = 0.999999$ when the greatest stress is used in the coupling equation. This assumes, of course, that the compressive yield strength of the material is equivalent to the tensile yield strength which is the case for most ductile materials.

4.8.6 Bimetallic strip deflection

Bimetallic elements are widely used in instruments such as thermostats to sense or control temperatures. There are several bimetallic element types available, such as straight strips, coils and discs, but all rely on the same working principle. In its most basic form, the bimetallic strip comprises of two dissimilar metal strips bonded together, usually of the same surface area, but not necessarily of the same thickness thermostat. The composite metal strip is clamped at one end to act as a cantilever beam, and is horizontal at a particular temperature. When the temperature is increased, the strip deflects in the direction of the metal with the least coefficient of linear expansion. Its working principle relies on the fact that the metals will expand at different rates as the strip is heated. The purpose of this deflection is to typically cause the strip to make contact with a switch or complete an electric circuit at a particular setpoint temperature above the ambient.

Figure 4.64 shows a thermostat comprising a bimetallic strip formed by bonding a strip of cold rolled 60/40 brass with a similar size strip of cold rolled mild steel. Both the brass and steel strips are exactly the same width of 15 mm. The bimetallic strip is precisely located and rigidly clamped giving a length of 70 ± 0.05 mm. When heated from an ambient or datum temperature, $T_0 = 15°C$, when the strip is horizontal, to its setpoint temperature, T, the strip deflects downwards and completes an electric circuit. It is required to determine the setpoint temperature for the bimetallic strip for a given deflection of 2 ± 0.05 mm, which is the distance that separates the base

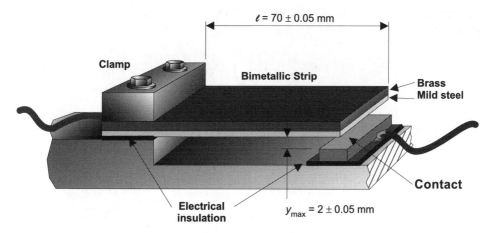

Figure 4.64 Thermostat arrangement

of the strip and the electrical contact. It is also required to determine the most critical variables involved in meeting this requirement through performing a sensitivity analysis. The thickness of the brass strip is 0.5 ± 0.02 mm, and the mild steel strip thickness is 0.4 ± 0.02 mm. The Modulus of Elasticity $E = 105$ GPa for 60/40 brass with a coefficient of variation $C_v \approx 0.02$, and $E = 208$ GPa for mild steel with $C_v \approx 0.03$. The coefficient of linear expansion $\alpha = 18.5\,(10^{-6}/°\text{C})$ for cold rolled brass, and for cold rolled mild steel, $\alpha = 12.7\,(10^{-6}/°\text{C})$, both with a coefficient of variation $C_v \approx 0.01$, estimated from engineering literature. It is assumed that the stresses set up in each strip on deflection are well below their yield strengths and that no residual stresses are acting in the individual strip material.

Determining the temperature variation

The maximum deflection at the free end of the bimetallic strip, y_{max}, due to a temperature increase from T_0 to T is given by (Young, 1989):

$$y_{max} = \frac{3\ell^2(T - T_0)(\alpha_S - \alpha_B)(d_B + d_S)}{d_S^2 \cdot \left[4 + 6\dfrac{d_B}{d_S} + 4\left(\dfrac{d_B}{d_S}\right)^2 + \dfrac{E_B}{E_S}\left(\dfrac{d_B}{d_S}\right)^3 + \dfrac{E_S \cdot d_S}{E_B \cdot d_B} \right]} \tag{4.112}$$

where:

$$\ell = \text{length of the bimetallic strip}$$

$$\alpha_S = \text{linear coefficient of expansion of mild steel}$$

$$\alpha_B = \text{linear coefficient of expansion of brass}$$

$$d_S = \text{thickness of mild steel strip}$$

$$d_B = \text{thickness of brass strip}$$

$$E_S = \text{Modulus of Elasticity for mild steel}$$

$$E_B = \text{Modulus of Elasticity for brass}.$$

Replacing the change in temperature $(T - T_0)$ with ΔT, which is a random variable itself, and rearranging for this term gives:

$$\Delta T = \frac{y_{max} \cdot d_S^2 \cdot \left[4 + 6\dfrac{d_B}{d_S} + 4\left(\dfrac{d_B}{d_S}\right)^2 + \dfrac{E_B}{E_S}\left(\dfrac{d_B}{d_S}\right)^3 + \dfrac{E_S \cdot d_S}{E_B \cdot d_B}\right]}{3\ell^2 \cdot (\alpha_S - \alpha_B)(d_B + d_S)} \tag{4.113}$$

The variance equation to determine the standard deviation of the change in temperature can be written as:

$$\sigma_{\Delta T} \approx \left[\begin{array}{l}\left(\dfrac{\partial \Delta T}{\partial y_{max}}\right)^2 \cdot \sigma_{y_{max}}^2 + \left(\dfrac{\partial \Delta T}{\partial d_S}\right)^2 \cdot \sigma_{d_S}^2 + \left(\dfrac{\partial \Delta T}{\partial d_B}\right)^2 \cdot \sigma_{d_B}^2 + \left(\dfrac{\partial \Delta T}{\partial E_S}\right)^2 \cdot \sigma_{E_S}^2 \\[4mm] + \left(\dfrac{\partial \Delta T}{\partial E_B}\right)^2 \cdot \sigma_{E_B}^2 + \left(\dfrac{\partial \Delta T}{\partial \alpha_S}\right)^2 \cdot \sigma_{\alpha_S}^2 + \left(\dfrac{\partial \Delta T}{\partial \alpha_B}\right)^2 \cdot \sigma_{\alpha_B}^2 + \left(\dfrac{\partial \Delta T}{\partial \ell}\right)^2 \cdot \sigma_{\ell}^2\end{array}\right]^{0.5} \tag{4.114}$$

The mean value, $\mu_{\Delta T}$, can be approximated by substituting the mean values for each variable into equation 4.114. All eight variables are assumed to be random in nature following the Normal distribution. A summary of the parameters for each variable is given below. The standard deviation of each has been derived, in the case of y_{max}, and ℓ, from assuming that 6 standard deviations cover the tolerance range given, in the case of the individual strip thicknesses, from an appraisal using CA, and in the case of the elastic constants and coefficients of linear expansion, directly from the coefficients of variation provided.

$$y_{max} \sim N(0.002, 0.000017) \, \text{m}$$
$$d_S \sim N(0.0003, 0.000004) \, \text{m}$$
$$d_B \sim N(0.0005, 0.000002) \, \text{m}$$
$$E_S \sim N(208, 6.24) \, \text{GPa}$$
$$E_B \sim N(105, 2.1) \, \text{GPa}$$
$$\alpha_S \sim N(12.7, 0.127) \, 10^{-6}/°\text{C}$$
$$\alpha_B \sim N(18.5, 0.185) \, 10^{-6}/°\text{C}$$
$$\ell \sim N(0.07, 0.000017) \, \text{m}$$

With reference to Appendix XI, we can solve each partial derivative term in equation 4.114 using the Finite Difference method to give:

$$\sigma_{\Delta T} = \left[\begin{array}{l}(2.648 \times 10^4)^2 \times (1.7 \times 10^{-5})^2 + (1.067 \times 10^5)^2 \times (4 \times 10^{-6})^2 \\[2mm] +(8.41 \times 10^4)^2 \times (2 \times 10^{-6})^2 + (1.172 \times 10^{-11})^2 \times (6.24 \times 10^9)^2 \\[2mm] +(-2.419 \times 10^{-11})^2 \times (2.1 \times 10^9)^2 + (1.472 \times 10^7)^2 \times (0.127 \times 10^{-6})^2 \\[2mm] +(-1.485 \times 10^7)^2 \times (0.185 \times 10^{-6})^2 + (-2.421 \times 10^4)^2 \times (1.7 \times 10^{-5})^2\end{array}\right]^{0.5}$$

and

$$\sigma_{\Delta T} = 3.41°C$$

and the mean $\mu_{\Delta T} = 84.74°C$. Therefore, the setpoint temperature is given by:

$$T = \Delta T + T_0 = 84.74 + 15 = 99.74°C$$

Sensitivity analysis

For the purposes of meeting a customer specification, a tolerance for the thermostat setpoint temperature can be estimated at $\pm 3\sigma_{\Delta T}$, from which the approximate thermostat specification becomes:

$$T \approx 100 \pm 11°C$$

From the Finite Difference method results above, the contribution of the variance of each variable to the temperature variance can be estimated to focus in on the key variables bounding the problem. This is an attempt to reduce the thermostat specification tolerance to around half its current design value.

$$\sigma_{\Delta T}^2 = (0.203) + (0.182) + (0.028) + (5.345 \times 10^{-3}) + (2.579 \times 10^{-3})$$
$$+ (3.497) + (7.550) + (0.169)$$
$$= 11.637$$

For the first term, the variance contribution of the deflection variable, y_{max}, to the variance of the temperature as a percentage becomes:

$$\frac{0.203}{11.637} \times 100 = 1.74\%$$

Repeating the above for each variable and ranking the percentage values in descending order gives the Pareto chart in Figure 4.65. It is evident that over 90% of the variance in the temperature is due to the variance of the expansion coefficients for the metals used. These are the most critical variables in the problem of setting the tolerance specification. A better understanding and control of these properties of the metals, or even the use of more exotic materials than those first chosen, will lead to a reduction in the achievable variation. This is because the other variables in the problem, such as the elastic constants of the metals, have a minimal contribution.

This case study has highlighted the use of probabilistic design principles other than for the purpose of SSI analysis. For nearly every engineering problem, the development of an answer using probabilistic techniques relies on manipulating the basic governing function for the parameter of interest, as would be done in a deterministic approach. A development from first principles is not necessary, and it is commonly the determination of the random variables involved in the problem which requires further and a more thorough investigation. For the situation above, a standard formulation is available, which is the case for many engineering problems. Although the mathematics needed for a probabilistic

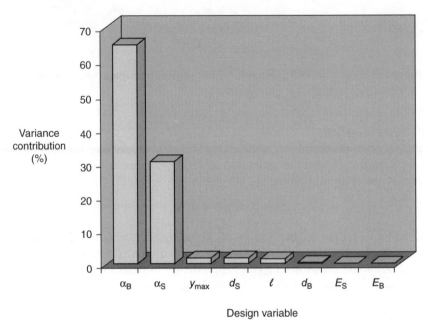

Figure 4.65 Sensitivity analysis for the temperature variance

analysis is more complex than a deterministic one, any engineering problem can be approached in a similar mechanistic fashion using the techniques and methodology discussed.

4.8.7 Design of a con-rod and pin

This case study discusses the design of a reciprocating mechanical press for the manufacture of can lids drawn from sheet steel material. The authors were involved in the early stages of the product development process to advise the company designing the press in choosing between a number of design alternatives with the goal of ensuring its reliability. The authors used a probabilistic approach to the problem to provide the necessary degree of clarity between the competing solutions.

The press had been designed with a capacity to deliver 280 kN press force and to work at a production rate of 40 lids per minute. Calculations to determine the distribution of forming loads required indicated that the press capacity was adequate to form the family of steel lids to be produced on the machine. One of the major areas of interest in the design was the con-rod and pin (see Figure 4.66). The first option considered was based on a previous design where the con-rod was manufactured from cast iron with phosphor bronze bearings at the big and small ends. However, weaknesses in this approach necessitated the consideration of other options. The case study presents the analysis of the pin and con-rod using simple probabilistic techniques in an attempt to provide in-service reliable press operation. The way a weak link was introduced to ensure ease of maintenance and repair in the event

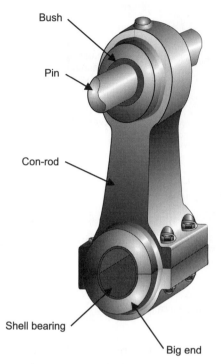

Bush

Pin

Con-rod

Shell bearing

Big end

Figure 4.66 General arrangement of the con-rod and pin

of an overload situation is also covered. Overloads typically occur if a pressed part is badly orientated within the press dies or if a foreign body is caught in the die set, literally described as 'a spanner in the works'. Analysis of the bolts is not covered.

Design strategy

In production, the press will be operating at millions of cycles *per annum*. Therefore, it should be designed against fatigue failure, and the con-rod and pin must be engineered in the light of the distribution of the associated endurance strength in shear. For the purposes of the analysis it was assumed that the applied stress would have an extreme value corresponding to the application of the 280 kN load. Also, in carrying out the analysis, approximate statistical models were needed for the material's endurance strength based on the available data. As mentioned earlier, the system was to be designed with a weak link. To satisfy this requirement the pin was designed such that it would fail in an overload situation.

The material selected for the pin was 070M20 normalized mild steel. The pin was to be manufactured by machining from bar and was assumed to have non-critical dimensional variation in terms of the stress distribution, and therefore the overload stress could be represented by a unique value. The pin size would be determined based on the −3 standard deviation limit of the material's endurance strength in shear. This infers that the probability of failure of the con-rod system due to fatigue would be very low, around 1350 ppm assuming a Normal distribution for the endurance strength in shear. This relates to a reliability $R \approx 0.999$ which is adequate for the

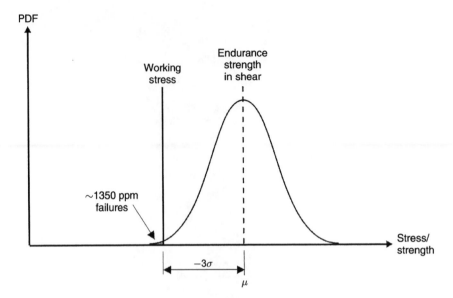

Figure 4.67 Stress–strength interference for the pin

non-safety critical nature of the failure mode. The situation is represented in the stress–strength interference diagram given in Figure 4.67.

Determination of the pin diameter and con-rod section size

There is no data available on the endurance strength in shear for the material chosen for the pin. An approximate method for determining the parameters of this material property for low carbon steels is given next. The pin steel for the approximate section size has the following Normal distribution parameters for the ultimate tensile strength, Su:

$$Su \sim N(505.9, 25.3)\,\text{MPa}$$

The endurance strength in bending, Se, is commonly found by multiplying Su by an empirical factor, typically 0.5 for steels. For mild steel, the relationship is (Waterman and Ashby, 1991):

$$Se \approx 0.47Su \qquad (4.115)$$

The relationship between the endurance strength in shear to that in bending is given by (Haugen, 1980):

$$\tau_e \approx 0.577Se \qquad (4.116)$$

Therefore, substituting equation 4.115 into equation 4.116 gives:

$$\mu_{\tau_e} = 0.27\mu_{Su} = 0.27(505.9) = 136.6\,\text{MPa}$$

Typically for the variation of the endurance strength in bending, at 10^6 cycles of operation (Furman, 1981):

$$\sigma_{Se} \approx 0.08\mu_{Se}$$

Substituting equation 4.115 into the above gives:

$$\sigma_{Se} = 0.038\mu_{Su}$$

Finally, substituting equation 4.116 into the above gives the standard deviation of the endurance strength in shear to be:

$$\sigma_{\tau_e} = 0.022\mu_{Su} = 0.022(505.9) = 11.1\,\text{MPa}$$

Summarizing, the parameters for the endurance strength for 070M20 normalized mild steel in shear are:

$$\tau_e \sim N(136.6, 11.1)\,\text{MPa}$$

The pin is in double shear in service. The diameter, d, is determined by:

$$d = \sqrt{\frac{0.63662F}{\tau_{e\,min}}} \tag{4.117}$$

where:

$$F = \text{shear force}$$

$$\tau_{e\,min} = \text{minimum endurance strength in shear.}$$

The minimum endurance strength for the problem stated earlier is set at the -3 standard deviations limit, therefore:

$$\tau_{e\,min} = \mu_{\tau_e} - 3\sigma_{\tau_e} = 136.6 - 3(11,1) = 103.6\,\text{MPa}$$

Therefore:

$$d = \sqrt{\frac{0.63662 \times 280 \times 10^3}{103.6 \times 10^6}} = 0.0415\,\text{m} = 41.5\,\text{mm}$$

The final selection of pin diameter based on preferred numbers gives $d = \varnothing 42\,\text{mm}$.

In designing the con-rod, we wish to ensure that the pin will fail, in the case of an overload, in preference to the con-rod. To realize this, the mean values of their individual strength distributions are to be set apart by a margin to ensure this requirement. In this way, the probability of con-rod failure will become insignificant to that of the pin. The force to shear the pin in an overload situation is a function of the ultimate shear strength, τ_u, of the material. The relationship between the ultimate tensile and shear properties for steel is (Green, 1992):

$$\tau_u = 0.75Su$$

Therefore:

$$\tau_u \sim N(379.4, 19)\,\text{MPa}$$

We assume that the maximum ultimate shear strength, $\tau_{u\,max}$, of the pin is $+3$ standard deviations from the mean value, therefore:

$$\tau_{u\,max} = \mu_{\tau_u} + 3\sigma_{\tau_u} = 379.4 + 3(19) = 436.4\,\text{MPa}$$

$$F = 1.5708\tau_{u\,max}d^2 = 1.5708 \times 436.4 \times 10^6 \times 0.042^2 = 1.21\,\text{MN}$$

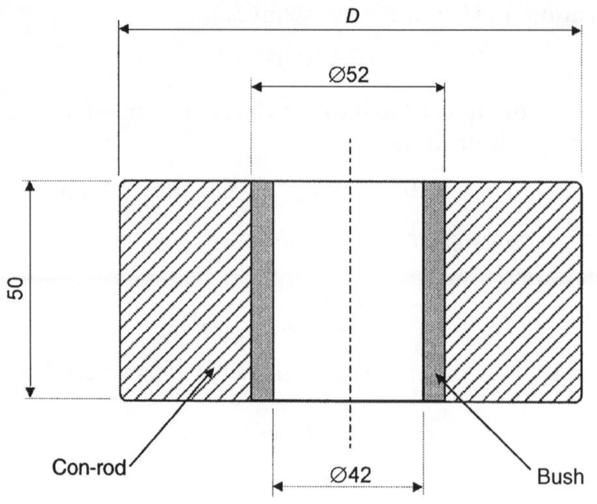

Figure 4.68 Basic dimensions of the con-rod small end cross-section

The next stage is to calculate the con-rod size based on this force transmitted from the pin in the overload situation. The basic dimensions of small end cross-section incorporating a phosphor bronze bearing are shown in Figure 4.68.

Given the cost constraints imposed on the press frame design, and the desire to minimize dimensional variation, the same material was selected for the con-rod cross-section. To determine the dimension, D, using the same steel for the con-rod:

$$Su_{min} = \mu_{Su} - 3\sigma_{Su} = 505.9 - 3(25.3) = 430 \text{ MPa}$$

$$A = \frac{F}{Su_{min}} = 0.05(D - 0.052)$$

where:

$$A = \text{area of the section}$$

$$F = \text{tensile force.}$$

Solving for D gives:

$$D = \frac{1.21 \times 10^6}{0.05(430 \times 10^6)} + 0.052 = 0.1083 \text{ m} = 108.3 \text{ mm}$$

Again, a preferred value for the actual section width would be $D = 110$ mm. In a more simplified way than that presented in Section 4.8.4, we have separated the failure of the pin from the con-rod by approximately $3 + 3$ strength standard deviations. The actual separation can be modelled for the distribution of the shear force in the pin and tensile force in the con-rod, as illustrated in Figure 4.69. The safety margin, SM, is calculated to be 4.61, or defined another way, the reliability $R = 0.999998$ which is adequate for the application to avoid overdesign of the con-rod.

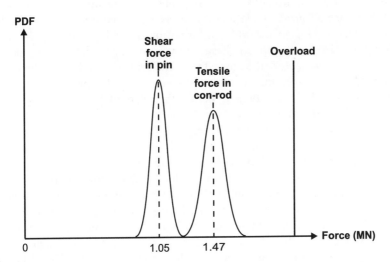

Figure 4.69 Separation of the pin and con-rod forces

Observations

An important aspect of the simple probabilistic approach used above was that it provided a transparent means of explaining to the company the reasons behind the design decisions. It gave a degree of clarity not provided by a deterministic approach and ultimately gave the engineers more confidence in their designs.

In this the final case study we have touched on a probabilistic approach in support of designing against fatigue failure, a topic which is actually outside the scope of the book. A fatigue analysis for the con-rod would need to take into account all factors affecting the fatigue life, such as stress concentrations and surface finish. However, it has indicated that a probabilistic design approach has a useful role in such a setting. Readers interested in more on stress concentrations and probabilistic fatigue design are directed to Carter (1986), Haugen (1980) and Mischke (1992).

4.9 Summary

Product life-time prediction, cost and weight optimization have enormous implications on the business of engineering manufacture. Deterministic design fails to provide the necessary understanding of the nature of manufacture, in-service loading and its variability. It is, however, appealing because of its simplicity in form and application, but since factors of safety are not performance related measures there is no way by which an engineer can know whether their design is near optimum or overconservative in critical applications. Probabilistic approaches offer much potential in this connection, but have yet to be taken up widely by manufacturing industry.

Virtually all design parameters such as tolerances, material properties and service loads exhibit some statistical variability and uncertainty that influence the adequacy of the design. A key requirement in the probabilistic approach is detailed knowledge about the distributions involved, to enable plausible results to be produced. The amount of information available at these early stages is limited, and the designer makes experienced judgements where information is lacking. This is why the deterministic approach is still popular, because many of the variables are taken under the 'umbrella' of one factor. If knowledge of the critical variables in the design can be estimated within a certain confidence level, then the probabilistic approach becomes more suitable. Probabilistic design then provides a more realistic way of thinking about the design problem.

This chapter has outlined the main concepts and techniques associated with designing for reliability that have been developed with regard to a probabilistic design. Past work on Conformability Analysis (CA) has made it possible to estimate dimensional variation, a key component in the probabilistic calculations used as part of the methodology. A key problem in probabilistic design is the generation of the characterizing distributions for experimental data. The effective statistical modelling of material property data and service loads is fundamental to the probabilistic approach. Where this data needs to be modelled, the most efficient techniques make it possible to estimate the parameters for several important distributions.

We need a special algebra to operate on the engineering stress equations as part of probabilistic design. Through the use of the variance equation, a means of estimating the stress variable is provided by relating geometric and load distributions with the failure governing stress equation. Methods to solve the variance relationships for near-Normal conditions have been discussed and examples given for more complex cases using the Finite Difference Method and Monte Carlo simulation. The variance equation also provides a valuable tool with which to draw sensitivity inferences to give the contribution of each variable to the overall variability in the problem. Sensitivity analysis is part of the standard reliability analysis and through its use probabilistic methods provide a more effective way to determine key design parameters in a design. From this and other information in Pareto chart form, the designer can quickly focus on the dominant variables for redesign purposes.

Stress–strength interference analysis offers a practical engineering approach for designing and quantitatively predicting the reliability of components subjected to mechanical loading and has been described as a simulative model of failure. The probability of failure, and hence the reliability, can be estimated as the area of interference between the stress and strength distributions. However, the analysis of reliability using this approach is more often than not incorrectly performed and a thorough understanding of the loading type is required by the practitioner. This chapter has reviewed the application of stress–strength interference analysis to some important cases in static design and has provided methods to solve the problem for any combination of stress and strength using closed-form equations or numerical techniques.

When designing a product it is useful to have a reliability target, in order to attain customer satisfaction and reduce levels of non-conformance and attendant failure costs. A clear and concise approach to failure mode descriptions and reliability targets determination is also crucial in the development of reliable designs, and the integrated use of FMEA in setting reliability target levels is a key benefit of the approach in this respect. A key objective of the methodology is to provide the designer with a deeper understanding of these critical design parameters and how they influence the adequacy of the design in its operating environment. The design intent must be to produce detailed designs that reflect a high reliability when in service. It is appreciated that there is not always published data on engineering variables. However, much can be done by approaching engineering problems with a probabilistic 'mindset'.

Probabilistic design provides a transparent means of explaining to a business more about the safety aspects of engineering design decisions with a degree of clarity not provided by the 'factor of safety' approach. The measures of performance determined using a probabilistic approach give the designers more confidence in their designs by providing better understanding of the variables involved and quantitative estimates for reliability.

5

Effective product development

5.1 Introduction

Effective product development can be the single most important driving force behind creating successful products. The objective is to develop a product that has been systematically optimized to meet the customers' needs as early as possible (Dertouzos *et al.*, 1989). Fierce competition and higher customer expectations are forcing manufacturing businesses to improve quality, reduce costs, and shorten time to market and this places new pressures on the product development process. In today's globally competitive world, successful product development means achieving a level of excellence that goes far beyond the traditional notions of a quality product and the added dimension of speed to market is seen as the key to success (Barclay and Poolton, 1994; Walsh, 1992).

The importance of time to market has recently been shown to be responsible for over 30% of the total profit to be made from a product during its life-cycle. However, it has been found that nearly 30% of product development programmes overran their planned times (Maylor, 1996; Nichols *et al.*, 1993). The reduction in profit due to late delivery of the product to market is shown in Figure 5.1 for a sample of businesses surveyed.

Protracted lead times were particularly the case in Western automotive companies in the 1980s. Compared to their Japanese counterparts, the number of engineering changes experienced were much higher which resulted in protracted lead times, as shown in Figure 5.2 (Clark and Fujimoto, 1991). The differences were due to the different design philosophies of the companies. Japanese companies tend to assign a large engineering staff to the project early in product development and encourage the engineers to utilize the latest techniques and to explore all the options early to preclude the need for changes later on. The use of quality tools and techniques by the Japanese, such as QFD, in the development of the product design was seen as key to reducing lead times.

Typically, fixing errors and redesign account for around 30% of product development time, as shown in Figure 5.3, and improving this position provides an opportunity for lead time reduction. This means doing more work early in the process when (Parker, 1997):

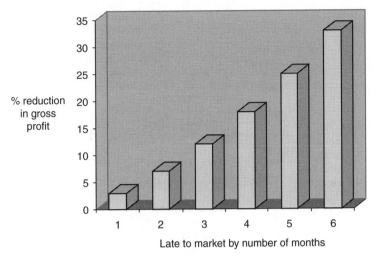

Figure 5.1 Reduction of profit due to late delivery of the product to market (Ostrowski, 1992)

- It is easy to influence the customer
- Design changes are easy
- The cost of change is low
- Management involvement is more cost effective.

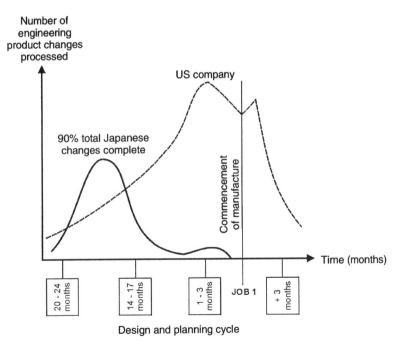

Figure 5.2 Engineering changes in the design and planning cycle of a motor vehicle under Japanese and Western practices (Sullivan, 1987)

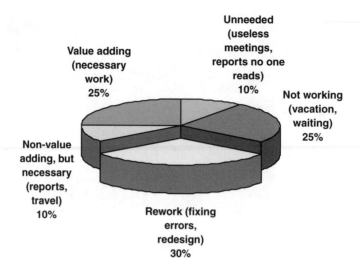

Figure 5.3 Lead time reduction opportunities in product development (Parker, 1997)

This puts the emphasis on design to improve the systems used in developing the products, and to detect and correct potential problems very early on in the development process.

One of the most important factors that influences the reduction of errors and defects is the number of quality tools and techniques used (Radovilsky *et al.*, 1996). Tools and techniques contribute to the engineering knowledge of the product design in the early stages of product development and consequently may decrease the numbers of errors and design changes (Norell and Andersson, 1996). The general recommendation is to use the methods as early as possible in the product development process because the greatest opportunity in reducing design changes and rework occurs here.

The elimination of rework not only includes the requirement to use robust technologies, but to have design information available at the right time (Clausing, 1998). It is recognized that increased communication is one of the most important effects of the usage of tools and techniques. Given two thirds of all technical modifications could be avoided by better communication, the need is clear (Clark and Fujimoto, 1992). The use of tools and techniques can make the difference when in the early stages of product development. They provide this necessary information at each critical stage from which informed decisions can be made together as a team. In essence, tools and techniques help anticipate problems upstream by considering downstream activities – visualizing other people's problems.

It has also become increasingly important for companies to develop their own product development process as a basis for competitive advantage (Jenkins *et al.*, 1997b). An effective product development process with the use of complementing processes should be used (Klit *et al.*, 1993). However, a recent survey shows that over 30% of companies surveyed in the UK do not employ a formal strategy for product development (Araujo *et al.*, 1996). It has been found that companies that use a strategy for product development have been consistently more successful than those who do not (Cooper and Kleinschmidt, 1993).

The design and development of a new product requires creativity which is often unplanned and unstructured; however, new product development can still be encouraged, managed and controlled proactively. Therefore, we also see that a company's success in launching new products is dependent upon the management of the product development process (Jenkins *et al.*, 1997b; Young and Guess, 1994). Merely having a new product development process has no impact on the business, but successfully managing it is, to a great extent, the process of separating the winners from the losers (Cooper and Kleinschmidt, 1995).

Product design has become crucially important for corporate competitiveness and long-term survival (Araujo *et al.*, 1996). Design methods or 'philosophies' have been extensively researched and documented, but they are far from a product design panacea. The primary purpose of these methods is to formalize the design process and externalize design thinking (Prasad *et al.*, 1993). Design methods have the effect of aiding the creative process by ensuring that all avenues are fully explored in a systematic way (Pahl and Beitz, 1989). Attempts at generalizing the product design process have yielded inconclusive results and their direct affect on improving product quality and reducing time to market have also been inconclusive. Despite the large number of design methods available, and the importance credited to them in some circles, the adoption of these methods by industry is not widespread or always successful (Andreasen, 1991; Gill, 1990; Prasad *et al.*, 1993).

Similarly, the achievement of quality assurance certification, such as BS EN ISO 9000 (1994), is also not a guarantee of good quality products. The quality assurance model recommends quality 'system requirements' for the purpose of a supplier demonstrating its capability, and for assessment of the capability in terms of design, development, production, installation and servicing (BS EN ISO 9001, 1994). The adoption of quality standards is only a first step in the realization of quality products as is the acceptance and implementation of a fundamental design method. Tools and techniques also provide useful aids in the process of quality improvement, but they do not ensure product quality (Andersson, 1994). The implementation of an effective product development process is the pivotal means of driving all the requirements and processes involved in producing a 'quality' product.

It is the objective of this chapter to present a model for effective product development. It is a generic framework to apply the tools and techniques seen as the most beneficial in the design of capable and reliable products. Given the diversity of companies and industries, as well as the complexity of the product development process, no single set of activities or steps can be defined that will be appropriate to all due to the dependence on so many subjective variables (Calantone *et al.*, 1995; Elliot *et al.*, 1998). However, for the demonstration of the effective placement of the tools and techniques, it is necessary that a generic process with defined functional stages be presented. Also discussed are some of the important peripheral issues with regard to the product development process and designing capable and reliable products that have been highlighted through industrial collaboration.

5.2 Product development models

5.2.1 Overview of product development models

Product development models are the driving force for delivering the product to market on time and at the right cost. In general, the models in the literature can be divided into just two types: sequential and concurrent. Each has its own characteristics, but there are several requirements that a new product development model should fulfil (Sum, 1992):

- It must be scaleable as organizations change size constantly
- The model will probably be introduced incrementally, perhaps into one team, and then over time it will spread to larger sections of the company
- It must be extensible, because not all features of product development will be foreseen at the time that the model is developed, for example new tools and techniques might emerge
- It must be adaptable as the situations within every organization are different and a uniform product development process will not capitalize on the strengths of individual enterprises or address the weaknesses.

Traditionally, product development has been viewed as an organizational activity, which is the result of various functional activities performed in stages, such as design followed by manufacture. The sequential operation of these functional stages resulted in long development times and many quality problems due to the lack of communication and understanding of the different design, manufacturing, and above all, customer requirements (Haque and Pawar, 1998). This is shown diagrammatically in Figure 5.4. A popular product development process reference (Dale and Oakland, 1994; Evbuomwan *et al.*, 1996) is taken from the British Standards (BS 7000, 1997) as shown in Figure 5.5. It is termed an 'idealized' product evolution cycle, and is clearly sequential in nature.

This series of walls between the functional areas are eventually being broken down and replaced with new alliances and modes of interaction. This has been accomplished by (Russell and Taylor, 1995):

- Establishing multifunctional design teams
- Making product and process design decisions concurrently instead of sequentially

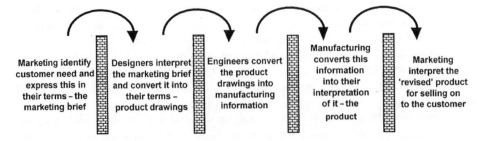

Figure 5.4 Traditional approach to product development (Maylor, 1996)

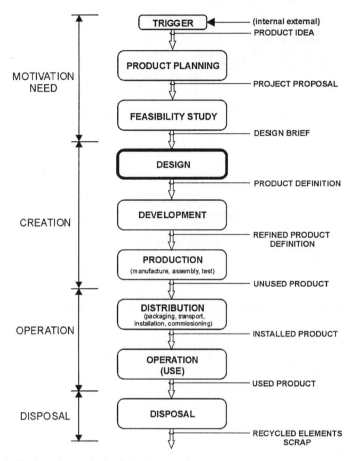

Figure 5.5 Idealized product evolution (BS 7000, 1997)

- Changing the role of design engineers
- Utilizing new and existing tools and techniques
- Measuring design quality.

The traditional sequential product development process is being replaced by a faster and far more effective team-based concurrent engineering approach, also called simultaneous engineering (Miles and Swift, 1992). A key feature is the involvement of several engineers at each stage of the process implying a high degree of collaboration and intrinsic overlap of the product development functions (Prasad *et al.*, 1993). The arrangement of product development into a process stream, with all the necessary parties involved at all stages to prevent the rework of ideas, has the natural effect of allowing activities to run along side one another (concurrently) as opposed to one after the other (sequentially). This is shown in Figure 5.6 (Maylor, 1996). The immediate benefit of the concurrent model over the sequential model is the saving in time during the product development process. The use of concurrent engineering has resulted in companies making new products better and faster (Haque and Pawar, 1998).

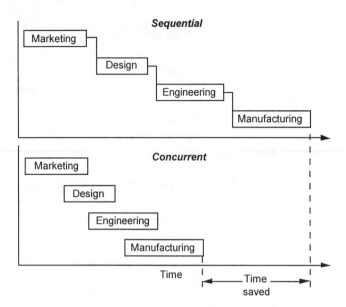

Figure 5.6 Sequential versus concurrent model for product development (Maylor, 1996)

Additionally, a structured concurrent product development process eventually allows a project to be split up into logical phases rather than functional steps, and allows decision points or gates to be inserted at the appropriate points. This gives senior management the opportunity to review a project's progress against an agreed set of deliverables, and to be involved in the project at appropriate points in the programme, rather than to attempt to micro-manage the project (Jenkins *et al.*, 1997b). This is demonstrated in the industrial models described later.

The automotive sector's quality assurance standard QS 9000 (1998) suggests a concurrent high level model, as opposed to the sequential model from BS 7000 (1997). This is shown in Figure 5.7. The automotive industry in particular has embraced the use of concurrent engineering models for product development, and this is reflected in the standards which facilitate their quality assurance programmes. A concurrent industrial model from the automotive sector will be discussed later.

An important benefit with regard to designing for quality in particular is that concurrent engineering processes are more conducive to the free flow of information between the design/manufacture interface. Companies that commit to this new way of working are more likely to gain the early commitment of production for this to be achieved. In contrast, those who do not embrace concurrent processes, typically express greater problems in gaining the commitment of production (Barclay and Poolton, 1994). The benefits of concurrent engineering can be summarized as the potential for the reduction in long-term costs, in terms of (Albin and Crefield, 1994; Maylor, 1996):

- Reduced time to market
- Reduced engineering costs due to the reduction in reworking of designs
- Better responsiveness to market needs
- Reduced manufacturing costs.

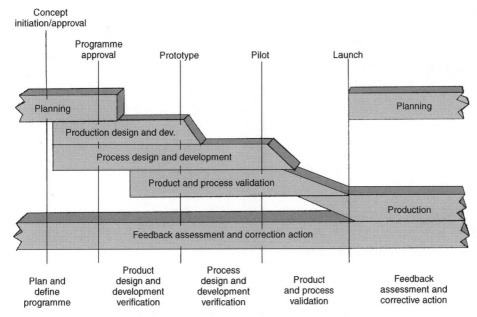

Figure 5.7 Product quality planning timing chart (QS 9000, 1998)

However, there are also disadvantages when applying the concurrent approach, which include (Maylor, 1996):

- Increased overheads – the teams require their own administration support
- Costs of co-location – people being relocated away from their functions to be with the team
- Cultural resistance
- Inappropriate application – it is not a panacea for development problems as poor conceptual designs will not be improved by using concurrent methods.

As mentioned, concurrent engineering principles should result in a better quality of the final design; however, it may increase the complexity of the design process and make it more difficult to manage (Kusiak and Wang, 1993). The need for effective management and commitment is again emphasized as having a key role in the product development process.

5.2.2 Industrial models

A few companies are currently making dramatic changes to the way in which products are brought to market, developing their own new product development processes which are employed and supported on site. Current industrial models which are referred to in the literature include:

- Lucas Industry's (now TRW) Product Introduction Management (PIM) process (Lucas, 1993)

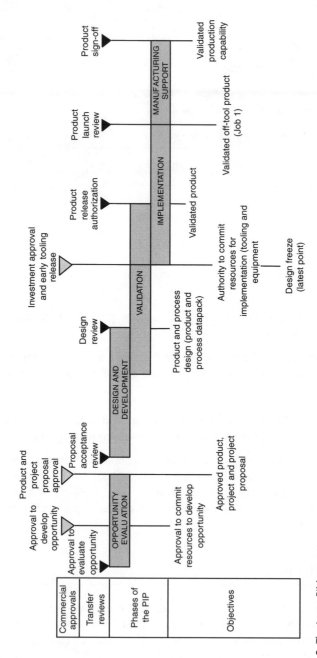

Figure 5.8 The Lucas PIM process

- British Aerospace Dynamics' Design-Make-Prove (DMP) (Ellis *et al.*, 1996)
- Toyota's X300 process (Kurogane, 1993)
- General Electric's 'Tollgate Process' (Wheelwright and Clark, 1992).

The Toyota X300 fork lift truck project design cycle is concurrent in nature spanning all the major disciplines in the process with quality assurance reviews, stipulating the use of appropriate tools and techniques at certain points. The product development process produced by General Electric is called the 'Tollgate Process'. Again, it is concurrent in nature and includes ten review points.

DMP was developed with the intention of realizing the power of concurrent engineering by providing a structured, process orientated view of product development. It was seen as a mechanism for embedding best practice at BAe Aerospace Dynamics in order to reduce lead times, reduce costs and improve quality. DMP makes use of concurrent principles due to the early multi-disciplinary input to the process and completion of tasks in parallel where possible. Key to the model is that risk management can be used to remove some of the iteration in a traditional product development process. Fundamental to DMP are the four 'risk gates'. They are intended to ensure that work in each phase is completed with the risk being assessed and alleviated to an acceptable, well-understood level, before moving onto the next phase. They are equivalent to design reviews as all issues relevant to the gate should be worked during the proceeding phase with the aim of alleviating risk. Many designers consider risk assessment to be an intuitive element of their work. However, a formal process model that makes risks explicit, and stresses accountability, can only be an advantage.

Lucas PIM

The need for change was recognized in Lucas Industries (now TRW) and led to the development of a PIM process for use in all Lucas businesses with the declared objectives of reducing time to market, reducing product costs and reducing project costs. The PIM model also reflects the requirements of the automotive industry standard QS 9000 (1998). The generic process is characterized by five phases and nine reviews as indicated in Figure 5.8. Each review has a relevant set of commercial, technical and project criteria for sign-off and hand-over to the next stage. To be effective, Lucas have found that PIM requires the collaborative use of (Miles and Swift, 1992):

- Team work
- Concurrent engineering
- Project management
- Tools and techniques.

The linkage between the above elements and PIM is represented diagrammatically in Figure 5.9.

DFA is one of the main tools and techniques prescribed by the PIM process. Other tools and techniques currently specified include QFD, FMEA and DOE. Significant benefits have been obtained through the use of the tools and techniques in a team-based concurrent engineering environment. They inject method, objectivity

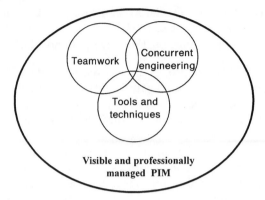

Figure 5.9 Key elements of successful PIM (Miles and Swift, 1992)

and structured continuous feedback, and improve a team's communication and understanding.

5.3 Tools and techniques in product development

5.3.1 Overview of tools and techniques

A summary of each of the key tools and techniques considered to be important in the product development process is given in Appendix III. This covers such techniques as FMEA, QFD, DFA/DFM and DOE. Included for each is a description of the tool or technique, placement issues in product development, key issues with regard to implementation, and the benefits that can accrue from their use, and finally a case study. It would be advantageous next, however, to determine exactly what a tool or technique does. In general, the main engineering activities that should be facilitated by their use are (Huang, 1996):

- Gather and present facts about products and processes
- Clarify and analyse relationships between products and processes
- Measure performance
- Highlight strengths and weaknesses and compare alternatives
- Diagnose why an area is strong or weak
- Provide redesign advice on how a design can be improved
- Predict what-if effects
- Carry out improvements
- Allow iteration to take place.

The decision to develop in-house or buy a tool or technique may arise as awareness of the opportunities in a business increases. Off-the-shelf products are more popular as the development of tools and techniques is very time consuming and labour intensive. A generally expressed desire is to use computerized techniques for easier documentation and design reuse. However, it should be noted that a paper-based analysis is especially advantageous for team-building (Norell and Andersson, 1996). For example, around 90% of companies that implemented DFA on a paper-based system perceived a significant contribution to product quality. Of the companies that implemented DFA on a computer-based system, 73% perceived that it provided a significant contribution to product quality (Araujo *et al.*, 1996). One explanation of this is that the use of the paper-based version of DFA in the initial stages deepens the understanding of the problem (Leaney, 1996b). However, there is a tendency for respondents to believe that methods implemented on a computer have a more positive influence on product quality than implementations of the same methods on paper based systems. This may suggest there is a need to provide more extensive computer implementation of methods (Araujo *et al.*, 1996), but also to make the methods more interactive and have greater networking capabilities.

The benefits of using tools and techniques in general can be grouped into three categories (Huang, 1996):

- Related to competitiveness measures – improved quality, compressed lead time, reduced life-cycle costs, increased flexibility, improved productivity, more satisfied customers
- Benefits including improved and rationalized decisions in designing products
- Far-reaching effects on operational efficiency in product development.

A recent survey of companies in the automotive and aerospace industry found that many companies are unaware of the benefits that can be gained from the utilization of quality tools and techniques. The adoption of BS EN ISO 9000 (1994) and Total Quality Management (TQM) strategies might be expected to increase the utilization of methods. However, the extent to which companies utilize methods is more strongly related to annual turnover than employee count, therefore the use of tools and techniques is dominated by large companies (Araujo *et al.*, 1996).

Practitioners tend to be overwhelmed by the wide spectrum and the diverse nature of the available tools and techniques. Some have been promoted as panaceas for solving all sorts of problems, rightly so or not (Huang, 1996). It has been argued that their application actually constrains the design solution, and that if numerous techniques are utilized, the constraints are often contradictory in nature (Meerkamm, 1994). Sophisticated techniques may also lose the advantage of being focused and pragmatic. Subsequently, practitioners become increasingly sceptical and gradually lose interest and commitment (Huang, 1996). An important requirement is that the tool or technique used should operate at a level of abstraction so that it does not inhibit the design development itself (Mørup, 1993).

There may also be an element of ignorance about a particular tool or technique's function and level of commitment, be it time, staffing or financial, to be effective in product development. Designers and managers alike need to be educated about the implementation issues, function and benefits of tools and techniques because there may be a significant difference between understanding the processes involved in an overview sense, and understanding it from the perspective of the person actually doing the work (Hughes, 1995). This can lead to serious implementation problems. A number of common difficulties encountered in tools and techniques implementation have been stated as (McQuater and Dale, 1995):

- Poorly designed training and support
- Not applying what has been learnt
- Inappropriate use
- Resistance to use
- Failure to lead by example
- Poor measurement and data handling
- Not sharing or communicating the benefits achieved.

Reflecting on the development of a new technique such as CA, for example, it was found that its full acceptance by the product team relied on several factors:

- **Understanding** – It should not be assumed that all team members understand the concepts of variability in manufacturing processes, the measurement and ideas of process capability indices, C_p and C_{pk} and SPC. The team must be brought to understand these ideas and that while processes can always be improved by special controls, there are boundaries which can be moved only by expensive and lengthy process development programmes.
- **Mindset** – Designers readily accept that their design must deliver the required level of functionality and within product cost limits. They do not always put the same emphasis on ensuring that they specify a design that can be manufactured capably. The responsibility for ensuring quality must be accepted by the whole team.
- **Focus** – Engineers usually like to realize their designs in hardware as soon as possible and to test prototypes and debug problems. With the ever increasing time pressures on teams, it is even more important to detect and eliminate as many problems in the design as early as possible, because they take much longer and cost much more to fix later.
- **Acceptance** – Deadlines put pressure on the team, and CA may be seen as yet another hurdle for the team to jump, or as a criticism of the designer's work. Techniques must be accepted at all levels in the organization as measures of performance on the design, not the designers, and as positive initiatives to promote creative ideas to modify designs and eliminate risk.

The final point requires that the implementation of new techniques must be management led, usually, as has been observed through the authors' industrial collaboration, by a 'Champion' within the business who has the necessary vision and influence to change opinion. Tools and techniques, therefore, rely on 'critical success factors' to make their application effective. Some of these are (McQuater and Dale, 1995):

- Full management support and commitment
- Effective and timely training
- A need to use the tool or technique
- Defined aims and objectives for use
- Back-up and support from facilitators.

It must be remembered that quality training is a long-term process and should be systematically delivered in a top-down manner until everyone is equipped with the necessary philosophies, attitudes and techniques. A quality transformation will simply not work without effective empowerment and practice (Kolarik, 1995).

5.3.2 Utilization of tools and techniques

As part of a concurrent engineering framework, four formal methods, FMEA, QFD, DFA/DFM and DOE, have been identified as complementing concurrent engineering working and their application in general shortens the total product development time (Norell, 1992; Poolton and Barclay, 1996). Although methods like QFD and Taguchi's Robust Design have been stated to yield good results by users in Japan and the USA, surveys have shown that their utilization is not as high as might be expected (Araujo *et al.*, 1996). Possible reasons for this are:

- Companies unaware of their existence
- Companies unaware of the quality related benefits that may accrue from their use
- There are differences in companies and/or their products that reduce their usefulness under certain circumstances
- Effective utilization requires experienced or trained staff who may not be available in some companies.

The popularity of some of the key tools and techniques mentioned is shown in Figure 5.10, taken from two different surveys of UK companies. The importance of FMEA, DFM and DFA is apparent, although the survey conducted in 1994 did not ask for the usage of DFM or DFA. Equally, the 1996 survey did not include Poka Yoke. The minor popularity of techniques such as QFD, DOE and Robust

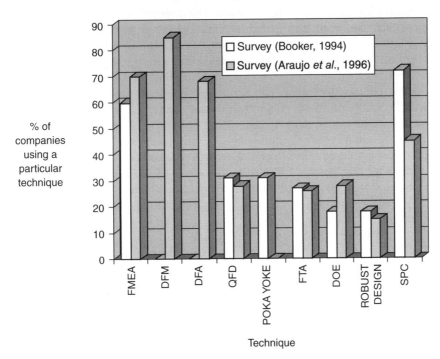

Figure 5.10 Percentage of UK companies surveyed using a particular tool or technique (Araujo *et al.*, 1996; Booker, 1994)

Design is a reflection on the comments made earlier. The popularity of SPC is shown for comparison to gauge the philosophies of 'off-line' and 'on-line' quality.

5.3.3 The integration of tools and techniques in the product development process

To achieve successful concurrent engineering design, one needs an integrated framework, a well-organized design team, and adequate tools and techniques (Jin *et al.*, 1995). Some companies in various countries have had great success in improving the quality and reliability of manufactured products by applying tools and techniques. However, without a proper understanding of the underlying processes they will not blend together to give an effective sequence of actions (Andersson, 1994). It is possible and quite common to perform them individually, but it is their joint use that gives the greatest value, through the increased co-operation from their use in cross-functional teams (Norell and Andersson, 1996).

When a single technique is employed only local life-cycle cost minimization is achieved. If the global life-cycle cost is to be minimized, a number of techniques have to be applied (Watson *et al.*, 1996). In this case, tools and techniques shouldn't compete with each other, but be complementary in the product development process. The correct positioning of the various off-line tools and techniques in the product development process, therefore, becomes an important consideration in their effective usage. Patterns of application have been proposed by a number of workers over several years (Brown *et al.*, 1989; Jakobsen, 1993; Norell, 1993) and the importance of concurrency has been highlighted as a critical factor in their use (Poolton and Barclay, 1996).

Before setting about the task of developing such a model, the product development process requires definition along with an indication of its key stages, this is so the appropriate tools and techniques can be applied (Booker *et al.*, 1997). In the approach presented here in Figure 5.11, the product development phases are activities generally defined in the automotive industry (Clark and Fujimoto, 1991). QFD Phase 1 is used to understand and quantify the importance of customer needs and requirements, and to support the definition of product and process requirements. The FMEA process is used to explore any potential failure modes, their likely Occurrence, Severity and Detectability. DFA/DFM techniques are used to minimize part count, facilitate ease of assembly and project component manufacturing and assembly costs, and are primarily aimed at cost reduction.

To be effective, we see the need for CA, which comes under the heading of *CAPRA* or **CA**pabilty and **PR**obabilistic Design Analysis, to be integrated into the early stages of design too, if it is to have its full impact. This will require the designer to define acceptable system performance more precisely than is presently the case. Design information from other techniques such as FMEA are crucial in this connection, as discussed in Chapter 1. CA is employed to provide a measure of potential process capability in manufacture and assembly, and to ensure robustness against failure and its associated failure costs. When used in conjunction with DFA/DFM, CA helps to assure the right part count and design of the assembly so that the required level of conformance can be achieved at the lowest cost. The integrated application

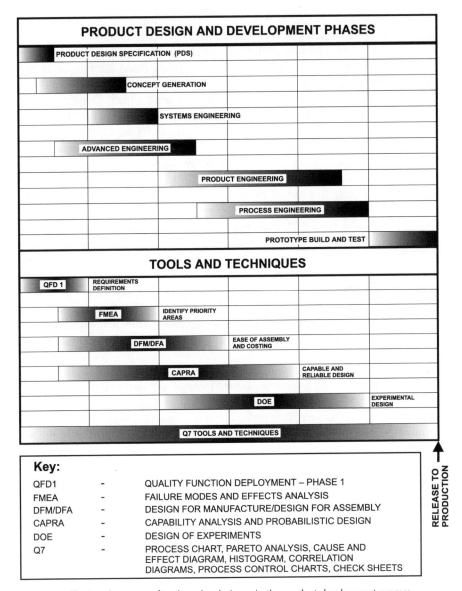

Figure 5.11 Effective placement of tools and techniques in the product development process

of FMEA, DFA/DFM and CA has many benefits including common data sharing and assembly sequence declaration. The outputs of FMEA and CA in terms of critical characteristics and any potential out of tolerance problems focuses attention on those areas where DOE is needed.

The process capability requirements for component characteristics determined by CA are used to support the supplier development process. In the approach, new designs of 'bought-in' parts such as castings, plastic mouldings and assembly work are discussed with process capability requirements as the first priority. Where a potential problem with the tolerance on a characteristic has been identified, this will be discussed and examined with the supplier. The supplier will be encouraged to provide evidence that they can meet the capability requirements or otherwise by reference to performance on similar part characteristics. The approach is supported by the application of SPC in the factory and the encouragement and facilitation of workforce involvement in a process of continuous improvement.

The so-called Q7 tools and techniques, Cause and Effect Diagrams, Pareto Analysis, etc. (Bicheno, 1994; Dale and McQuater, 1998; Straker, 1995), are applicable to any stage of the product development process. Indeed they support the working of some of the techniques mentioned, for example using a Pareto chart for prioritizing the potential risks in terms of the RPN index for a design as determined in FMEA (see Appendix III).

A particular difficulty in-product development is to properly and efficiently tie in reliability prediction methods with the design process so as to gain maximum return on investment (Klit *et al.*, 1993). Businesses need to develop economical and timely methods for obtaining the information needed to meet overall reliability goals at each step of the product design and development process (Meeker and Hamada, 1995). Also under the *CAPRA* heading is the design for reliability methodology, *CAPRAstress*, described in Chapter 4 of this book. Data relevant to CA is applied here also, therefore *CAPRAstress* should ideally be performed at the end of this phase of the product development process, as capability knowledge and knowledge of the service conditions accumulates, together with qualitative data already available from an FMEA.

A proposed product development process that facilitates designing capable and reliable products has been outlined above. It must be stressed that the product development process itself will not produce quality products, and consideration of many issues are crucial to success, such as company strategy, management structure, commitment, sufficient resources, communication, and most importantly proficient engineering practices, such as the following.

5.4 Supporting issues in effective product development

Tools and techniques can enhance the success of a product, but alone they will not solve all product development issues (Jenkins *et al.*, 1997a). Any implementation of tools and techniques within the product development process must take the following into account if the outcome is to be effective at all:

- Team approach to engineering design
- The company's quality philosophy
- Adequate Product Design Specification (PDS)
- Assessment of external supplier quality
- Adequate number of design solutions
- Adequate design reviews
- Adequate configuration control of product design and processes issues
- Adequate Research and Development (RandD) should be completed before commencement of the project.

Several of these issues will be discussed in detail next.

5.4.1 Team approach to engineering design

Even with the aid of tools and techniques, engineering is still a task that requires creative solutions (Urban and Hauser, 1993). Companies recognizing the importance of product development have searched to resolve this problem, with most opting for some kind of 'team approach', involving a multitude of persons supposedly providing the necessary breadth of experience in order to obtain 'production friendly products'.

Research (Urban and Hauser, 1993) has shown that teams produce better engineering solutions. Team sizes should be kept to three or four for best performance; however, up to nine or ten can work effectively together (BS 7000, 1997; Straker, 1995). The use of multi-functional teams also enables concurrent engineering, since all the team members are aware of the effect of their functional input to the project on other areas, as discussed earlier. The team are able to plan activities that may be carried out in parallel. While sometimes obtaining reasonable results, this approach often faces a number of obstacles (Towner, 1994):

- Assembling the persons with the relevant experience
- Lack of formal structure. Typically such meetings tend to be unstructured and often *ad hoc* attacks on various 'pet' themes
- The location of the persons required in the team can also present problems. Not only are designers and production engineers found in different functional departments, but they can frequently be on different sites and are in the case of subcontractors in different companies.

In addition, the chances are that the expertise in the team will only cover the primary activities of the business and hence opportunity to exploit any benefits from alternative processes may be lost. The ideal concurrent engineering scenario would involve a single product development team whose co-location would enable close communication and mutual understanding between team members (McCord and Eppinger, 1993). An important way of encouraging multi-functional team working, however, is to use tools and techniques that provide structured ways of achieving important objectives, such as (Parker, 1997):

- Understanding what the customer wants
- Design products for ease of manufacture and assembly
- Making sure products and processes are safe
- Improving process capabilities.

Furthermore, in order more effectively to create quality, the people involved must be empowered to make changes in the products and processes they are involved with (Kolarik, 1995).

5.4.2 Quality philosophy of the company

It is quite possible for a company to be registered under BS EN ISO 9000 or QS 9000 and still be producing products which are defective and not to customer requirements.

Product quality may not necessarily be better than that in a non-registered company. To some, especially small companies, registration is an unnecessary bureaucracy. However, to many it is a way of demonstrating, internationally, that the company takes quality seriously and has thought through its quality system. BS EN ISO 9000 registration is seen as a marketing advantage and a trade facilitator (Bicheno, 1994). However, it does reveal bottlenecks in the organization and in the handling of projects. The general opinion throughout companies surveyed is that once you are registered, as far as quality is concerned it made little or no difference. In fact, recent articles on the quality standards will, in general, reveal an anti-trend. More specifically, the main criticisms are (McLachlan, 1996):

- It is too expensive
- It does not address the needs of small businesses
- It is unduly biased towards manufacturing
- Is it relevant?
- You can still make and sell rubbish.

Traditionalists say that you cannot write a standard to achieve quality. Quality depends on people doing their job properly having made sure that they know what their job is, that it has been correctly defined and that allowance has been made for continuous improvement as defined by TQM. In view of the above, BS EN ISO 9000 registration and TQM complement each other (McLachlan, 1996) and the quality movement has become a driver for change in product development (Rosenau and Moran, 1993). Creating an environment which complements successful product development also involves changing the culture of the organization and the core beliefs of the people who form it (Parnaby, 1995). BS EN ISO 9000 registration and TQM support this process.

TQM affects three areas of the product development process (Rosenau and Moran, 1993):

- Strategic – external and internal management of processes
- Cultural – empowerment and teamwork
- Technical – thought of as a toolbox, techniques used to facilitate TQM and the product development process.

TQM involves all the organizations, all the functions, the external suppliers, the external customers and involves the quality policy. Similarly, TQM cannot be achieved without good Quality Management Systems (QMS) which bring together all functions relevant to the product, providing policies, procedures and documentation. The elements of a quality organization consist of these three mutually dependent items (Field and Swift, 1996):

- The culture – TQM
- Registration (BS EN ISO 9000)
- Quality Management Systems (QMS).

The importance of TQM is now recognized nationally with the issue of two British Standards (BS 7850, 1992; BS 7850, 1994). However, while the advantages of TQM seem obvious, implementing TQM systems has not produced equally good results in all cases. Several researchers report that TQM programmes have produced improvements in quality, productivity and competitiveness in only 20 to 30% of the companies that have implemented such initiatives (Benson, 1993; Schonberger, 1992).

5.4.3 Product design specifications

The PDS or design brief should be a definitive statement or instruction of what is required. It should define all the requirements and constraints, for example standards and regulations that the designer has to observe, but should not impose design solutions. It should receive contributions from many sources and should evolve through a series of iterations (BS 7000, 1997). The PDS produced from the initial market analysis is the basic reference for the new product. The PDS is important since it encourages quality in up-front stages of new product design, and reference to the PDS, throughout the product development process, ensuring that a final product is developed which will satisfy the customers' needs (Jenkins et al., 1997a). PDS is one of the key factors in successful quality engineering, since it clarifies the design objectives. Important items in its construction are shown in Figure 5.12 (Andersson, 1994).

5.4.4 External supplier quality

A key success factor for reducing the costs and lead times for vehicle manufacturers, for example, is the degree of integration of the suppliers within the product development process. This is seen as a natural extension to concurrent engineering principles (Wyatt et al., 1998). For many years, in engineering companies, a substantial proportion of the finished product, typically two thirds, consists of components or subassemblies produced by suppliers (Noori and Radford, 1995).

An effective programme for product quality must, therefore, include a means of assuring supplier quality and reliability (Nelson, 1996). This means that the customer must get much closer to the supplier's operations. Many companies have failed to develop a supplier strategy and traditionally have used an almost gladiatorial and hostile approach to their suppliers, for example to drive them to impossibly low prices regardless of quality, or to terminate trading with those who fail to perform. Most buyers are only interested in price and delivery and not the quality of the supplier's goods and services. The more enlightened companies have now established a procurement strategy that controls the quality of a supplier, but does not take from a supplier the responsibility for their quality. The company collaborates with the supplier through the project ensuring standards are maintained. To effectively manage suppliers, a number of guidelines are proposed:

- Choose suppliers who produce the required quality, not who offer the lowest price
- Reduce the number of suppliers
- Build up close working relationships

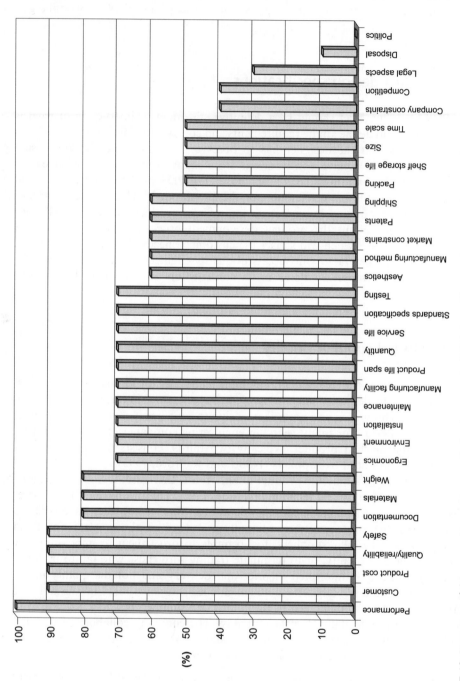

Figure 5.12 Percentage of the companies applying a particular element in the PDS (adapted from Andersson, 1994)

- Collaboration and mutual trust
- Operational integration at all stages of design and production process.

This approach has demonstrated a marked reduction in supplier prices, improved quality and delivery, and external customer satisfaction. The supplier must be regarded as part of the team and their full commitment to the project will ensure minimum inspection on receipt and the implementation of joint improvement programmes. For example, the car manufacturer Nissan has been involved in a process to improve supplier quality and productivity. During a two year period they claim that 47% of their suppliers were producing 10 ppm or less, failing specification (Greenfield, 1996). More on the benefits of the approach are discussed by Galt and Dale (1990) and Lloyd (1994).

A company will typically categorize its suppliers according to their type of operation, for example:

- Fully design, develop and manufacture their own products
- Manufacture major components from supplier designs
- Manufacture simple components from third party designs.

In every case the company may demand the supplier be registered to BS EN ISO 9000 or conform to their company requirements for the adequacy of their quality systems. However, as previously stated these registrations do not control the quality of the product itself. Any new suppliers must show evidence of their capability with other contracts before being awarded any orders. Any current supplier who fails to meet the standards may be removed from the bidding list. The quality assurance model represents quality system requirements for the purpose of a supplier demonstrating its capability (BS EN ISO 9001, 1994). Evaluating supplier capability at the production level involves two parts (Gryna, 1988):

- Qualifying the supplier's design through evaluation of product samples
- Qualifying the supplier's capability to meet quality requirements on production lots.

The results from CA, described in Chapter 2, serve as a good basis for supplier dialogue very early in the development process. Problem areas in the design are systematically identified and discussed openly with suppliers. The necessary process or design changes are then effectively communicated in order to meet the customer's requirements leading to a process capable design.

In summary, an effective supplier development process should contain the following elements (Gryna, 1988):

- Define a product and programme quality requirements
- Evaluate alternative suppliers
- Select suppliers
- Conduct joint quality planning
- Co-operate with the supplier during contact
- Obtain proof of conformance
- Certify qualified supplier
- Conduct quality improvement programmes as required
- Create and utilize supplier quality ratings.

5.4.5 Design scheme generation

Designing for quality methods have the objective of selecting the 'technically perfect one' from a number of alternative solutions that have been arrived at systematically and not the first satisfactory solution (Braunsperger, 1996). Developing more than one design scheme is therefore a key issue, the highest percentages of companies generating on average two design schemes at system level and three or greater subsystem schemes, as shown in Figure 5.13.

In general, the more schemes generated, the more likely that an effective design solution can be isolated through the determined performance measures. The process of evaluating and comparing the design alternatives is an important mode of application of tools and techniques and replaces the aspect of exactitude which designers sometimes seek through their use.

5.4.6 Design reviews

Design reviews are a formal procedure to establish the total understanding and acceptance of a design task. The purpose of reviews is to (Parker, 1997):

- Monitor progress
- Review quality
- Identify issues
- Ensure level of support
- Grant permission to proceed early in the process.

A design review must not consider cost and time scales; it forms a traceable record of technical decisions and current performance against requirements. Design reviews are a crucial part of product development and are an important factor in enhancing product performance (Dieter, 1986). They should also take place at the critical stages of the product development process (as highlighted in the industrial models of PIM and DMP). The participants at the design review generally include representatives

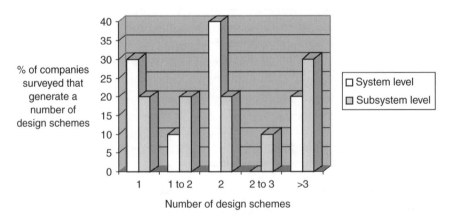

Figure 5.13 Number of design schemes generated at system and subsystem level (Andersson, 1994)

	Product	Process
Eliminate	Can any of the components be eliminated?	Can any of the activities be eliminated?
Integrate	Can one component be integrated with another component?	Can one activity be integrated with another activity?
Combine	Can the given components be combined in a better way?	Can a better sequence of activities be followed?
Simplify	Can components be simplified?	Can activities be simplified?
Standardize	Can components be standardized into one?	Can activities be standardized into one?
Substitute	Can any component be replaced?	Can any activity be replaced?
Revise	Can any component be revised?	Can any activity be revised?

Figure 5.14 Techniques for redesigning products and processes (Huang, 1996)

of all functions affecting quality as appropriate to the phase being reviewed (Dale and Oakland, 1994).

Following the completion of a design review, it is usual for a final report to be submitted to the responsible party to summarize the recommendations made and any modifications subsequently incorporated in a design (Dale and Oakland, 1994). Feedback from various tools and techniques is an important part of the design review process. At the end of the day, the design review is probably the only means of communicating the current design to all concerned. It should be readily interpreted by all in a simple language. Some guidelines for presenting a design review are given below (Ullman, 1992):

- Make it understandable
- Carefully consider the order of presentations
- Be prepared with quality material.

Redesign activities based on feedback from the review are the next logical stage for the progression of the design. Redesign offers the opportunity of functional substitution where the objective is not simply to replace, but to find a new and better way of achieving the same function (Dieter, 1986). The elementary processes of redesign are shown in Figure 5.14.

5.5 Summary

This chapter has overviewed various product development models found in the literature and several of those employed by industry. When designing a new product, it is essential to facilitate and drive the operation by an adequate product development process. The application of concurrent engineering, rather than sequential activities, has many benefits, in particular giving a reduction in the number of design changes and a reduction in the time it takes to bring the product to market. Tools and techniques cannot be employed in isolation and the integration of several tools and techniques has been found to effectively support the process of designing capable and reliable products. For example, an important aspect of the use of CA is its reliance on information from an FMEA, specifically the Severity Rating and the potential failure modes associated with the design.

Additionally, information from the PDS or QFD is critical in the development of the design principles, as the final design solution should satisfy the customer/functional requirements stipulated. Tools and techniques can supply the required design information and design performance measures through the early stages of the product development process and it is this combined use which has been shown to provide the most benefit in reducing costs and shortening lead times.

A concurrent engineering framework allows a more efficient flow of information from the various tools and techniques used and effectively communicates the design through requirements based performance measures. The primary advantage of employing concurrent principles in terms of the use of tools and techniques is that the overlap of the engineering activities, which is natural in any case, enhances a team-based approach. The application of the tools and techniques in practice has also been discussed together with a review of each, including their effective positioning in the product development process, implementation and management issues and likely benefits from their usage.

Important issues to support the development of capable and reliable products have also been discussed. These include a team approach to engineering design, the generation of an adequate number of design schemes from which to determine the most effective design solution, assessment of external supplier capability and the importance of the design review process in forming a traceable record of technical performance versus customer requirements. The vital role that quality philosophies, such as TQM, have on improving and driving the product development process has been emphasized.

It is hoped that the reader using the approaches described in this book will find great benefit when designing new products for high quality and reliability. Within the framework for application detailed above, the techniques discussed combine to provide an effective process for the development of new products that can be implemented in a manufacturing company. Finally, design has a key role in the competitiveness of a manufacturing company. Designers trained in the use of appropriate tools and techniques, and with the correct mindset, can create capable and reliable designs of high value to the company and customer. Manufacturing businesses must strive to reduce the costs of failure using such approaches.

Appendix I

Introductory statistics

Statistical representation of data

The random nature of most physical properties, such as dimensions, strength and loads, is well known to statisticians. Engineers too are familiar with the typical appearance of sets of tensile strength data in which most of the individuals congregate around mid-range and fewer further out to either side. Statisticians use the *mean* to identify the location of a set of data on the scale of measurement and the *variance* (or *standard deviation*) to measure the dispersion about the mean. In a variable '*x*', the symbols used to represent the mean are μ and $\bar{x}$ for a population and sample respectively. The symbol for variance is V. The symbols for standard deviation are σ and s respectively, although σ is often used for both. In this book we will always use the notation μ for mean and σ for the standard deviation.

In a set of '*N*' individual values, $x_1, x_2, x_3, \ldots, x_i, \ldots, x_N$, the statistical measures are:

$$\mu = \frac{\sum_{i=1}^{N} x_i}{N} \tag{1}$$

$$V = \frac{\sum_{i=1}^{N} (x_i - \mu)^2}{N} \tag{2}$$

where:

$$x = \text{variable}$$

$$N = \text{population (size of data set)}$$

$$\mu = \text{mean}$$

$$V = \text{variance.}$$

These equations are called the *moment equations*, because we are effectively taking moments of the data about a point to measure the dispersion over the whole set of data. Note that in the variance, the positive and negative deviates when squared do not cancel each other out but provide a powerful measure of dispersion which takes account of every single individual in the data set. The importance of variance is that when independent sources of variability combine in a complex situation, the variances usually combine additively.

To express the measure of dispersion in the original scale of measurement, it is usual to take the square root of the variance to give the standard deviation:

$$\sigma = \sqrt{V} \tag{3}$$

It follows that the variance can also be given as:

$$V = \sigma^2 \tag{4}$$

and the standard deviation is commonly given as:

$$\sigma = \sqrt{\frac{\sum_{i=1}^{N}(x_i - \mu)^2}{N}} \tag{5}$$

The statistical measures can be calculated using most scientific calculators, but confusion can arise if the calculator offers the choice between dividing the sum of squares by 'N' or by '$N-1$'. If the object is to simply calculate the variance of a set of data, divide by 'N'. If, on the other hand, a sample set of data is being used to estimate the properties of a supposed population, division of the sum of squares by '$N-1$' gives a better estimate of the population variance. The reason is that the sample mean is unlikely to coincide exactly with the (unknown) true population mean and so the sum of squares about the sample mean will be less than the true sum of squares about the population mean. This is compensated for by using the divisor '$N-1$'. Obviously, this becomes important with smaller samples.

In general, the larger the sample size, the more accurate will be the experimental calculations of the population parameters. A sample size $N \geq 30$ is typical, but sometimes 100 is preferred.

Representing data using histograms

An effective way of grouping large amounts of data is by using bar charts or *histograms*. The simplest method by which to make a crude hypothesis about the type of distribution the data could follow is by inspecting the histogram, although for small samples they are meaningless. The first type is shown in Figure 1 and is commonly referred to as *discontinuous* or discrete data representation. The data is collated and any having the same value for the variable has its frequency recorded, which is proportional to the height of the bar.

The use of grouped or classed data in histogram form may be even more efficient for large amounts of data, as shown in Figure 2. Here, the data is analysed and any falling within the same class limits are grouped together giving a frequency distribution, where the height of the plotted bars is proportional to the number of data falling within each class this time.

To efficiently construct a histogram for grouped data firstly requires the determination of the optimum number of classes the data range is separated into. A suitable

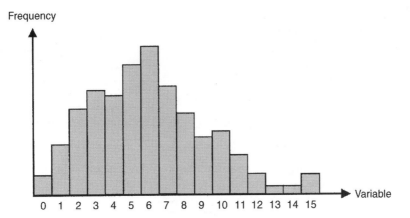

Figure 1 Histogram showing discontinuous or discrete frequency distribution

number of data classes are selected depending on the population size. Care should be taken in the selection of the number of class intervals. Too few will cause omission of some important features of the data; too many will not give a clear overall picture because there may be wild fluctuations in the frequencies. A number of commonly used formulations are given below:

$$k = 1 + 3.22 \log_{10}(N) \tag{6}$$

$$k = \sqrt{N} \tag{7}$$

$$k = \frac{(\text{max} - \text{min})N^{1/3}}{2(Q_3 - Q_1)} \tag{8}$$

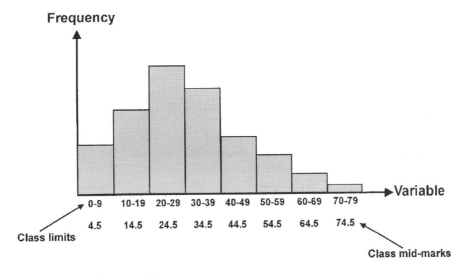

Figure 2 Histogram for grouped discrete data

where:

$$k = \text{class interval}$$
$$N = \text{population}$$
$$\text{max} = \text{maximum value in population}$$
$$\text{min} = \text{minimum value in population}$$
$$Q_3 = \text{median of upper half of data when in ascending order}$$
$$Q_1 = \text{median of lower half of data when in ascending order.}$$

From the data range (maximum minus minimum value) the optimum class width, w, can then be determined from:

$$w = \frac{\text{max} - \text{min}}{k} \tag{9}$$

Data can then be grouped between the various class limits, the nominal value of each group being known as the class mid-mark and the number of occurrences in each group being the frequency, as shown in Figure 2. The mean, variance and standard deviation of the data can then be determined using the equations below:

$$\mu = \frac{\sum_{i=1}^{k} f_i x_i}{N} \tag{10}$$

$$V = \frac{\sum_{i=1}^{k} f_i (x_i - \mu)^2}{N} \tag{11}$$

where:

$$f = \text{frequency}$$
$$x = \text{class mid-mark}$$
$$k = \text{number of classes.}$$

Again, the standard deviation can be given by:

$$\sigma = \sqrt{\frac{\sum_{i=1}^{k} f_i (x_i - \mu)^2}{N}} \tag{12}$$

Properties of the Normal distribution

For mathematical tractability, the experimental data can be modelled with a *continuous distribution* which will adequately describe the pattern of the data using just a single equation and its related parameters.

In the above calculations of the mean, variance and standard deviation, we make no prior assumption about the shape of the population distribution. Many of the data distributions encountered in engineering have a bell-shaped form similar to that showed in Figure 1. In such cases, the Normal or *Gaussian* continuous distribution can be used to model the data using the mean and standard deviation properties.

The Normal distribution is the most widely used of all distributions and through empirical evidence provides a good representation of many engineering variables

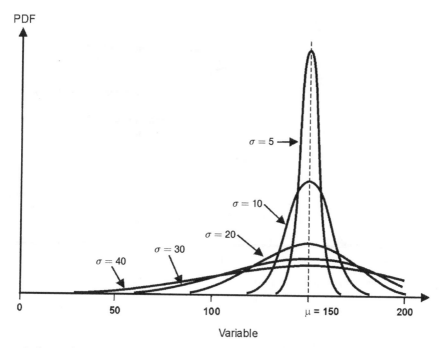

PDF

$\sigma = 5$ →

$\sigma = 10$

$\sigma = 20$

$\sigma = 30$

$\sigma = 40$

0   50   100 $\mu = 150$ 200

Variable

Figure 3 Shape of the probability density function (PDF) for a normal distribution with varying standard deviation, σ, and mean, $\mu = 150$

and is easily tractable mathematically. Typical applications have been cited, for example dimensional tolerances, ultimate tensile strength and yield strength of some metallic alloys.

A useful measure of the Normal distribution is derived from its parameters, and is called the *coefficient of variation*, C_v:

$$C_v = \frac{\sigma}{\mu} \qquad (13)$$

The shape of the Normal distribution is shown in Figure 3 for an arbitrary mean, $\mu = 150$ and varying standard deviation, σ. Notice it is symmetrical about the mean and that the area under each curve is equal representing a probability of one. The equation which describes the shape of a Normal distribution is called the Probability Density Function (PDF) and is usually represented by the term $f(x)$, or the function of 'x', where 'x' is the variable of interest or *variate*.

$$f(x) = \frac{1}{\sigma\sqrt{2\pi}} \exp\left(-\frac{(x-\mu)^2}{2\sigma^2}\right) \quad \text{for } -\infty < x < \infty \qquad (14)$$

Once the mean and standard deviation have been determined, the frequency distribution determined from the PDF can be compared to the original histogram, if one was constructed, by using a scaling factor in the PDF equation. For example, the expected frequency for the Normal distribution is given by:

$$y = \frac{Nw}{\sigma\sqrt{2\pi}} \exp\left(-\frac{(x-\mu)^2}{2\sigma^2}\right) \qquad (15)$$

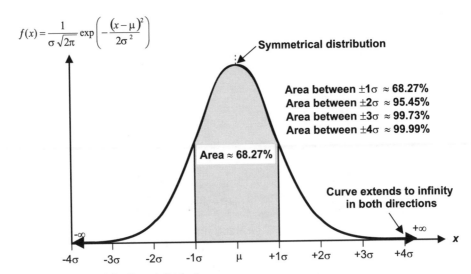

Figure 4 Properties of the Normal distribution

where:

y = expected frequency

Nw = scaling factor (population, N, multiplied by the class width, w).

If we plot a Normal distribution for an arbitrary mean and standard deviation, as shown in Figure 4, it can be shown that at $\pm 1\sigma$ about the mean value, the area under the frequency curve is approximately 68.27% of the total, and at $\pm 2\sigma$, the area is 95.45% of the total under the curve, and so on. This property of the Normal distribution then becomes useful in estimating the proportion of individuals within prescribed limits.

The Standard Normal distribution

The Normal distribution can be fitted to a set of data by 'standardizing' (i.e. adjusting the mean to zero and rescaling the standard deviation to unity). The engineering utility of this procedure is that the proportion of individuals to be expected on either side of a given ordinate, called the *Standard Normal variate*, z, can be read from a table, which makes up the *Standard Normal Distribution* (SND). Table 1 at the end of this Appendix provides values for the area under the cumulative SND from $z = \pm 5$ in steps on 0.01. In engineering design, this then gives a connotation of probability, where the area under the SND curve represents the probability of failure, P, determined from:

$$z = \left(\frac{x - \mu}{\sigma} \right) \tag{16}$$

$$P = \Phi_{\text{SND}}(z) \tag{17}$$

where:

$$x = \text{variable of interest}$$

$$\Phi_{SND} = \text{function of the Standard Normal distribution.}$$

Example

The following set of data represents the outcome of a tensile test experiment to determine the yield strength in MPa of a metal. There are 50 individual results and they are displayed in the order they were recorded. It is required to find the mean and standard deviation when the data is represented by a histogram. It is also required to find the strength at -3σ from the mean for the metal and the proportion of individuals that could be expected to have a strength greater than 500 MPa.

482, 470, 469, 490, 471, 442, 480, 472, 453, 450, 415, 433, 420, 459, 424, 444, 455, 462, 495, 426, 502, 447, 438, 436, 498, 474, 515, 483, 454, 453, 447, 485, 456, 443, 449, 466, 445, 452, 478, 407, 433, 466, 428, 452, 440, 450, 460, 451, 496, 477.

Solution

Sorting the data into ascending order gives:

407, 415, 420, 424, 426, 428, 433, 433, 436, 438, 440, 442, 443, 444, 445, 447, 447, 449, 450, 450, 451, 452, 452, 453, 453, 454, 455, 456, 459, 460, 462, 466, 466, 469, 470, 471, 472, 474, 477, 478, 480, 482, 483, 485, 490, 495, 496, 498, 502, 515.

From equation 6, we can determine the optimum number of class intervals to be:

$$k = 1 + 3.22 \log_{10}(N) = 1 + 3.22 \log_{10}(50) \approx 6$$

The class width is given by equation 9 as:

$$w = \frac{\text{max} - \text{min}}{k} = \frac{515 - 407}{6} = 18$$

The class limits become:

$$\text{minimum} = 407$$

$$415 + 18 = 425$$

$$425 + 18 = 443$$

$$443 + 18 = 461$$

$$461 + 18 = 479$$

$$479 + 18 = 497$$

$$\text{maximum} = 515$$

Table 1 Area under the cumulative Standard Normal Distribution (SND)

$$z = \left(\frac{x-\mu}{\sigma}\right)$$

z	Φ(z)	z	Φ(z)	z	Φ(z)	z	Φ(z)	z	Φ(z)	z	Φ(z)	z	Φ(z)	z	Φ(z)	z	Φ(z)
-5.00	0.0000003	-3.32	0.000450	-2.36	0.009137	-1.40	0.080757	-0.44	0.329968	0.45	0.673645	1.41	0.920730	2.37	0.991106	3.33	0.999566
-4.90	0.0000005	-3.31	0.000467	-2.35	0.009387	-1.39	0.082265	-0.43	0.333598	0.46	0.677242	1.42	0.922196	2.38	0.991344	3.34	0.999581
-4.80	0.0000009	-3.30	0.000484	-2.34	0.009642	-1.38	0.083793	-0.42	0.337243	0.47	0.680823	1.43	0.923641	2.39	0.991576	3.35	0.999596
-4.70	0.000001	-3.29	0.000501	-2.33	0.009903	-1.37	0.085344	-0.41	0.340903	0.48	0.684387	1.44	0.925066	2.40	0.991802	3.36	0.999610
-4.60	0.000002	-3.28	0.000519	-2.32	0.010170	-1.36	0.086915	-0.40	0.344578	0.49	0.687933	1.45	0.926471	2.41	0.992024	3.37	0.999624
-4.50	0.000003	-3.27	0.000538	-2.31	0.010444	-1.35	0.088508	-0.39	0.348268	0.5	0.691462	1.46	0.927855	2.42	0.992240	3.38	0.999637
-4.40	0.000005	-3.26	0.000557	-2.30	0.010724	-1.34	0.090123	-0.38	0.351973	0.51	0.694975	1.47	0.929219	2.43	0.992451	3.39	0.999650
-4.30	0.000009	-3.25	0.000577	-2.29	0.011011	-1.33	0.091759	-0.37	0.355694	0.52	0.698468	1.48	0.930563	2.44	0.992656	3.40	0.999663
-4.20	0.000013	-3.24	0.000598	-2.28	0.011304	-1.32	0.093417	-0.36	0.359424	0.53	0.701944	1.49	0.931888	2.45	0.992857	3.41	0.999675
-4.19	0.000014	-3.23	0.000619	-2.27	0.011604	-1.31	0.095098	-0.35	0.363169	0.54	0.705401	1.50	0.933193	2.46	0.993053	3.42	0.999687
-4.18	0.000015	-3.22	0.000641	-2.26	0.011911	-1.30	0.096801	-0.34	0.366928	0.55	0.708840	1.51	0.934478	2.47	0.993244	3.43	0.999698
-4.17	0.000015	-3.21	0.000664	-2.25	0.012224	-1.29	0.098525	-0.33	0.370699	0.56	0.712260	1.52	0.935744	2.48	0.993431	3.44	0.999709
-4.16	0.000016	-3.20	0.000687	-2.24	0.012545	-1.28	0.100273	-0.32	0.374483	0.57	0.715661	1.53	0.936992	2.49	0.993613	3.45	0.999720
-4.15	0.000017	-3.19	0.000711	-2.23	0.012874	-1.27	0.102042	-0.31	0.378280	0.58	0.719043	1.54	0.938220	2.50	0.993790	3.46	0.999730
-4.14	0.000018	-3.18	0.000736	-2.22	0.013209	-1.26	0.103835	-0.30	0.382088	0.59	0.722405	1.55	0.939429	2.51	0.993963	3.47	0.999740
-4.13	0.000018	-3.17	0.000762	-2.21	0.013553	-1.25	0.105650	-0.29	0.385908	0.60	0.725747	1.56	0.940620	2.52	0.994132	3.48	0.999749
-4.12	0.000019	-3.16	0.000789	-2.20	0.013903	-1.24	0.107488	-0.28	0.389738	0.61	0.729069	1.57	0.941792	2.53	0.994297	3.49	0.999758
-4.11	0.000020	-3.15	0.000816	-2.19	0.014262	-1.23	0.109349	-0.27	0.393580	0.62	0.732371	1.58	0.942947	2.54	0.994457	3.50	0.999767
-4.10	0.000021	-3.14	0.000845	-2.18	0.014629	-1.22	0.111233	-0.26	0.397432	0.63	0.735653	1.59	0.944083	2.55	0.994614	3.51	0.999776
-4.09	0.000022	-3.13	0.000874	-2.17	0.015003	-1.21	0.113140	-0.25	0.401294	0.64	0.738914	1.60	0.945201	2.56	0.994766	3.52	0.999784
-4.08	0.000022	-3.12	0.000904	-2.16	0.015386	-1.20	0.115070	-0.24	0.405165	0.65	0.742154	1.61	0.946301	2.57	0.994915	3.53	0.999793
-4.07	0.000023	-3.11	0.000935	-2.15	0.015777	-1.19	0.117023	-0.23	0.409046	0.66	0.745374	1.62	0.947384	2.58	0.995060	3.54	0.999800
-4.06	0.000024	-3.10	0.000968	-2.14	0.016177	-1.18	0.119000	-0.22	0.412936	0.67	0.748572	1.63	0.948449	2.59	0.995201	3.55	0.999807
-4.05	0.000025	-3.09	0.001001	-2.13	0.016586	-1.17	0.121001	-0.21	0.416834	0.68	0.751748	1.64	0.949497	2.60	0.995339	3.56	0.999815
-4.04	0.000027	-3.08	0.001035	-2.12	0.017003	-1.16	0.123024	-0.20	0.420740	0.69	0.754903	1.65	0.950529	2.61	0.995473	3.57	0.999821
-4.03	0.000028	-3.07	0.001070	-2.11	0.017429	-1.15	0.125072	-0.19	0.424655	0.70	0.758036	1.66	0.951543	2.62	0.995604	3.58	0.999828
-4.02	0.000029	-3.06	0.001107	-2.10	0.017864	-1.14	0.127143	-0.18	0.428577	0.71	0.761148	1.67	0.952540	2.63	0.995731	3.59	0.999835
-4.01	0.000031	-3.05	0.001144	-2.09	0.018309	-1.13	0.129138	-0.17	0.432506	0.72	0.764238	1.68	0.953521	2.64	0.995855	3.60	0.999841
-4.00	0.000032	-3.04	0.001183	-2.08	0.018763	-1.12	0.131357	-0.16	0.436440	0.73	0.767305	1.69	0.954486	2.65	0.995975	3.61	0.999847
-3.99	0.000033	-3.03	0.001264	-2.07	0.019226	-1.11	0.133500	-0.15	0.440382	0.74	0.770350	1.70	0.955435	2.66	0.996093	3.62	0.999853
-3.98	0.000034	-3.02	0.001264	-2.06	0.019699	-1.10	0.135666	-0.14	0.444329	0.75	0.773373	1.71	0.956367	2.67	0.996207	3.63	0.999858
-3.97	0.000036	-3.01	0.001306	-2.05	0.020182	-1.09	0.137857	-0.13	0.448283	0.76	0.776373	1.72	0.957284	2.68	0.996319	3.64	0.999864
-3.96	0.000037	-3.00	0.001350	-2.04	0.020675	-1.08	0.140080	-0.12	0.452241	0.77	0.779350	1.73	0.958185	2.69	0.996427	3.65	0.999869
-3.95	0.000039	-2.99	0.001395	-2.03	0.021178	-1.07	0.142310	-0.11	0.456204	0.78	0.782305	1.74	0.959071	2.70	0.996533	3.66	0.999874

z	Φ(z)
-3.94	0.000041
-3.93	0.000042
-3.92	0.000044
-3.91	0.000046
-3.90	0.000048
-3.89	0.000050
-3.88	0.000052
-3.87	0.000054
-3.86	0.000057
-3.85	0.000059
-3.84	0.000062
-3.83	0.000064
-3.82	0.000067
-3.81	0.000069
-3.80	0.000072
-3.79	0.000075
-3.78	0.000078
-3.77	0.000082
-3.76	0.000085
-3.75	0.000088
-3.74	0.000092
-3.73	0.000096
-3.72	0.000100
-3.71	0.000104
-3.70	0.000108
-3.69	0.000112
-3.68	0.000117
-3.67	0.000121
-3.66	0.000126
-3.65	0.000131
-3.64	0.000136
-3.63	0.000142
-3.62	0.000147
-3.61	0.000153
-3.60	0.000159
-3.59	0.000165
-3.58	0.000172
-3.57	0.000179
-3.56	0.000185
-3.55	0.000193
-3.54	0.000200
-3.53	0.000207
-3.52	0.000216
-3.51	0.000224
-3.50	0.000233
-3.49	0.000242
-3.48	0.000251
-3.47	0.000260
-3.46	0.000270
-3.45	0.000280
-3.44	0.000291
-3.43	0.000302
-3.42	0.000313
-3.41	0.000325
-3.40	0.000337
-3.39	0.000350
-3.38	0.000363
-3.37	0.000376
-3.36	0.000390
-3.35	0.000404
-3.34	0.000419
-3.33	0.000434

z	Φ(z)
-2.98	0.001441
-2.97	0.001489
-2.96	0.001538
-2.95	0.001589
-2.94	0.001641
-2.93	0.001695
-2.92	0.001750
-2.91	0.001807
-2.90	0.001866
-2.89	0.001926
-2.88	0.001988
-2.87	0.002052
-2.86	0.002118
-2.85	0.002186
-2.84	0.002256
-2.83	0.002327
-2.82	0.002401
-2.81	0.002477
-2.80	0.002555
-2.79	0.002635
-2.78	0.002718
-2.77	0.002803
-2.76	0.002890
-2.75	0.002980
-2.74	0.003072
-2.73	0.003167
-2.72	0.003264
-2.71	0.003364
-2.70	0.003467
-2.69	0.003573
-2.68	0.003681
-2.67	0.003793
-2.66	0.003907
-2.65	0.004025
-2.64	0.004145
-2.63	0.004269
-2.62	0.004396
-2.61	0.004527
-2.60	0.004661
-2.59	0.004799
-2.58	0.004940
-2.57	0.005085
-2.56	0.005234
-2.55	0.005386
-2.54	0.005543
-2.53	0.005703
-2.52	0.005868
-2.51	0.006037
-2.50	0.006210
-2.49	0.006387
-2.48	0.006569
-2.47	0.006756
-2.46	0.006947
-2.45	0.007143
-2.44	0.007344
-2.43	0.007549
-2.42	0.007760
-2.41	0.007976
-2.40	0.008198
-2.39	0.008424
-2.38	0.008656
-2.37	0.008894

z	Φ(z)
-2.02	0.021692
-2.01	0.022216
-2.00	0.022750
-1.99	0.023295
-1.98	0.023852
-1.97	0.024419
-1.96	0.024998
-1.95	0.025588
-1.94	0.026190
-1.93	0.026803
-1.92	0.027429
-1.91	0.028067
-1.90	0.028716
-1.89	0.029379
-1.88	0.030054
-1.87	0.030742
-1.86	0.031443
-1.85	0.032157
-1.84	0.032884
-1.83	0.033625
-1.82	0.034379
-1.81	0.035148
-1.80	0.035930
-1.79	0.036727
-1.78	0.037538
-1.77	0.038364
-1.76	0.039204
-1.75	0.040059
-1.74	0.040929
-1.73	0.041815
-1.72	0.042716
-1.71	0.043633
-1.70	0.044565
-1.69	0.045514
-1.68	0.046479
-1.67	0.047460
-1.66	0.048457
-1.65	0.049471
-1.64	0.050503
-1.63	0.051551
-1.62	0.052616
-1.61	0.053699
-1.60	0.054799
-1.59	0.055917
-1.58	0.057053
-1.57	0.058208
-1.56	0.059380
-1.55	0.060571
-1.54	0.061780
-1.53	0.063008
-1.52	0.064256
-1.51	0.065522
-1.50	0.066807
-1.49	0.068112
-1.48	0.069437
-1.47	0.070781
-1.46	0.072145
-1.45	0.073529
-1.44	0.074934
-1.43	0.076359
-1.42	0.077804
-1.41	0.079270

z	Φ(z)
-1.06	0.144572
-1.05	0.146859
-1.04	0.149170
-1.03	0.151505
-1.02	0.153864
-1.01	0.156248
-1.00	0.158655
-0.99	0.161087
-0.98	0.163543
-0.97	0.166023
-0.96	0.168527
-0.95	0.171056
-0.94	0.173609
-0.93	0.176185
-0.92	0.178786
-0.91	0.181411
-0.90	0.184060
-0.89	0.186733
-0.88	0.189430
-0.87	0.192150
-0.86	0.194895
-0.85	0.197663
-0.84	0.200454
-0.83	0.203269
-0.82	0.206108
-0.81	0.208970
-0.80	0.211855
-0.79	0.214764
-0.78	0.217695
-0.77	0.220650
-0.76	0.223627
-0.75	0.226627
-0.74	0.229650
-0.73	0.232695
-0.72	0.235762
-0.71	0.238852
-0.70	0.241964
-0.69	0.245097
-0.68	0.248252
-0.67	0.251428
-0.66	0.254626
-0.65	0.257846
-0.64	0.261086
-0.63	0.264347
-0.62	0.267629
-0.61	0.270931
-0.60	0.274253
-0.59	0.277595
-0.58	0.280957
-0.57	0.284339
-0.56	0.287740
-0.55	0.291160
-0.54	0.294599
-0.53	0.298056
-0.52	0.301532
-0.51	0.305025
-0.50	0.308538
-0.49	0.312067
-0.48	0.315613
-0.47	0.319117
-0.46	0.322758
-0.45	0.326355

z	Φ(z)
-0.10	0.460171
-0.09	0.464143
-0.08	0.468118
-0.07	0.472096
-0.06	0.476078
-0.05	0.480061
-0.04	0.484046
-0.03	0.488034
-0.02	0.492022
-0.01	0.496011
0.00	0.500000
0.01	0.503989
0.02	0.507798
0.03	0.511966
0.04	0.515954
0.05	0.519939
0.06	0.523922
0.07	0.527904
0.08	0.531882
0.09	0.535857
0.10	0.539829
0.11	0.543796
0.12	0.547759
0.13	0.551717
0.14	0.555671
0.15	0.559618
0.16	0.563560
0.17	0.567494
0.18	0.571423
0.19	0.575345
0.20	0.579260
0.21	0.583166
0.22	0.587064
0.23	0.590954
0.24	0.594835
0.25	0.598706
0.26	0.602568
0.27	0.606420
0.28	0.610262
0.29	0.614092
0.30	0.617912
0.31	0.621720
0.32	0.625517
0.33	0.629301
0.34	0.633072
0.35	0.636831
0.36	0.640576
0.37	0.644306
0.38	0.648027
0.39	0.651732
0.40	0.655422
0.41	0.659097
0.42	0.662757
0.43	0.666402
0.44	0.670032

z	Φ(z)
0.79	0.785236
0.80	0.788145
0.81	0.791030
0.82	0.793892
0.83	0.796731
0.84	0.799546
0.85	0.802337
0.86	0.805105
0.87	0.807850
0.88	0.810570
0.89	0.813267
0.90	0.815940
0.91	0.818589
0.92	0.821214
0.93	0.823815
0.94	0.826391
0.95	0.828944
0.96	0.831473
0.97	0.833977
0.98	0.836457
0.99	0.838913
1.00	0.841345
1.01	0.843752
1.02	0.846136
1.03	0.848495
1.04	0.850830
1.05	0.853141
1.06	0.855428
1.07	0.857690
1.08	0.859929
1.09	0.862143
1.10	0.864334
1.11	0.866500
1.12	0.868643
1.13	0.870762
1.14	0.872857
1.15	0.874928
1.16	0.876976
1.17	0.878999
1.18	0.881000
1.19	0.882977
1.20	0.884930
1.21	0.886860
1.22	0.888767
1.23	0.890651
1.24	0.892512
1.25	0.894350
1.26	0.896165
1.27	0.897958
1.28	0.899727
1.29	0.901475
1.30	0.903199
1.31	0.904902
1.32	0.906583
1.33	0.908241
1.34	0.909877
1.35	0.911492
1.36	0.913085
1.37	0.914656
1.38	0.916207
1.39	0.917735
1.40	0.919243

z	Φ(z)
1.75	0.959941
1.76	0.960796
1.77	0.961636
1.78	0.962462
1.79	0.963273
1.80	0.964070
1.81	0.964852
1.82	0.965621
1.83	0.966375
1.84	0.967116
1.85	0.967843
1.86	0.968557
1.87	0.969258
1.88	0.969946
1.89	0.970621
1.90	0.971284
1.91	0.971933
1.92	0.972571
1.93	0.973197
1.94	0.973810
1.95	0.974412
1.96	0.975002
1.97	0.975581
1.98	0.976148
1.99	0.976705
2.00	0.977250
2.01	0.977784
2.02	0.978308
2.03	0.978822
2.04	0.979325
2.05	0.979818
2.06	0.980301
2.07	0.980774
2.08	0.981237
2.09	0.981691
2.10	0.982136
2.11	0.982571
2.12	0.982997
2.13	0.983414
2.14	0.983823
2.15	0.984223
2.16	0.984614
2.17	0.984997
2.18	0.985371
2.19	0.985738
2.20	0.986097
2.21	0.986447
2.22	0.986791
2.23	0.987126
2.24	0.987455
2.25	0.987776
2.26	0.988089
2.27	0.988396
2.28	0.988696
2.29	0.988989
2.30	0.989276
2.31	0.989556
2.32	0.989830
2.33	0.990097
2.34	0.990358
2.35	0.990613
2.36	0.990863

z	Φ(z)
2.71	0.996636
2.72	0.996736
2.73	0.996833
2.74	0.996928
2.75	0.997020
2.76	0.997110
2.77	0.997197
2.78	0.997282
2.79	0.997365
2.80	0.997445
2.81	0.997523
2.82	0.997599
2.83	0.997673
2.84	0.997744
2.85	0.997814
2.86	0.997882
2.87	0.997948
2.88	0.998012
2.89	0.998074
2.90	0.998134
2.91	0.998193
2.92	0.998250
2.93	0.998305
2.94	0.998359
2.95	0.998411
2.96	0.998462
2.97	0.998511
2.98	0.998559
2.99	0.998605
3.00	0.998650
3.01	0.998694
3.02	0.998736
3.03	0.998777
3.04	0.998817
3.05	0.998856
3.06	0.998893
3.07	0.998930
3.08	0.998965
3.09	0.998999
3.10	0.999032
3.11	0.999065
3.12	0.999096
3.13	0.999126
3.14	0.999155
3.15	0.999184
3.16	0.999211
3.17	0.999238
3.18	0.999264
3.19	0.999289
3.20	0.999313
3.21	0.999336
3.22	0.999359
3.23	0.999381
3.24	0.999402
3.25	0.999423
3.26	0.999443
3.27	0.999462
3.28	0.999481
3.29	0.999499
3.30	0.999516
3.31	0.999533
3.32	0.999550

z	Φ(z)
3.67	0.999879
3.68	0.999883
3.69	0.999888
3.70	0.999892
3.71	0.999896
3.72	0.999900
3.73	0.999904
3.74	0.999908
3.75	0.999912
3.76	0.999915
3.77	0.999918
3.78	0.999922
3.79	0.999925
3.80	0.999928
3.81	0.999931
3.82	0.999933
3.83	0.999936
3.84	0.999938
3.85	0.999941
3.86	0.999943
3.87	0.999946
3.88	0.999948
3.89	0.999950
3.90	0.999952
3.91	0.999954
3.92	0.999956
3.93	0.999958
3.94	0.999959
3.95	0.999961
3.96	0.999963
3.97	0.999964
3.98	0.999966
3.99	0.999967
4.00	0.999968
4.01	0.999969
4.02	0.999971
4.03	0.999972
4.04	0.999973
4.05	0.999974
4.06	0.999975
4.07	0.999976
4.08	0.999977
4.09	0.999978
4.10	0.999979
4.11	0.999980
4.12	0.999981
4.13	0.999982
4.14	0.999983
4.15	0.999984
4.16	0.999985
4.17	0.999985
4.18	0.999986
4.19	0.999987
4.20	0.999987
4.30	0.999991
4.40	0.999995
4.50	0.999997
4.60	0.999998
4.70	0.999999
4.80	0.9999991
4.90	0.9999995
5.00	0.9999997

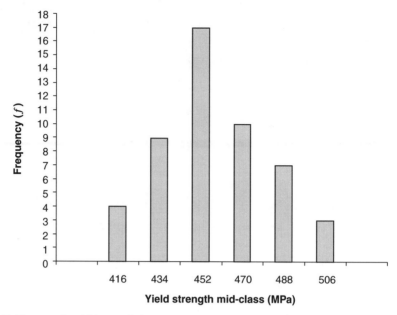

From the original data, the number of yield strength values falling between these class limits is recorded to give the frequency and a histogram can be generated, as shown in Figure 5. The mid-class values are determined by taking the mid-point between each pair of class limits.

The mean and standard deviation are given by equations 10 and 12 respectively:

$$\mu = \frac{\sum_{i=1}^{k} f_i x_i}{N}$$

$$= \frac{(4 \times 416) + (9 \times 434) + (17 \times 452) + (10 \times 470) + (7 \times 488) + (3 \times 506)}{50}$$

$$= 457.76 \, \text{MPa}$$

$$\sigma = \sqrt{\frac{\sum_{i=1}^{k} f_i (x_i - \mu)^2}{N}}$$

$$= \sqrt{\frac{\begin{aligned}&4 \times (416 - 457.76)^2 + 9 \times (434 - 457.76)^2 + 17 \times (452 - 457.76)^2 \\ &+ 10 \times (470 - 457.76)^2 + 7 \times (488 - 457.76)^2 + 3 \times (506 - 457.76)^2\end{aligned}}{50}}$$

$$= 23.45 \, \text{MPa}$$

We can now plot the Normal frequency distribution superimposed over the histogram bars for comparison. The curve is generated using equation 15, where the variables of interest, x, are values in steps of 10 on the x-axis from, say, 380 to 540. The Normal frequency equation is given below, and Figure 6 shows the histogram and the Normal

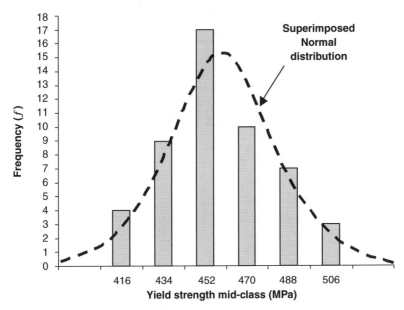

Figure 6 Histogram for yield strength data with superimposed Normal distribution

distribution for comparison:

$$y = \frac{Nw}{\sigma\sqrt{2\pi}} \exp\left(-\frac{(x-\mu)^2}{2\sigma^2}\right) = \frac{50 \times 18}{23.45\sqrt{2\pi}} \exp\left(-\frac{(x-457.76)^2}{2(23.45)^2}\right)$$

To find the strength at -3σ from the mean simply requires that we take three standard deviations away from the mean. Therefore, the strength at this point on the distribution is:

$$457.763(23.45) = 387.41 \text{ MPa}$$

The proportion of individuals that could be expected to have a strength greater than 500 MPa requires using SND theory. The variable of interest is 500 MPa, and so from Equation 16:

$$z = \left(\frac{x-\mu}{\sigma}\right) = \left(\frac{500 - 457.76}{23.45}\right) = 1.80$$

The area under the cumulative SND curve at $z = 1.80$ is equivalent to the probability that the yield strength is less than 500 MPa. Referring to Table 1 gives:

$$P = \Phi_{SND}(z) = \Phi_{SND}(1.80) = 0.964070$$

We require the proportion that has a strength greater than 500 MPa, therefore:

$$1 - 0.964070 = 0.035930$$

or, in other words, approximately 3.6% of the population can be expected to have a yield strength greater than 500 MPa.

Process capability studies

Process capability concepts

A capability study is a statistical tool which measures the variations within a manufacturing process. Samples of the product are taken, measured and the variation is compared with a tolerance. This comparison is used to establish how 'capable' the process is in producing the product. Process capability is attributable to a combination of the variability in all of the inputs. Machine capability is calculated when the rest of the inputs are fixed. This means that the process capability is not the same as machine capability. A capability study can be carried out on any of the inputs by fixing all the others. All processes can be described by Figure 1, where the distribution curve for a process shows the variability due to its particular elements.

There are five occasions when capability studies should be carried out, these are:

- Before the machine/process is bought (to see if it is capable of producing the components you require it to)
- When it is installed
- At regular intervals to check that the process is given the performance required
- If the operating conditions change (for example, materials, lubrication)
- As part of a process capability improvement.

The aim is to have a process where the product variability is sufficiently small so that all the products produced are within tolerance. Since variation can never be eliminated, the control of variation is the key to product quality and capability studies give us one tool to achieve this.

There are two main kinds of variability:

- Common-cause or inherent variability is due to the set of factors that are inherent in a machine/process by virtue of its design, construction and the nature of its operation, for example, positional repeatability, machine rigidity, which cannot be removed without undue expense and/or process redesign.
- Assignable-cause or special-cause variability is due to identifiable sources which can be systematically identified and eliminated.

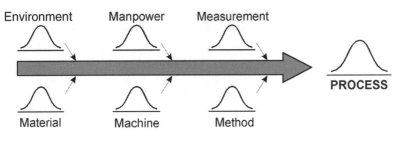

ORGANIZATIONAL

Environment   Manpower   Measurement

PROCESS

Material   Machine   Method

TECHNOLOGICAL

Figure 1 Factors affecting process capability

When only common-cause variability is present, the process is performing at its best possible level under the current process design. For a process to be capable of producing components to the specification, the sum of the common-cause and assignable-cause variability must be less than the tolerance.

The way of measuring capability is to carry out a capability study and calculate a capability index. There are two commonly used process capability indices, C_p and C_{pk}. In both cases, it is assumed the data is adequately represented by the Normal distribution (see Appendix I).

Process capability index, C_p

The process capability index is a means of quantifying a process to produce components within the tolerances of the specification. Its formulation is shown below:

$$C_p = \frac{U - L}{6\sigma} \tag{1}$$

where U = upper tolerance limit, L = lower tolerance limit, σ = standard deviation.

$U - L$ also equals the unilateral tolerance, T. Where a bilateral tolerance, t, is used (for example, ± 0.2 mm), equation (1) simplifies to:

$$C_p = \frac{t}{3\sigma} \tag{2}$$

where t = bilateral tolerance.

A value of $C_p = 1.33$ would indicate that the distribution of the product characteristics covers 75% of the tolerance. This would be sufficient to assume that the process is capable of producing an adequate proportion to specification. The numbers of failures falling out of specification for various values of C_p and C_{pk} can be determined from Standard Normal Distribution (SND) theory (see an example later for how to determine the failure in 'parts-per-million' or ppm). For example, at $C_p = 1.33$, the expected number of failures is 64 ppm in total.

The minimum level of capability set by Motorola in their 'Six Sigma' quality philosophy is $C_p = 2$ which equates to 0.002 ppm failures. Some companies set a

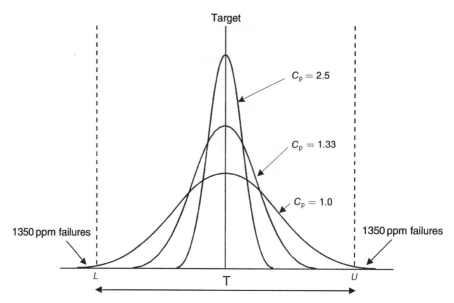

Figure 2 Process capability in terms of tolerance

capability level of $C_p = 1$ which relates to 2700 ppm failures in total. This may be adequate for some manufactured products, say a nail manufacturer, but not for safety critical products and applications where the characteristic controlled has been determined as critical. In general, the more severe the potential failure, the more capable the requirement must be to avoid it. General capability target values are given below.

Interpretation of process capability index, C_p:

Less than 1.33 → Process not capable.

Between 1.33 and 2.5 → Process capable.

Where a process is producing a characteristic with a capability index greater than 2.5 it should be noted that the unnecessary precision may be expensive. Figure 2 shows process capability in terms of the tolerance on a component. The area under each distribution is equal to unity representing the total probability, hence the varying heights and widths.

The variability or spread of the data does not always take the form of the true Normal distribution of course. There can be 'skewness' in the shape of the distribution curve, this means the distribution is not symmetrical, leading to the distribution appearing 'lopsided'. However, the approach is adequate for distributions which are fairly symmetrical about the tolerance limits. But what about when the distribution mean is not symmetrical about the tolerance limits? A second index, C_{pk}, is used to accommodate this 'shift' or 'drift' in the process. It has been estimated that over a very large number of lots produced, the mean could expect to drift about $\pm 1.5\sigma$ (standard deviations) from the target value or the centre of the tolerance limits and is caused by some problem in the process, for example tooling settings have been altered or a new supplier for the material being processed.

Process capability index, C_{pk}

By calculating where the process is centred (the mean value) and taking this, rather than the target value, it is possible to account for the shift of a distribution which would render C_p inaccurate (see Figure 3). C_{pk} is calculated using the following equation:

$$C_{pk} = \frac{|\mu - L_n|}{3\sigma} \tag{3}$$

where L_n = nearest tolerance limit and μ = mean.

Note, that the $|\mu - L_n|$ part of the equation means that the value is always positive.

By using the nearest tolerance limit, L_n, which is the tolerance limit physically closest to the distribution mean, the worst case scenario is being used ensuring that overoptimistic values of process capability are not employed. In Figure 3, a -1.5σ shift is shown from the target value for a $C_{pk} = 1.5$. C_{pk} is a much more valuable tool than C_p because it can be applied accurately to shifted distributions. As a large percentage of distributions are shifted, C_p is limited in its usefulness. If C_{pk} is applied to a non-shifted Normal distribution, by the nature of its formula it reverts to C_p.

Again, the minimum level of capability at Motorola using $C_{pk} = 1.5$ (or $\sim$3.4 ppm) at the nearest limit, where it is assumed the sample distribution is $\pm 1.5\sigma$ shifted from the target value. From Figure 3, it is evident that at $\pm 1.5\sigma$, the number of items falling out of specification on the opposite limit is negligible. However, more typical values are shown below. $C_{pk} = 1.33$ is regarded to be the absolute minimum by industry. This relates to 32 ppm failures, although it is commonly rounded down to 30 ppm.

C_{pk} is interpreted in the same way as C_p:

Less than 1.33 $\rightarrow$ Process not capable.

Between 1.33 and 2.5 $\rightarrow$ Process capable.

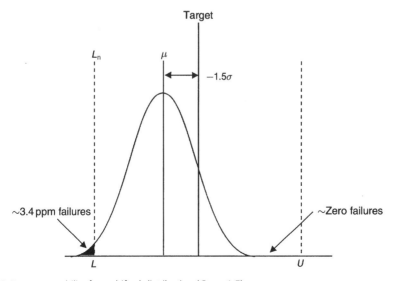

Figure 3 Process capability for a shifted distribution ($C_{pk} = 1.5$)

Again, for C_{pk} greater than 2.5, it should be noted that the unnecessary precision may be expensive.

Also note that

$$C_{pk} = C_p - 0.5 \qquad (4)$$

when the distribution is $\pm 1.5\sigma$ shifted from the target value. For a sample set of data, a C_p and C_{pk} value can be determined at the same time; however, the selection of which one best models the data is determined by the degree of shift. From equation (3), if $C_p - 0.5$ approaches the value of C_{pk} calculated, then a $\pm 1.5\sigma$ shift is evident in the sample distribution, and C_{pk} is a more suitable model.

Example – process capability and failure prediction

The component shown in Figure 4 is a spacer from a transmission system. The component is manufactured by turning/boring at the rate of 25 000 *per annum* and the component characteristic to be controlled, X, is an internal diameter. From the statistical data in the form of a histogram for 40 components manufactured, shown in Figure 5, we can calculate the process capability indices, C_p and C_{pk}. It is assumed that a Normal distribution adequately models the sample data.

The solution is as follows:

$$\mu = \frac{\displaystyle\sum_{i=1}^{k} f_i x_i}{N}$$

$$= \frac{\begin{array}{l} 2 \times 49.95 + 4 \times 49.96 + 7 \times 49.7 + 10 \times 49.8 \\ + 8 \times 49.9 + 6 \times 50 + 2 \times 50.01 + 1 \times 50.02 \end{array}}{40} = 49.98\,\text{mm}$$

$$\sigma = \sqrt{\frac{\displaystyle\sum_{i=1}^{k} f_i (x_i - \mu)^2}{N}}$$

$X = \varnothing 50 \pm 0.05\ \text{mm}$

Figure 4 Spacer component showing a critical characteristic, X

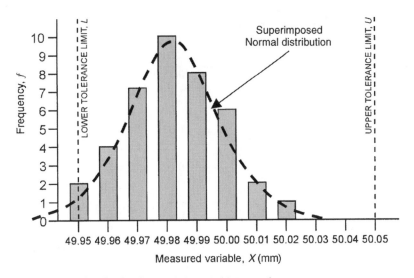

Figure 5 Measurement data for the characteristic, X, in histogram form

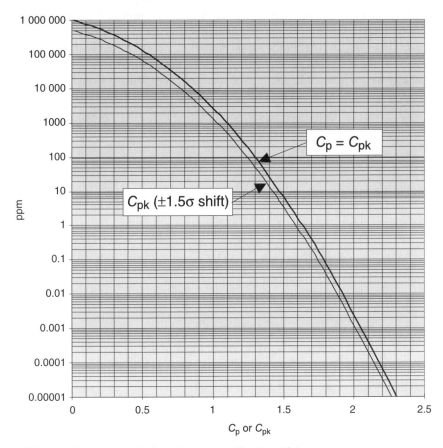

Figure 6 Relationship between C_p, C_{pk} and parts-per-million (ppm) failure

$$\sigma = \sqrt{\frac{\begin{array}{c} 2 \times (49.95 - 49.98)^2 + 4 \times (49.96 - 49.98)^2 + 7 \times (49.97 - 49.98)^2 \\ +10 \times (49.98 - 49.98)^2 + 8 \times (49.99 - 49.98)^2 + 6 \times (50 - 49.98)^2 \\ +2 \times (50.01 - 49.98)^2 + 1 \times (50.02 - 49.98)^2 \end{array}}{40}}$$

$$= 0.0162 \, \text{mm}$$

$$C_p = \frac{t}{3\sigma} \qquad = \frac{0.05}{3 \times 0.0162} \qquad = \quad 1.03$$

$$C_{pk} = \frac{|\mu - L_n|}{3\sigma} = \frac{49.9 - 49.95}{3 \times 0.0162} = \boxed{0.62} \qquad \text{Compare to see}$$
$$\updownarrow \qquad \text{if shift}$$
$$C_{pk} = C_p - 0.5 = 1.03 - 0.5 \quad = \boxed{0.53} \qquad \text{approaches } \pm 1.5\sigma$$

It is evident that an approximate -1.5σ shift can be determined from the data and so the C_{pk} value is more suitable as a model. Using the graph on Figure 6, which shows the relationship C_p, C_{pk} (at $\pm 1.5\sigma$ shift) and parts-per-million (ppm) failure at the nearest limit, the likely annual failure rate of the product can be calculated. The figure has been constructed using the Standard Normal Distribution (SND) for various limits. The number of components that would fall out of tolerance at the nearest limit, L_n, is potentially 30 000 ppm at $C_{pk} = 0.62$, that is, 750 components of the 25 000 manufactured *per annum*. Of course, action in the form of a process capability study would prevent further out of tolerance components from being produced and avoid this failure rate in the future and a target $C_{pk} = 1.33$ would be aimed for.

Appendix III

Overview of the key tools and techniques

A. Failure Mode and Effects Analysis (FMEA)

(BS 5760, 1991; Chrysler Corporation *et al.*, 1995; Ireson *et al.*, 1996; Kehoe, 1996)

Description

FMEA is a systematic element by element assessment to highlight the effects of a component, product, process or system failure to meet all the requirements of a customer specification, including safety. It helps to indicate by high point scores those elements of a component, product, process or system requiring priority action to reduce the likelihood of failure. This can be done through redesign, safety back-ups, design reviews, etc. It can be carried out at the design stage using experience or judgement, or integrated with existing data and knowledge on components and products.

FMEA was first used in the 1960s by the aerospace sector, but has since found applications in the nuclear, electronics, chemical and motor manufacturing sectors. FMEA can also apply to office processes as well as design and manufacturing processes, which are the main application areas.

Placement in product development

It should be started as early as possible once the concept designs have been generated from initial requirements, generating information on the critical elements of the design.

Application of the technique

The following factors are assessed in an FMEA:

- **Potential Failure Mode**. How could the component, product, process or system element fail to meet each aspect of the specification?

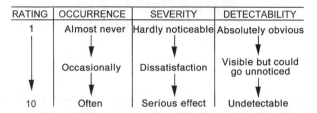

Figure 1 General ratings for FMEA Occurrence, Severity and Detectability

- **Potential Effects of Failure**. What would be the consequences of the component, product, process or system element failure?
- **Potential Causes of Failure**. What would make the component, product, process or system fail in the way suggested by the potential failure mode?
- **Current Controls**. What is done at present to reduce the chance of this failure occurring?
- **Occurrence (O)**. The probability that a failure will take place, given that there is a fault.
- **Severity (S)**. The effect the failure has on the user/environment, if the failure takes place.
- **Detectability (D)**. The probability that the fault will go undetected before the failure takes place. (An additional detectability rating is sometimes considered which relates to the probability that the failure will go undetected before having an effect.)

The Occurrence, Severity and Detectability Ratings are assessed on a scale of 1 to 10 and are illustrated in general terms in Figure 1.

Note, a comprehensive list of failure modes and causes of failure for mechanical components is provided by Dieter (1986). These tables are particularly useful when assessing the likely stress rupture failure mechanism for reliability work.

The Risk Priority Number (RPN) is the Occurrence (O), Severity (S) and Detectability (D) ratings multiplied together:

$$RPN = O \times S \times D \tag{1}$$

This number should be used as a guide to the most serious problems, with the highest numbers (typically greater than 100) requiring the most urgent action, particularly if they have scored high Severity Ratings.

For more comprehensive FMEA Occurrence, Severity and Detectability Ratings, see Figure 2. Note that Occurrence can be replaced by field reliability data in the form of failure rates scaled to the original ratings. This may be useful when assessing product families or new designs based on existing ones.

Benefits

In the case of performing a design FMEA, the RPN score can:

- Highlight the need for design improvement
- Highlight priority areas for focusing limited resources

Occurrence (O)

Rating	Guide	ppm
1	Remote possibility of failure occurring	0.1
2	Low possibility of occurrence	0.5
3		2
4	Moderate possibility of occurrence	10
5		50
6	Significant number of failures possible	200
7	High possibility of occurrence	1000
8		5000
9	Very high possibility of occurrence	20 000
10	Almost certain that many failures will occur	100 000

Severity (S)

Rating	Guide to effect on user
1*	No effect or minimal effect on customer
2*	Minor annoyance to customer
3*	Annoyance to customer but no loss of major function
4*	Possible return to manufacturer
5*	Definite return to manufacturer
6	Failure leading to violation of statutory requirement
7	Failure leading to injury or a more safety critical related problem with secondary back-up
8	Safety problem — degradation of function with possible severe injury
9	Complete failure with probable severe injury and/or loss of life
10	Catastrophic failure with high probability of loss of life

* Note: These failures are non-safety critical

Detectability (D)

Rating	Guide
1*	Always obvious — foolproof
2*	Obvious to human senses
3*	Inspection effort required
4*	Careful inspection by human senses
5*	Very careful inspection by human senses
6	Simple aids and/or disassembly required
7	Inspection aids and/or disassembly required
8	Complex inspection and/or disassembly required
9	Very high possibility of non-detection
10	Undetectable

* Note: Failure detection by human senses

Figure 2 Typical FMEA ratings for Occurrence, Severity and Detectability

- Highlight the need for Statistical Process Control (SPC), 100% inspection, no inspection
- Identify parts which have redundant function
- Prioritize those suppliers to which attention needs to be given
- Provide a basis for measures of performance.

Through its effective use, FMEA has been found to reduce:

- Customer complaints
- Late design changes
- Defects during manufacture and assembly
- Failures in the field
- Failure costs (rework, warranty claims).

Key issues

- Can be used in product design or process development
- Must be management led and have a strategy of use

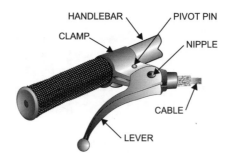

Product function	Potential failure mode	Potential effects of failure	Potential causes of failure	Current controls	Existing conditions			
					Occurrence (O)	Severity (S)	Detectability (D)	RPN = O×S×D
TO ACTIVATE THE REAR BRAKE ASSEMBLY OF A BICYCLE	REAR BRAKE DOES NOT WORK	CANNOT STOP SAFELY GOING DOWN HILL OR IN SUFFICIENT TIME	LEVER BENDS		1	6	4	24
			LEVER BREAKS		1	8	5	40
			CABLE SNAPS		2	8	8	128
			CABLE PULL OUT FROM NIPPLE	PERSONAL INSPECTION	3	8	10	240
			PIVOT PIN SHEARS		1	7	10	70
			CLAMP LOOSENS		4	6	2	48

Figure 3 Design FMEA for a bicycle rear brake lever

- Training required to use analysis initially
- Initial resources committed minimal
- Team-based application essential
- Can be subjective
- Performed too late in product development in many cases
- Links well with Fault Tree Analysis (FTA) where possible causes of failure are identified
- Support design FMEAs with failure data wherever possible
- Input from the customer and suppliers is important
- Should be reviewed at regular stages
- Can help to build up a knowledge base for product families.

Example – bicycle rear brake lever

The design FMEA process is shown in a case study on Figure 3. It highlights the areas that need special attention, reflected in the highest Risk Priority Numbers, when designing a bicycle rear brake lever. Using a Pareto chart, the RPN number, relating to the relative risk of each failure mode, can be displayed as shown in Figure 4. The analysis shows that design effort should be focused on the flexible element in the assembly, i.e. the brake cable. This perhaps supports the personal experiences of the reader, but the FMEA has shown this in a structured and rigorous appraisal of the concept design.

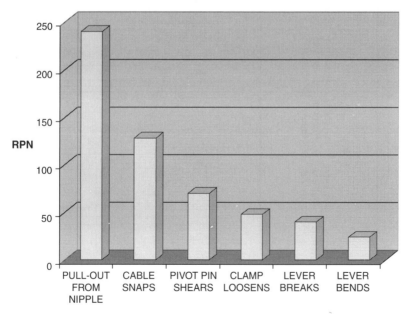

Figure 4 Pareto chart of RPN values against potential cause of failure for the rear brake lever design

PRODUCT NAME _____
PRODUCT CODE/ID _____
PRODUCT QUANTITY _____

SHEET No. _____
ANALYSIS REF. _____
ANALYSIS DATE _____
ENGINEER _____

Product function	Potential failure mode	Potential effects of failure	Potential causes of failure	Existing conditions					Recommended action and status	Resulting conditions				
				Current controls	Occurrence (O)	Severity (S)	Detectability (D)	RPN = O×S×D		Revised controls	Occurrence (O)	Severity (S)	Detectability (D)	RPN = O×S×D

Figure 5 Blank FMEA table

Figure 5 provides a blank table (complete with design revision section) used to perform a design FMEA. It forms a traceable record of the design and its failure modes and associated risks.

B. Quality Function Deployment (QFD)

(ASI, 1987; ASI, 1992; Clausing, 1994)

Description

In QFD customer requirements or 'the voice of the customer' are cascaded down through the product development process in four separate phases keeping the effort focused on the important issues, linking customer requirements directly with actual shop-floor operations/procedures.

In summary the objectives of QFD are to:

- Prioritize customer requirements and relate them to all stages of product development
- Focus resources on the aspects of the product that are important for customer satisfaction
- Provide a structured, team-driven product development process.

Placement in product development

QFD should be applied at the start of product development to help understand and quantify customer requirements and support the definition of product requirements, giving an overall picture of the requirements definition throughout the product development process. Other tools and techniques can be used in conjunction with it, for example Pareto chart, histogram, or data coming out of the FMEA can be fed back at an early stage.

Application of the technique

The four phases of QFD are described below and shown in Figure 6:

- **Phase 1 – Product Planning.** Customer requirements from market research information (including competitor analysis) and product specification are ranked in a matrix against the important design requirements, yielding a numerical quantification between the important customer requirements and product design issues. The quantification is performed by assessing whether the matrix elements have either a weak, medium or strong relationship. Together with the importance rating from the customer for a particular requirement, a points rating for the design requirements is then determined. A popular convention for QFD Phase 1,

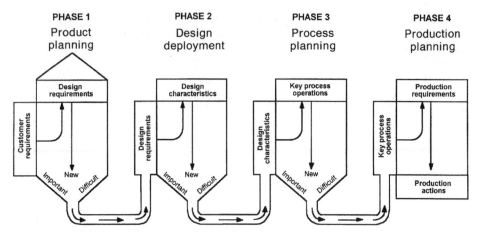

Figure 6 The four phases of QFD (ASI, 1987)

the product planning matrix, is shown in Figure 7. Critical design requirements that are considered to be either new to a product, important by the customer or difficult to produce are carried forward to the next phase.

- **Phase 2 – Design Deployment.** The critical design requirements from Phase 1 (those with a high points rating) are ranked with design characteristics where the relationship between them is again quantified, yielding important interface issues between design and manufacture, for example those characteristics that require close control during production. The critical design characteristics are then carried forward to the next phase.
- **Phase 3 – Process Planning.** The important design characteristics from Phase 2 are ranked with key process operations, where quantification yields actions to improve the understanding of the processes involved and gain the necessary expertise early on. The critical process operations highlighted are then carried forward to the next phase.
- **Phase 4 – Production Planning.** The critical process operations from Phase 3 are ranked with production requirement issues, ultimately translating the important customer requirements to production planning and establishing important actions to be taken.

Benefits

The potential benefits of QFD are:

- Increased customer satisfaction
- Improved product quality
- Reduced design changes and associated costs
- Reduced lead times
- Improved documentation and traceability
- Promotes team working

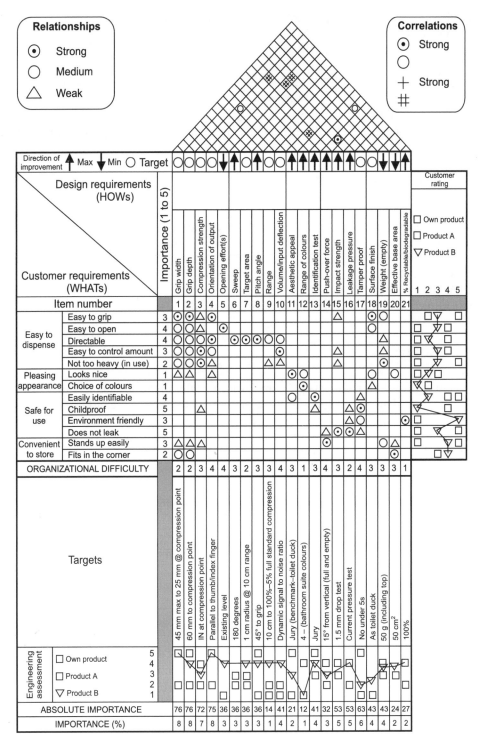

Figure 7 QFD phase 1 matrix example for a toilet bleach container (ASI, 1992)

- Provides a structured approach to product development
- Improves customer/supplier relationship.

Key issues

- Can be used on products, software or services
- Must be management led and have an overall strategy for implementation and application.
- Training required to use analysis initially
- Multi-disciplinary team-based application essential
✓ • Can be subjective and tedious
- Organizations do not extend the use of QFD past the first phase usually
- Involvement of customers and suppliers essential
- Review at regular intervals with customers and suppliers
- Identification of customer requirements difficult sometimes
- Support QFD with existing data wherever possible
✓ • Applied too late in many cases
- Can help to build up a knowledge base for product families.

Case study

An illustrative chart is shown in Figure 7, which relates to the design of a toilet bleach container. It does not include all the customer wants. Figure 7 can be regarded as Phase 1 product planning covering design requirements. The three remaining phases are design deployment, process planning and production planning. Similar charts are constructed for each of these phases (see Figure 6).

C. Design for Assembly/Design for Manufacture (DFA/DFM)

(Boothroyd *et al.*, 1994; CSC, 1995; Huang, 1996; Shimada *et al.*, 1992)

Description

DFA/DFM is a team-based product design evaluation tool which, through simple structured analysis, gives the information required by designers to achieve:

- Part count reduction
- Calculation of component manufacturing and assembly costs
- Ease of part handling
- Ease of assembly
- Ability to reproduce identically and without waste products which satisfy customer requirements.

Placement in product development

DFA/DFM should be performed during concept design leading into the detailed design stage to revise the product structure.

Application of the technique (CSC's DFA/MA methodology)

The analysis is carried out in four main stages (see Figure 8).

1. **Functional analysis.** The methodology enables each part to be classified as essential 'A' parts or non-essential 'B' parts. Given this awareness, redesigns can be evolved around the essential components, from which reduced part count normally results.
2. **Manufacturing analysis.** Component costs are calculated by considering materials, manufacturing processes and aspects such as complexity, volume and tolerance. This allows ideas for part count reduction to be tested since combining parts can lead to more complex components and changes to manufacturing processes.
3. **Handling analysis.** Components must be correctly orientated before assembly can take place. The difficulty achieving this by either manual or mechanical means can be assessed using the analysis. Components receiving high ratings should be modified to give an acceptable rating. Mistake proofing (or Poka Yoke) devices

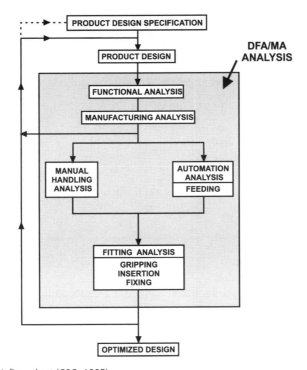

Figure 8 DFA/MA flow chart (CSC, 1995)

can be installed in the manufacturing and assembly processes to help ensure zero defects.

4. **Fitting analysis.** For this to be possible, an assembly sequence plan must be constructed and the difficulty of assembling each part in the sequence rated using the design for assembly analysis tables. Difficult assembly tasks and non-value added processes are revealed as candidates for correction by redesign. Simple concepts such as the ability to assemble in a layer fashion can result in major cost savings.

The assembly analysis stages 3 and 4 have the following measures of performance which are accepted as an indication of good design:

- Functional analysis gives a Design Efficiency = Essential Parts/Total Parts ≥ 60%
- Handling analysis gives a Handling Ratio = Total Ratings/Essential Parts ≤ 2.5
- Fitting analysis gives a Fitting Ratio = Total Ratings/Essential Parts ≤ 2.5.

Benefits

The potential benefits of DFA/DFM are:

- Reduced part count
- Fewer parts means: improved reliability, fewer stock costs, fewer invoices from fewer suppliers and possibly fewer quality problems
- Systematic component costing and process selection
- Improved yields
- Lower component and assembly costs
- Standardize components, assembly sequence and methods across product 'families' leading to improved reproducibility
- Faster product development and reduced time to market
- Lower level of engineering changes, modifications and concessions.

Typical results achieved by application of the DFA technique to date are approximately as follows:

- Parts count reduction = 50%
- Assembly cost reduction = 50%
- Product cost reduction = 30%.

Key issues

- Must be management led
- Training required before use
- Resources consumed in product development can be significant
- Team-based application and systematic approach is essential
- Many subjective analysis processes
- Manufacturing and technical feasibility of new component design solutions need to be validated
- Early life failures are caused by latent defects.

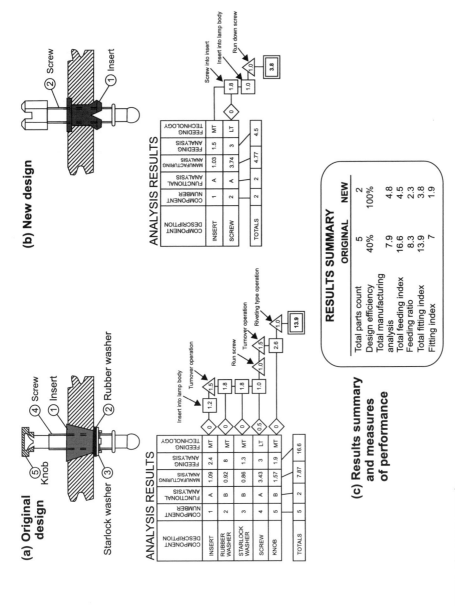

Figure 9 DFA/MA trim screw example

Case study

A feasibility study aimed at automating the assembly of trim screws on a car headlight design revealed difficulties in handling the components and the assembly processes – complex assembly structure, complex access, turnover operation and automation problems.

The trim screw assembly in its original form is shown in Figure 9(a). A product improvement team used the DFA/MA methodology to analyse the design. The results of the analysis are shown in the figure against a component list and a sequence of assembly. Having analysed the existing trim screw design, certain undesirable elements of the design were highlighted. This information together with original conceptual changes to suit the proposed automation programme were used to prompt a better design solution shown in Figure 9(b).

The number of parts in the redesign reduced from 5 to 2. A more simple product structure and increased assembly design efficiency resulted. Overall, component and assembly costs were significantly reduced. Figure 9(c) summarizes the results of the analysis.

D. Design of Experiments (DOE)

(Grove and Davis, 1992; Kapur, 1993; Taguchi *et al.*, 1989)

Description

DOE encompasses a range of techniques used to enable a business to understand the effects of important variables in product and process development. It is normally used when investigating a situation where there are several variables, one or more of which may result in a problem either singly or in combination.

In summary the objectives of DOE are to:

- Identify causes of variation in critical product or process parameters
- Limit the effects of the variations identified (where they can be controlled)
- Achieve reproducibility of best system performance in manufacture and use
- Optimize the product or process
- Reduce cost.

Placement in product development

QFD, FMEA and CA are useful in identifying critical characteristics early on in product development and the results from these can be fed into DOE. DOE is useful in investigating and validating these critical characteristics with respect to technical requirements and their influence on product and process quality.

Application of the technique

The systematic application of DOE should be based on three distinct phases:

Phase 1 – Preparation. This is often referred to as the pre-experimental stage. Experiments can take considerable time and resources, and good preparation is all important. The results from other techniques are important inputs to this phase of the methodology, providing focus and priority selection filtering. A summary of the steps that should be considered is given below:

- Define the problem to be solved
- Agree the objectives and prepare a project plan
- Examine and understand the situation. Obtain and study all available data related to the problem
- Define what needs to be measured to satisfy the project objectives
- Identify both the noise and factors to be controlled during the experiment and select the levels to be considered. These should be representative of normal operating range and sufficiently spaced to spot changes
- Establish an effective measuring system. Understand its variance and the likely effects of this apparent variation in output.

Phase 2 – Experimentation and analysis. Carrying out the planned trials and analytical work will include the steps below:

- Select the techniques to be used. Decisions on the number of trials is invariably coloured by cost, but always check they are balanced otherwise it may be false economy
- Plan the trials. Consecutive trials should not affect each other
- Conduct the trials as planned. Results should be carefully recorded in a table or machine along with any observations which may be regarded as potential errors
- Analysis and reporting of results, including a conformation run. The analysis of results should be transparent. Simple approaches should be used where possible to build confidence and provide clarity.

Phase 3 – Implementation. This phase is often called the post-experimental stage. It is about acting on the results and communicating the lessons learnt. Some points on the process are given below:

- Apply the findings to resolve the problem
- Adopt a procedure for measuring and monitoring results such as SPC to detect future changes that could influence quality
- Communicate lessons learnt through the business and included in training programmes.

Benefits

The potential benefits of DOE are:

- Improved product quality

- Focuses efforts on critical areas in the product or process development
- Identification of critical process parameters
- Reduced scrap or rework
- Reduced need for inspection
- Systematic approach to process development.

Key issues

- An overall strategy for its implementation and application should be in place with clear objectives
- Can be complex at first. Begin on small experiments and then expand to larger ones
- Poor understanding of the concepts and underlying methods can lead to poor results
- Problems associated with interpretation of the results and assessing their significance are common
- Team should involve those knowledgeable about the process
- Use software wherever possible to reduce complexity and tediousness.

Case study

Consider the example of fluid flow through a filter, where the objective is to maximize flow rate. Assume that the relevant control factors are filter, fluid viscosity and

(a)

Trial No.	A Filter density	B Fluid viscosity	C System temperature	Result*
1	coarse	low	low	8
2	coarse	low	high	10
3	coarse	high	low	4
4	coarse	high	high	6
5	fine	low	low	4
6	fine	low	high	6
7	fine	high	low	2
8	fine	high	high	4

Full factorial exploring all combinations of variables and their levels. This approach can prove to be costly and time consuming.

(b)

1	coarse	low	low	8
2	coarse	high	high	6
3	fine	low	high	6
4	fine	high	low	2

Orthogonal array involving 4 trials. Here the effect of different factors can be separated out and their effects estimated.

*Non-dimensional measure of fluid flow

Figure 10 Experimental variations for the fluid flow through a filter

temperature, and that each can have two settings, coarse/fine, high/low and high/low respectively. Figure 10(a) shows a full factorial experiment where all combinations of factors and levels are explored based on eight trials. As touched on previously, this approach can be costly. While we could try varying a single factor at a time, in a more effective experiment, the different levels of each factor would occur on the same number of occasions. Figure 10(b) illustrates an Orthogonal Array involving four trials. Note that between any two columns each combination level occurs equally often. Using this approach the effect of the different factors can be separated out and their effects estimated.

In many practical cases, useful experiments can be carried out using only two levels. Often this can be employed where a single change in level will illustrate whether a factor is likely to be significant or not. However, where a response is non-linear, three or more levels may be needed. When the experiment is complete it is time to investigate the significance of each factor. This can be done with the aid of straight-forward graphical or statistical analysis of variance techniques. The graphical approach is particularly simple and effective and the approach is outlined in Figure 11. In this figure, values from Figure 10(b) are used to generate plots of factors

(a) Individual results

The gradient of the graph indicates the significance of the factor.

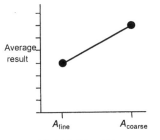
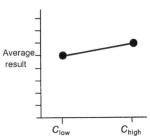

The high differential means highly significant.

A low differential means low significance. A zero differential means not significant at all.

(b) Combination of results

The relationship of the lines shows the significance of the combination.

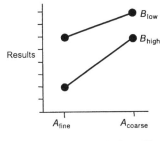
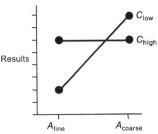

Significant combinations of factors will be indicated when the lines show a large difference in gradient. Strong significance is indicated where the lines cross. Parallel lines indicate little significance.

Figure 11 Graphical analysis of the experimental results

and levels. The plots are employed to study the effects of both individual results and combinations. Note that only four of the six graphs are included in Figure 11.

In the case of individual results in Figure 11(a), the gradient of the line drawn between average values of the results at the levels under consideration is a measure of the significance of the factor. A high gradient indicates a high significance, while a horizontal line infers no significance at all. Analysis of the effects of combinations is based on lines constructed as shown in Figure 11(b). The relationship between the lines indicates the significance of the results. Significant combinations are inferred when the lines show a large difference in gradient. Strong significance is indicated when lines cross. Parallel lines suggest no significance. Note that in fractional factorial experiments, the main effects are compounded by interaction effects.

Process capability maps

Index to maps

Cold forging
Swaging – All metals

Sheet F   **Cold drawing and rolling processes**
Cold drawing – Steel
Cold drawing – Copper and copper alloys
Cold rolling – Steel
Cold rolling – Copper and copper alloys
Cold rolling – Aluminium alloys

Sheet G   **Extrusion processes**
Hot extrusion – Aluminium and magnesium alloys
Extrusion – Thermoplastics
Extrusion – Elastomers
Cold extrusion

Sheet H   **Sheet metalworking processes**
Cutting
Punching
Fine blanking
Bending
Deep drawing and ironing
Roll forming

Sheet J   Spinning

Sheet K   **Machining processes**
Turning/boring
Diamond turning/boring
Cylindrical grinding
Drilling
Reaming
Honing

Sheet L   Planing/shaping
Milling
Surface grinding
Lapping
Broaching
Machining centre (positional) – Direct C_{pk} values

Sheet M   **Powder metallurgy processes**
Powder metal sintering (radial)
Powder metal sintering (axial)
Powder metal sizing (radial)

Sheet N   **Plastic moulding processes**
Injection/compression moulding – Thermoplastics and thermosets
Blow moulding – Thermoplastics
Rotational moulding – Thermoplastics

Sheet P   **Elastomer and composite moulding processes**
Injection moulding – Elastomers
Compression moulding – Composites
Hand/spray lay-up – Composites

Sheet Q   **Non-traditional machining processes**
Electrical discharge machining
Electrochemical milling
Electron beam machining
Laser beam machining
Ultrasonic machining

Sheet R   Chemical milling – Steel
Chemical milling – Aluminium alloys
Photochemical blanking – Steel
Photochemical blanking – Copper

Key to maps

1. To determine the risk value 'A', look along the horizontal axis until the characteristic dimension is found, and at the same time, locate the adjusted tolerance on the vertical axis. Read off the 'A' value in the zone at which these lines intersect on the map by interpolating as required between the zone bands, $A = 1$ to $A = 9$.
2. $A = 1$ corresponds to a tolerance that is easily achieved by the manufacturing process.
3. The intermediate values of 'A' ($A = 1.1$, $A = 1.3$, $A = 1.7$) correspond to increasing technical difficulty/cost to achieve the tolerance.
4. $A = 3$ to $A = 9$ correspond to a tolerance that is technically unattainable using the manufacturing process. A secondary (and possibly a tertiary) process may be required to achieve the tolerance.
5. The two bold lines at $A = 1$ and $A = 1.7$ represent the normal range of tolerance capability for the process under ideal material and component geometry conditions. This information can be used as knowledge in the redesign process if required.
6. All tolerances are plus or minus unless stated otherwise.
7. All dimensions and tolerances are in millimetres and both axis have log scales.

Note: The process capability maps for casting processes (excluding investment casting), closed die forging/stamping and plastic/rubber/composite moulding processes are for tolerances in one half of the die/mould only. Additional allowances will need to be added for tolerances that are across the die/mould parting line when the process capability maps are used in the redesign process as a knowledge base (see Swift and Booker (1997) for approximate parting line allowances).

The map for machining centre positional tolerances gives C_{pk} values directly, and no other analysis is required.

Material key for plastic moulding processes

Abbreviation	Name
ABS	Acrylonitrile butadiene styrene
CA	Cellulose acetate
CP	Cellulose propionate
PF	Phenolic
PA	Polyamide
PBTP	Polybutylene terephthalate
PC	Polycarbonate
PCTFE	Polychlorotrifluoroethylene
PE	Polyethylene
PESU	Polyethersulphone
PETP	Polyethyleneterephthalate
PMMA	Polymethylmethacrylate
POM	Polyoxymethylene
PPS	Polyphenylene sulphide
PP	Polypropylene
PS	Polystyrene
PSU	Polysulphone
PVC-U	Polyvinylchloride – unplasticized
SAN	Styrene acrylonitrile
UP	Polyester

Note: The injection/compression moulding process capability maps 1, 2 and 3 are used for large parts with a major dimension greater than 50 mm typically and/or for large production volumes. Map 4 is for injection moulded parts that have a major dimension less than 150 mm and which are produced in small volumes.

Sheet A Casting processes

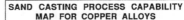

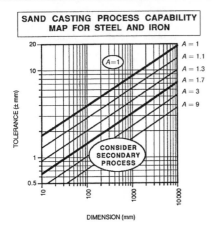

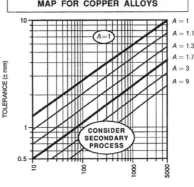

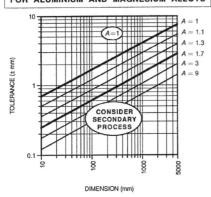

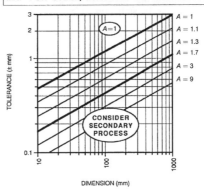

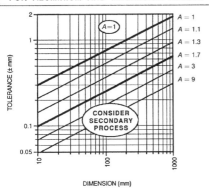

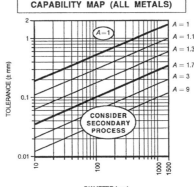

Sheet B   Casting processes (cont'd)

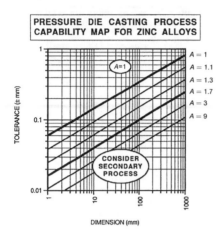

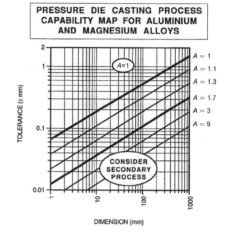

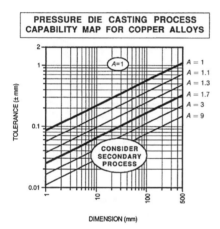

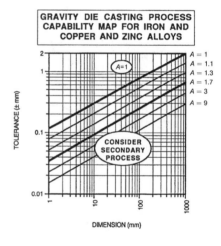

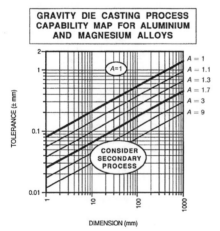

Sheet C Casting processes (cont'd)

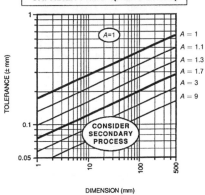

INVESTMENT CASTING PROCESS CAPABILITY MAP (ALL METALS)

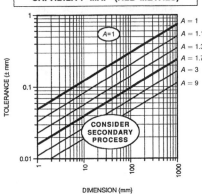

CERAMIC MOULD CASTING PROCESS CAPABILITY MAP (ALL METALS)

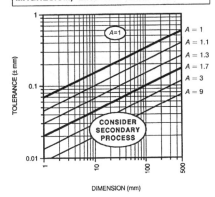

PLASTER MOULD CASTING PROCESS CAPABILITY MAP FOR ALUMINIUM, MAGNESIUM, ZINC AND COPPER ALLOYS

Sheet D　Hot forging processes

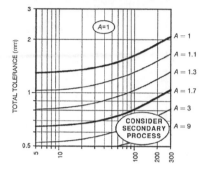

1. CLOSED DIE FORGING PROCESS CAPABILITY MAP FOR LOW TO MEDIUM CARBON AND LOW ALLOY STEELS
(WEIGHT UP TO 1 kg)

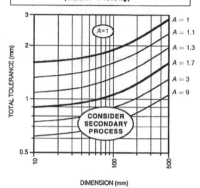

2. CLOSED DIE FORGING PROCESS CAPABILITY MAP FOR LOW TO MEDIUM CARBON AND LOW ALLOY STEELS
(WEIGHT 1→3.2 kg)

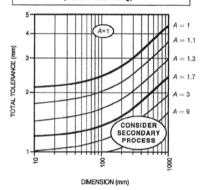

3. CLOSED DIE FORGING PROCESS CAPABILITY MAP FOR LOW TO MEDIUM CARBON AND LOW ALLOY STEELS
(WEIGHT 3.2→10 kg)

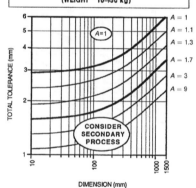

4. CLOSED DIE FORGING PROCESS CAPABILITY MAP FOR LOW TO MEDIUM CARBON AND LOW ALLOY STEELS
(WEIGHT 10→50 kg)

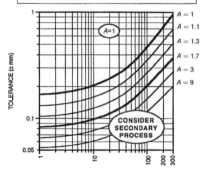

STAMPING PROCESS CAPABILITY MAP FOR COPPER ALLOYS

Sheet E   Cold forming processes

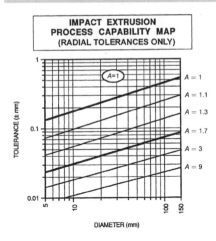

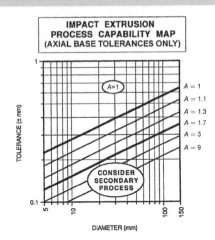

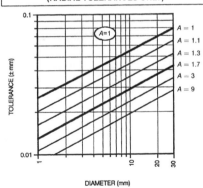

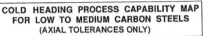

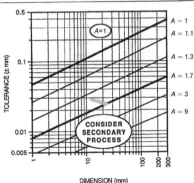

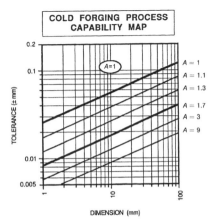

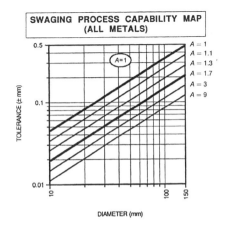

Sheet F   Cold drawing and rolling processes

COLD DRAWING PROCESS CAPABILITY MAP FOR LOW TO MEDIUM CARBON STEELS

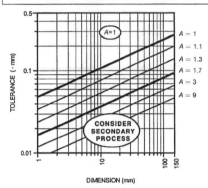

COLD DRAWING PROCESS CAPABILITY MAP FOR COPPER AND COPPER ALLOYS

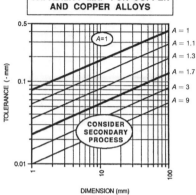

COLD ROLLING PROCESS CAPABILITY MAP FOR LOW TO MEDIUM CARBON AND LOW ALLOY STEELS

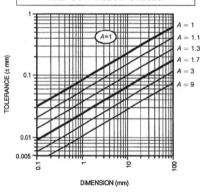

COLD ROLLING PROCESS CAPABILITY MAP FOR COPPER AND COPPER ALLOYS

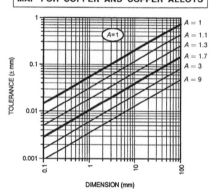

COLD ROLLING PROCESS CAPABILITY MAP FOR ALUMINIUM ALLOYS

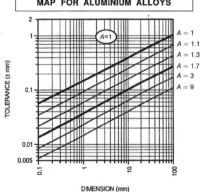

Sheet G   Extrusion processes

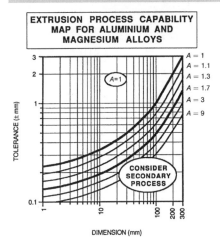

EXTRUSION PROCESS CAPABILITY
MAP FOR ALUMINIUM AND
MAGNESIUM ALLOYS

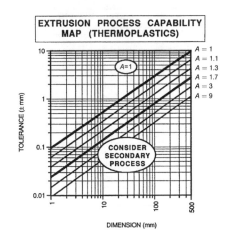

EXTRUSION PROCESS CAPABILITY
MAP (THERMOPLASTICS)

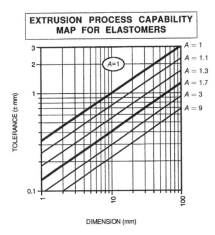

EXTRUSION PROCESS CAPABILITY
MAP FOR ELASTOMERS

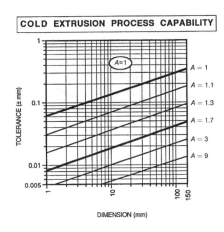

COLD EXTRUSION PROCESS CAPABILITY

Sheet H   Sheet metalworking processes

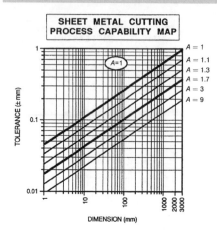

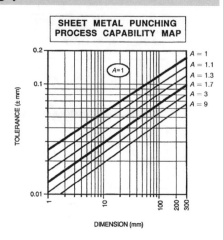

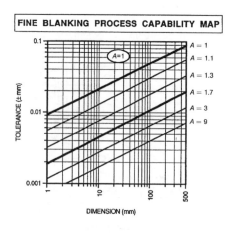

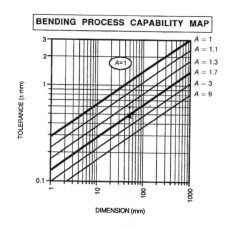

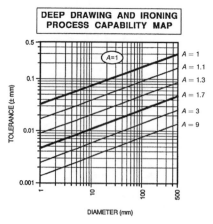

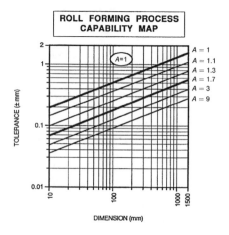

Sheet J   Sheet metalworking processes (cont'd)

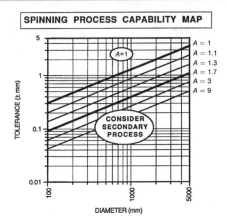

Sheet K   Machining processes

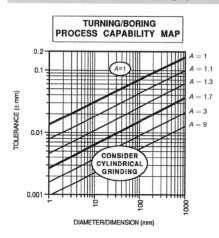

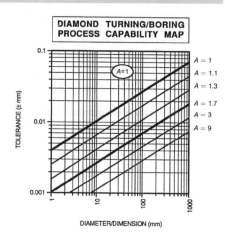

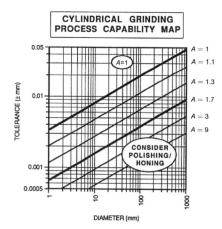

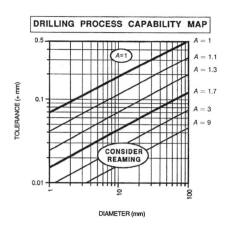

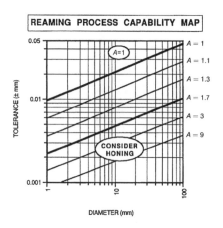

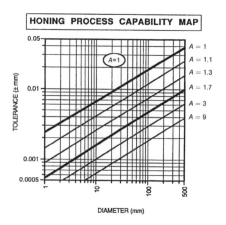

Sheet L   Machining processes (cont'd)

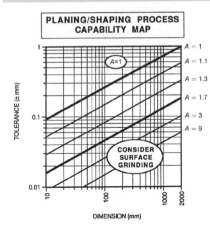

PLANING/SHAPING PROCESS CAPABILITY MAP

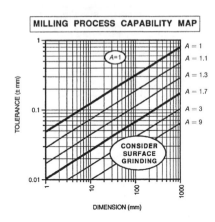

MILLING PROCESS CAPABILITY MAP

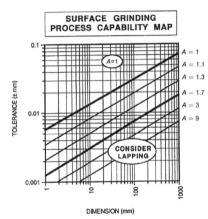

SURFACE GRINDING PROCESS CAPABILITY MAP

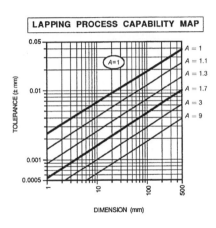

LAPPING PROCESS CAPABILITY MAP

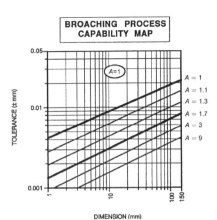

BROACHING PROCESS CAPABILITY MAP

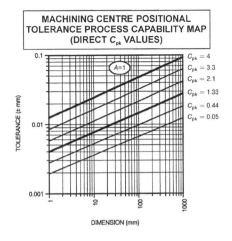

MACHINING CENTRE POSITIONAL TOLERANCE PROCESS CAPABILITY MAP (DIRECT C_{pk} VALUES)

Sheet M   Powder metallurgy processes

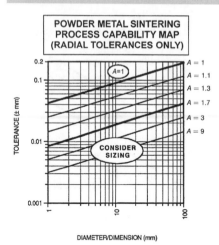

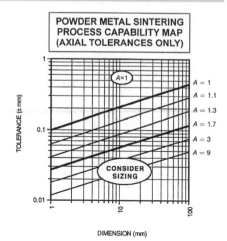

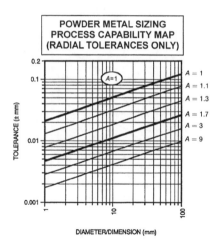

Sheet N Plastic moulding processes

See material key and notes.

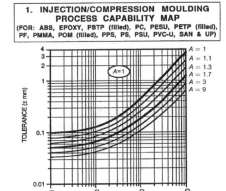

1. INJECTION/COMPRESSION MOULDING PROCESS CAPABILITY MAP
(FOR: ABS, EPOXY, PBTP (filled), PC, PESU, PETP (filled), PF, PMMA, POM (filled), PPS, PS, PSU, PVC-U, SAN & UP)

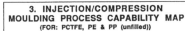

A = 1
A = 1.1
A = 1.3
A = 1.7
A = 3
A = 9

TOLERANCE (± mm) / DIMENSION (mm)

2. INJECTION/COMPRESSION MOULDING PROCESS CAPABILITY MAP
(FOR: CA, CP, PA, PBTP (unfilled) & PP (filled))

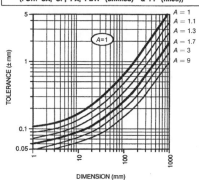

A = 1
A = 1.1
A = 1.3
A = 1.7
A = 3
A = 9

TOLERANCE (± mm) / DIMENSION (mm)

3. INJECTION/COMPRESSION MOULDING PROCESS CAPABILITY MAP
(FOR: PCTFE, PE & PP (unfilled))

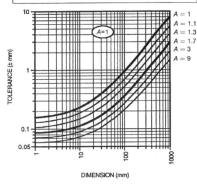

A = 1
A = 1.1
A = 1.3
A = 1.7
A = 3
A = 9

TOLERANCE (± mm) / DIMENSION (mm)

4. INJECTION MOULDING PROCESS CAPABILITY MAP FOR LIGHT ENGINEERING (ALL THERMOPLASTICS)

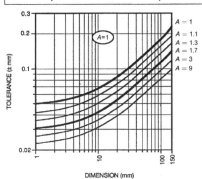

A = 1
A = 1.1
A = 1.3
A = 1.7
A = 3
A = 9

TOLERANCE (± mm) / DIMENSION (mm)

BLOW MOULDING PROCESS CAPABILITY MAP (THERMOPLASTICS)

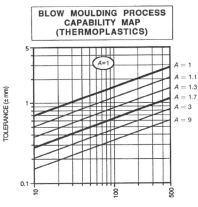

A = 1
A = 1.1
A = 1.3
A = 1.7
A = 3
A = 9

TOLERANCE (± mm) / DIMENSION (mm)

ROTATIONAL MOULDING PROCESS CAPABILITY MAP (THERMOPLASTICS)

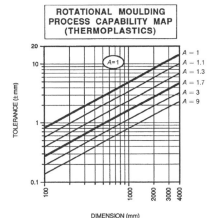

A = 1
A = 1.1
A = 1.3
A = 1.7
A = 3
A = 9

TOLERANCE (± mm) / DIMENSION (mm)

Sheet P   Elastomer and composite moulding processes

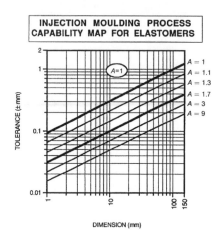

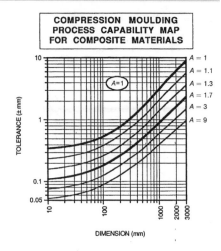

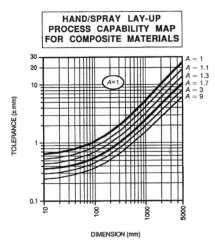

Sheet Q   Non-traditional machining processes

These processes are not strictly dimensional dependent.

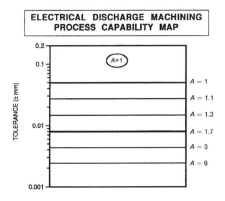

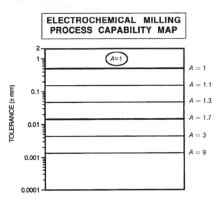

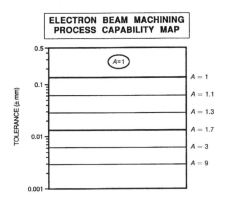

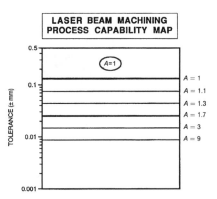

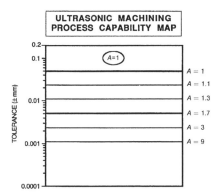

Sheet R Non-traditional machining processes (cont'd)

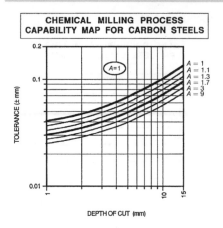

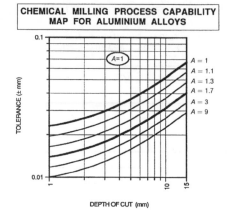

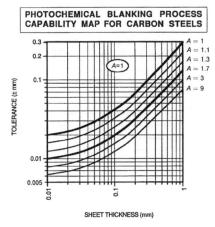

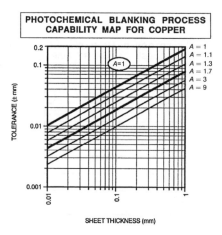

Appendix V

Sample case studies used in validation

Component name	Cylinder housing	Calculation of q_m
Drawing number	CH1878-1018	$m_p = 1.3 \times 1.5 = 1.95$
Material	Alloy steel	$g_p = 1 \times 1.1 \times 1 \times 1 \times 1 \times 1 = 1.1$
Manufacturing process	Turning/boring	
Characteristic description	Internal core length	Adjusted tolerance $= \dfrac{\text{Design tolerance}}{m_p \times g_p}$
Characteristic dimension	122.09 (mm)	
Design tolerance	± 0.05 (mm)	$= \dfrac{\pm 0.05}{1.95 \times 1.1} = \pm 0.023$
Surface roughness	0.6 (μm Ra)	

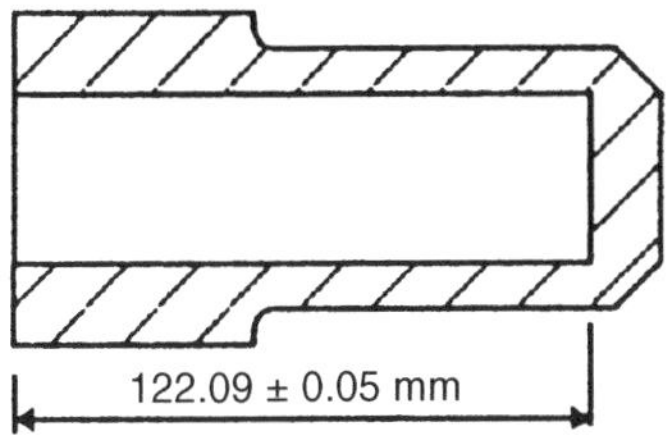

122.09 ± 0.05 mm

$t_p = 1.4 \times 1 = 1.4$

$s_p = 1 \times 1 = 1$

$q_m = t_p \times s_p = 1.4 \times 1 = 1.4$

Process capability index $C_{pk} = 1.79$ Process capability index $C_p = 2.1$	Manufacturing variability risk $q_m = 1.4$

Component name	Cover support leg	Calculation of q_m
Drawing number	N/A	$m_p = 1.7 \times 1.5 = 2.55$
Material	0.6 mm spring steel	$g_p = 1 \times 1 \times 1 \times 1.7 \times 1 \times 1.1 = 1.87$
Manufacturing process	Sheet metal bending	
Characteristic description	Top face to bottom lip	
Characteristic dimension	43.6 (mm)	$\text{Adjusted tolerance} = \dfrac{\text{Design tolerance}}{m_p \times g_p}$
Design tolerance	±0.2 (mm)	$= \dfrac{\pm 0.2}{2.55 \times 1.87} = \pm 0.042$
Surface roughness	– (µm Ra)	

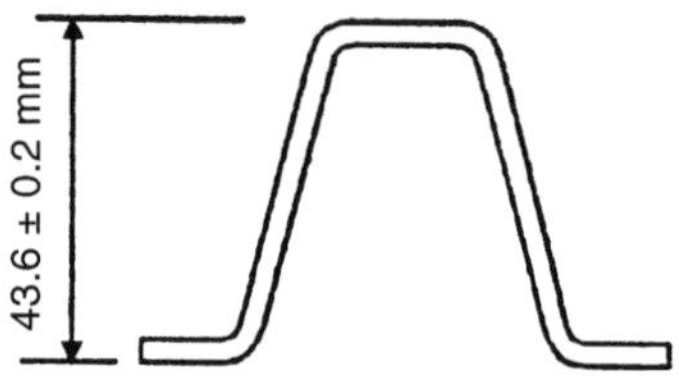

$t_p = 9 \times 1 = 9$

$s_p = 1 \times 1 = 1$

$q_m = t_p \times s_p = 9 \times 1 = 9$

Process capability index $C_{pk} = 0$ Process capability index $C_p = 0$	Manufacturing variability risk $q_m = 9$

Component name	End cap	Calculation of q_m
Drawing number	N/A	$m_p = 1 \times 1 = 1$
Material	Phenolic	$g_p = 1 \times 1.1 \times 1 \times 1 \times 1.1 \times 1 = 1.21$
Manufacturing process	Injection moulding	
Characteristic description	Clip length	
Characteristic dimension	55.0 (mm)	$\text{Adjusted tolerance} = \dfrac{\text{Design tolerance}}{m_p \times g_p}$
Design tolerance	±0.2 (mm)	$= \dfrac{\pm 0.2}{1 \times 1.21} = \pm 0.165$
Surface roughness	– (µm Ra)	

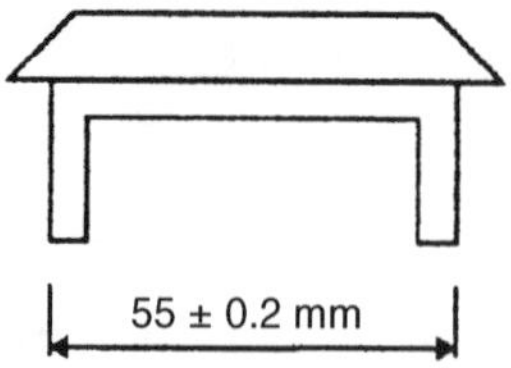

55 ± 0.2 mm

$t_p = 1.3 \times 1 = 1.3$

$s_p = 1 \times 1 = 1$

$q_m = t_p \times s_p = 1.3 \times 1 = 1.3$

Process capability index $C_{pk} = 2.18$ Process capability index $C_p = 2.67$	Manufacturing variability risk $q_m = 1.3$

		Calculation of q_m
Component name	Channel	$m_p = 1 \times 1.2 = 1.2$
Drawing number	FF476	
Material	0.55 mm mild steel	$g_p = 1 \times 1 \times 1 \times 1 \times 1 \times 1.1 = 1.1$
Manufacturing process	Sheet metal bending	
Characteristic description	Bottom width	
Characteristic dimension	57.1 (mm)	$\text{Adjusted tolerance} = \dfrac{\text{Design tolerance}}{m_p \times g_p}$
Design tolerance	±0.5 (mm)	$= \dfrac{\pm 0.5}{1.2 \times 1.1} = \pm 0.379$
Surface roughness	– (μm Ra)	

$t_p = 3 \times 1 = 3$

$s_p = 1 \times 1 = 1$

$q_m = t_p \times s_p = 3 \times 1 = 3$

57.1 ± 0.5 mm

Process capability index $C_{pk} = 0.85$ Process capability index $C_p = 1.21$	Manufacturing variability risk $q_m = 3$

		Calculation of q_m
Component name	End cap	$m_p = 1 \times 1 = 1$
Drawing number	N/A	
Material	Phenolic	$g_p = 1.7 \times 1.1 \times 1 \times 1 \times 1.1 \times 1 = 2.06$
Manufacturing process	Injection moulding	
Characteristic description	Length	
Characteristic dimension	6.5 (mm)	$\text{Adjusted tolerance} = \dfrac{\text{Design tolerance}}{m_p \times g_p}$
Design tolerance	±0.2 (mm)	$= \dfrac{\pm 0.2}{1 \times 2.06} = \pm 0.097$
Surface roughness	– (μm Ra)	

6.5 ± 0.2 mm

$t_p = 1.1 \times 1 = 1.1$

$s_p = 1 \times 1 = 1$

$q_m = t_p \times s_p = 1.1 \times 1 = 1.1$

Process capability index $C_{pk} = 3.88$ Process capability index $C_p = 4.48$	Manufacturing variability risk $q_m = 1.1$

Component name	Channel	Calculation of q_m
Drawing number	FF476	$m_p = 1 \times 1.2 = 1.2$
Material	0.55 mm mild steel	$g_p = 1 \times 1 \times 1 \times 1.7 \times 1 \times 1.1 = 1.87$
Manufacturing process	Sheet metal bending	
Characteristic description	Top width	
Characteristic dimension	57.1 (mm)	Adjusted tolerance $= \dfrac{\text{Design tolerance}}{m_p \times g_p}$
Design tolerance	±0.5 (mm)	$= \dfrac{\pm 0.5}{1.2 \times 1.87} = \pm 0.223$
Surface roughness	– (μm Ra)	

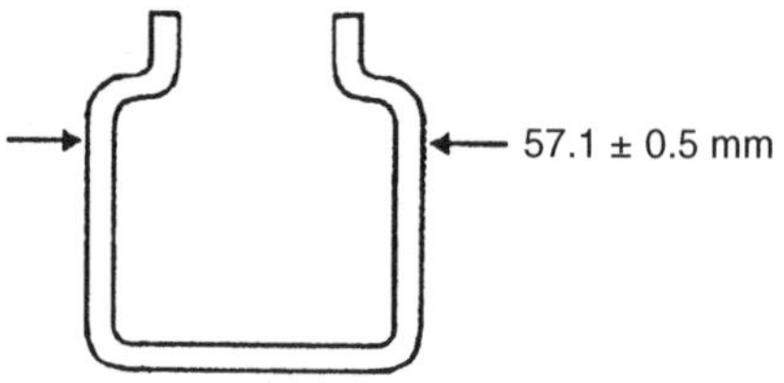

57.1 ± 0.5 mm

$t_p = 9 \times 1 = 9$

$s_p = 1 \times 1 = 1$

$q_m = t_p \times s_p = 9 \times 1 = 9$

Process capability index $C_{pk} = 0$ Process capability index $C_p = 0$	Manufacturing variability risk $q_m = 9$

Component name	Cylinder housing	Calculation of q_m
Drawing number	CH1878-1018	$m_p = 1.3 \times 1.5 = 1.95$
Material	Alloy steel	$g_p = 1 \times 1.1 \times 1 \times 1 \times 1 \times 1 = 1.1$
Manufacturing process	Turning/boring	
Characteristic description	Outside diameter	
Characteristic dimension	⌀50.9 (mm)	Adjusted tolerance $= \dfrac{\text{Design tolerance}}{m_p \times g_p}$
Design tolerance	±0.1 (mm)	$= \dfrac{\pm 0.1}{1.95 \times 1.1} = \pm 0.047$
Surface roughness	0.8 (μm Ra)	

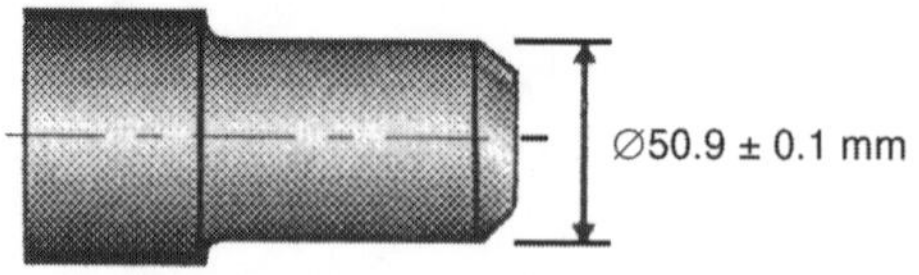

⌀50.9 ± 0.1 mm

$t_p = 1.02 \times 1 = 1.02$

$s_p = 1 \times 1 = 1$

$q_m = t_p \times s_p = 1.02 \times 1 = 1.02$

Process capability index $C_{pk} = 4.31$ Process capability index $C_p = 4.33$	Manufacturing variability risk $q_m = 1.02$

Component name	Washer	Calculation of q_m
Drawing number	N/A	$m_p = 1 \times 1.4 = 1.4$
Material	1mm mild steel	$g_p = 1 \times 1 \times 1 \times 1 \times 1 \times 1 = 1$
Manufacturing process	Sheet metal punching	
Characteristic description	Internal bore length	$\text{Adjusted tolerance} = \dfrac{\text{Design tolerance}}{m_p \times g_p}$
Characteristic dimension	⌀5.4 (mm)	
Design tolerance	±0.1 (mm)	$= \dfrac{\pm 0.1}{1.4 \times 1} = \pm 0.071$
Surface roughness	– (µm Ra)	

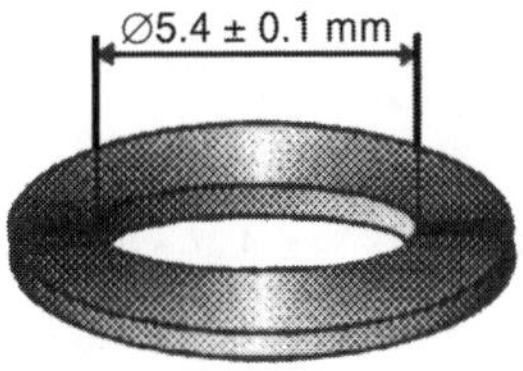

⌀5.4 ± 0.1 mm

$t_p = 1 \times 1 = 1$

$s_p = 1 \times 1 = 1$

$q_m = t_p \times s_p = 1 \times 1 = 1$

Process capability index $C_{pk} = 2.63$ Process capability index $C_p = 3.56$	Manufacturing variability risk $q_m = 1$

Component name	Bush	Calculation of q_m
Drawing number	ERC4215	$m_p = 1 \times 1.7 = 1.7$
Material	Bronze	$g_p = 1.7 \times 1.1 \times 1 \times 1 \times 1 \times 1 = 1.87$
Manufacturing process	Sintering/sizing	
Characteristic description	Length	$\text{Adjusted tolerance} = \dfrac{\text{Design tolerance}}{m_p \times g_p}$
Characteristic dimension	12.7 (mm)	
Design tolerance	±0.076 (mm)	$= \dfrac{\pm 0.076}{1.7 \times 1.87} = \pm 0.024$
Surface roughness	– (µm Ra)	

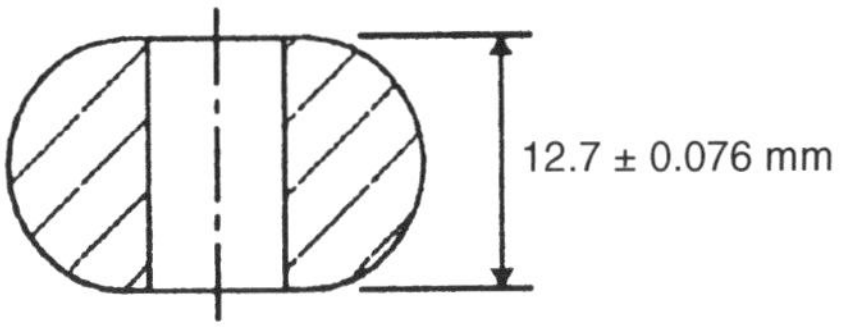

12.7 ± 0.076 mm

$t_p = 1.25 \times 1.1 = 1.4$

$s_p = 1 \times 1 = 1$

$q_m = t_p \times s_p = 1.4 \times 1 = 1.4$

Process capability index $C_{pk} = 1.88$ Process capability index $C_p = 2.15$	Manufacturing variability risk $q_m = 1.4$

Component name	Bush	Calculation of q_m
Drawing number	ERC4215	$m_p = 1 \times 1.7 = 1.7$
Material	Bronze	$g_p = 1 \times 1.1 \times 1 \times 1 \times 1 \times 1 = 1.1$
Manufacturing process	Sintering/sizing	
Characteristic description	Spherical diameter	Adjusted tolerance $= \dfrac{\text{Design tolerance}}{m_p \times g_p}$
Characteristic dimension	$\varnothing16.64$ (mm)	
Design tolerance	±0.025 (mm)	$= \dfrac{\pm0.025}{1.7 \times 1.1} = \pm0.013$
Surface roughness	$-$ (μm Ra)	

$\varnothing16.64 \pm 0.025$ mm

$t_p = 1.7 \times 1.1 = 1.87$

$s_p = 1 \times 1 = 1$

$q_m = t_p \times s_p = 1.87 \times 1 = 1.87$

Process capability index $C_{pk} = 1.2$ Process capability index $C_p = 1.33$	Manufacturing variability risk $q_m = 1.87$

Component name	Plenum	Calculation of q_m
Drawing number	N/A	$m_p = 1 \times 1 = 1$
Material	Aluminium alloy	$g_p = 1 \times 1 \times 1 \times 1 \times 1 \times 1 = 1$
Manufacturing process	Internal grinding	
Characteristic description	Internal diameter	Adjusted tolerance $= \dfrac{\text{Design tolerance}}{m_p \times g_p}$
Characteristic dimension	$\varnothing65.025$ (mm)	
Design tolerance	±0.025 (mm)	$= \dfrac{\pm0.025}{1 \times 1} = \pm0.025$
Surface roughness	1.6 (μm Ra)	

$\varnothing65.025 \pm 0.025$ mm

$t_p = 1 \times 1.1 = 1.1$

$s_p = 1 \times 1 = 1$

$q_m = t_p \times s_p = 1.1 \times 1 = 1.1$

Process capability index $C_{pk} = 4.53$ Process capability index $C_p = 4.7$	Manufacturing variability risk $q_m = 1.1$

Component name	Plunger housing	Calculation of q_m
Drawing number	CH1878-1021	$m_p = 1 \times 1 = 1$
Material	Aluminium bronze	$g_p = 1 \times 1 \times 1 \times 1 \times 1 \times 1 = 1$
Manufacturing process	Turning/boring	
Characteristic description	Internal length	Adjusted tolerance $= \dfrac{\text{Design tolerance}}{m_p \times g_p}$
Characteristic dimension	21.46 (mm)	
Design tolerance	± 0.05 (mm)	$= \dfrac{\pm 0.05}{1 \times 1} = \pm 0.05$
Surface roughness	0.8 (μm Ra)	

$t_p = 1 \times 1 = 1$

$s_p = 1 \times 1 = 1$

$q_m = t_p \times s_p = 1 \times 1 = 1$

21.46 ± 0.05 mm

Process capability index $C_{pk} = 3.98$ Process capability index $C_p = 4.96$	Manufacturing variability risk $q_m = 1$

Component name	Channel	Calculation of q_m
Drawing number	FF476	$m_p = 1 \times 1.2 = 1.2$
Material	0.55 mm mild steel	$g_p = 1 \times 1 \times 1 \times 1 \times 1 \times 1.3 = 1.3$
Manufacturing process	Punching/bending	
Characteristic description	D-slot centre	Adjusted tolerance $= \dfrac{\text{Design tolerance}}{m_p \times g_p}$
Characteristic dimension	42.05 (mm)	
Design tolerance	± 0.5 (mm)	$= \dfrac{\pm 0.5}{1.2 \times 1.3} = \pm 0.321$
Surface roughness	– (μm Ra)	

$t_p = 4 \times 1 = 4$

$s_p = 1 \times 1 = 1$

42.05 ± 0.5 mm

$q_m = t_p \times s_p = 4 \times 1 = 4$

Process capability index $C_{pk} = 0.28$ Process capability index $C_p = 0.42$	Manufacturing variability risk $q_m = 4$

Component name	Plunger housing	Calculation of q_m
Drawing number	CH1878-1021	$m_p = 1 \times 1 = 1$
Material	Aluminium bronze	$g_p = 1 \times 1 \times 1 \times 1 \times 1 \times 1.1 = 1.1$
Manufacturing process	Turning	
Characteristic description	Overall length	Adjusted tolerance $= \dfrac{\text{Design tolerance}}{m_p \times g_p}$
Characteristic dimension	33.99 (mm)	
Design tolerance	±0.05 (mm)	$= \dfrac{\pm 0.05}{1 \times 1.1} = \pm 0.045$
Surface roughness	1.6 (µm Ra)	

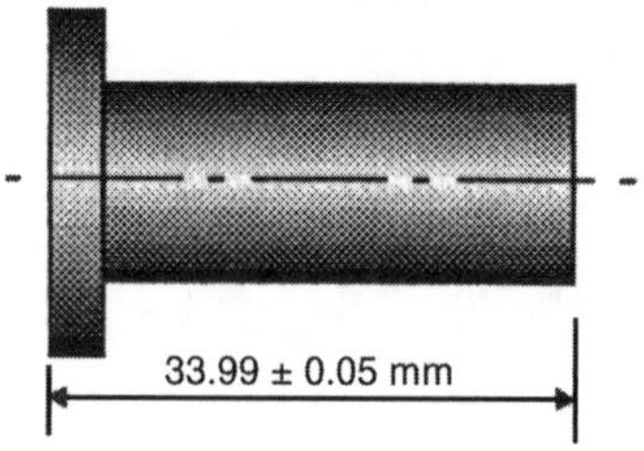

33.99 ± 0.05 mm

$t_p = 1.02 \times 1 = 1.02$

$s_p = 1 \times 1 = 1$

$q_m = t_p \times s_p = 1.02 \times 1 = 1.02$

Process capability index $C_{pk} = 2.99$ Process capability index $C_p = 3.22$	Manufacturing variability risk $q_m = 1.02$

Component name	Channel	Calculation of q_m
Drawing number	FF476	$m_p = 1 \times 1.4 = 1.4$
Material	0.55 mm mild steel	$g_p = 1 \times 1.1 \times 1 \times 1 \times 1 \times 1 = 1.1$
Manufacturing process	Punching	
Characteristic description	Profile length	Adjusted tolerance $= \dfrac{\text{Design tolerance}}{m_p \times g_p}$
Characteristic dimension	38.25 (mm)	
Design tolerance	±0.006 (mm)	$= \dfrac{\pm 0.006}{1.4 \times 1.1} = \pm 0.004$
Surface roughness	– (µm Ra)	

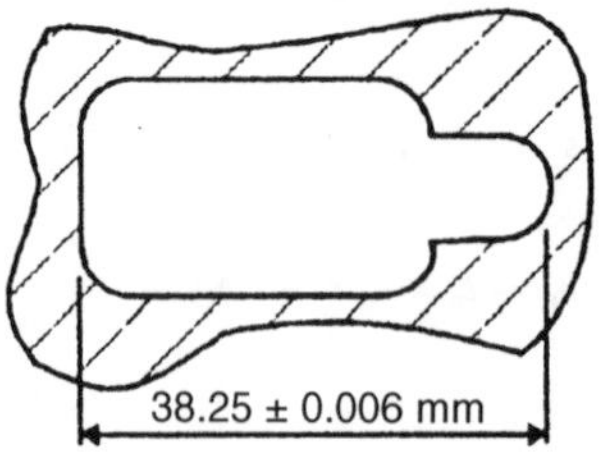

38.25 ± 0.006 mm

$t_p = 9 \times 1 = 9$

$s_p = 1 \times 1 = 1$

$q_m = t_p \times s_p = 9 \times 1 = 9$

Process capability index $C_{pk} = 0$ Process capability index $C_p = 0$	Manufacturing variability risk $q_m = 9$

Component name	Lamp holder bracket
Drawing number	FF476
Material	0.9 mm mild steel
Manufacturing process	Sheet metal bending
Characteristic description	Bottom leg length
Characteristic dimension	8.4 (mm)
Design tolerance	±0.25 (mm)
Surface roughness	– (µm Ra)

8.4 ± 0.25 mm

Process capability index $C_{pk} = 0.14$
Process capability index $C_p = 0.49$

Calculation of q_m

$m_p = 1 \times 1.2 = 1.2$

$g_p = 1 \times 1 \times 1 \times 1 \times 1 \times 1.2 = 1.2$

$$\text{Adjusted tolerance} = \frac{\text{Design tolerance}}{m_p \times g_p}$$

$$= \frac{\pm 0.25}{1.2 \times 1.2} = \pm 0.174$$

$t_p = 6 \times 1 = 6$

$s_p = 1 \times 1 = 1$

$q_m = t_p \times s_p = 6 \times 1 = 6$

Manufacturing variability risk $q_m = 6$

Component name	Lamp holder bracket
Drawing number	FQ2186
Material	0.9 mm mild steel
Manufacturing process	Sheet metal bending
Characteristic description	Leg clearance
Characteristic dimension	2.95 (mm)
Design tolerance	±0.25 (mm)
Surface roughness	– (µm Ra)

2.95 ± 0.25 mm

Process capability index $C_{pk} = 0.03$
Process capability index $C_p = 0$

Calculation of q_m

$m_p = 1 \times 1.2 = 1.2$

$g_p = 1 \times 1 \times 1.7 \times 1 \times 1 \times 1.2 = 2.04$

$$\text{Adjusted tolerance} = \frac{\text{Design tolerance}}{m_p \times g_p}$$

$$= \frac{\pm 0.25}{1.2 \times 2.04} = \pm 0.102$$

$t_p = 9 \times 1 = 9$

$s_p = 1 \times 1 = 1$

$q_m = t_p \times s_p = 9 \times 1 = 9$

Manufacturing variability risk $q_m = 9$

Component name	Lamp holder bracket	Calculation of q_m
Drawing number	FQ2186	$m_p = 1 \times 1.2 = 1.2$
Material	0.9 mm mild steel	$g_p = 1 \times 1 \times 1.7 \times 1 \times 1 \times 1.2 = 2.04$
Manufacturing process	Sheet metal bending	
Characteristic description	Top width	
Characteristic dimension	50.75 (mm)	Adjusted tolerance $= \dfrac{\text{Design tolerance}}{m_p \times g_p}$
Design tolerance	±0.25 (mm)	$= \dfrac{\pm0.25}{1.2 \times 2.04} = \pm0.102$
Surface roughness	– (μm Ra)	

50.75 ± 0.25 mm

$t_p = 9 \times 1 = 9$

$s_p = 1 \times 1 = 1$

$q_m = t_p \times s_p = 9 \times 1 = 9$

Process capability index $C_{pk} = 0$
Process capability index $C_p = 0$

Manufacturing variability risk $q_m = 9$

Component name	Capacitor clip	Calculation of q_m
Drawing number	SKSK7	$m_p = 1.7 \times 1.5 = 2.55$
Material	0.4 mm spring steel	$g_p = 1 \times 1 \times 1 \times 1.7 \times 1 \times 1.1 = 1.87$
Manufacturing process	Sheet metal bending	
Characteristic description	Overall height	
Characteristic dimension	49 (mm)	Adjusted tolerance $= \dfrac{\text{Design tolerance}}{m_p \times g_p}$
Design tolerance	±0.2 (mm)	$= \dfrac{\pm0.2}{2.55 \times 1.87} = \pm0.042$
Surface roughness	– (μm Ra)	

49 ± 0.2 mm

$t_p = 9 \times 1 = 9$

$s_p = 1 \times 1 = 1$

$q_m = t_p \times s_p = 9 \times 1 = 9$

Process capability index $C_{pk} = 0$
Process capability index $C_p = 0$

Manufacturing variability risk $q_m = 9$

Component name	Capacitor clip	**Calculation of q_m**
Drawing number	SKSK7	$m_p = 1.7 \times 1.5 = 2.55$
Material	0.45 mm spring clip	$g_p = 1 \times 1 \times 1 \times 1 \times 1 \times 1 = 1$
Manufacturing process	Sheet metal cutting	
Characteristic description	Width	Adjusted tolerance $= \dfrac{\text{Design tolerance}}{m_p \times g_p}$
Characteristic dimension	18 (mm)	
Design tolerance	± 0.2 (mm)	$= \dfrac{\pm 0.2}{2.55 \times 1} = \pm 0.078$
Surface roughness	– (µm Ra)	

$t_p = 1.3 \times 1 = 1.3$

$s_p = 1 \times 1 = 1$

$q_m = t_p \times s_p = 1.3 \times 1 = 1.3$

18 ± 0.2 mm

Process capability index $C_{pk} = 2.24$ Process capability index $C_p = 2.64$	Manufacturing variability risk $q_m = 1.3$

Component name	Lamp holder bracket	**Calculation of q_m**
Drawing number	FQ2186	$m_p = 1 \times 1.2 = 1.2$
Material	0.9 mm mild steel	$g_p = 1 \times 1 \times 1 \times 1 \times 1 \times 1.3 = 1.3$
Manufacturing process	Sheet metal bending	
Characteristic description	Bottom leg length	Adjusted tolerance $= \dfrac{\text{Design tolerance}}{m_p \times g_p}$
Characteristic dimension	10.9 (mm)	
Design tolerance	± 0.25 (mm)	$= \dfrac{\pm 0.25}{1.2 \times 1.3} = \pm 0.160$
Surface roughness	– (µm Ra)	

10.9 ± 0.25 mm

$t_p = 9 \times 1 = 9$

$s_p = 1 \times 1 = 1$

$q_m = t_p \times s_p = 9 \times 1 = 9$

Process capability index $C_{pk} = 0.03$ Process capability index $C_p = 0$	Manufacturing variability risk $q_m = 9$

Additional assembly process risk charts

A. Miscellaneous operations

Key to values: AUTOMATED (MANUAL)

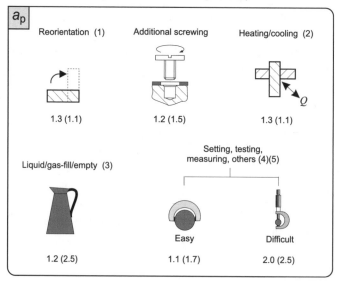

Notes

① Part may require reorientating for further assembly to take place or adjusted to achieve the required performance.

② Heating/cooling implies the application of heat, or the part is cooled to allow subsequent operations to be performed, for example heat shrinking.

③ Refers to the filling or emptying of fluids during the assembly process, for example lubricating oil.

④ These non-assembly processes can range from simple measurement tasks to complex setting operations. The user should select an appropriate value to suit the process.

⑤ Disassembly and reassembly processes (e.g. selective assembly of shims) are not covered in process considerations because they are non-value added processes and should be targeted for elimination.

B. Later mechanical deformation

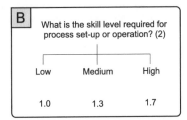

A — Is the process automated or performed manually? (1)

| Automated | Manual |
| 1.0 | 1.3 |

B — What is the skill level required for process set-up or operation? (2)

| Low | Medium | High |
| 1.0 | 1.3 | 1.7 |

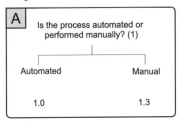

C — Is the joint easily accessible to the process? (3)

| Yes | No |
| 1.0 | 1.6 |

D — What is the type of tool motion used for deformation? (4)

| Vibration/Ultrasonic | Pressure/Orbital | Impact |
| 1.0 | 1.2 | 1.7 |

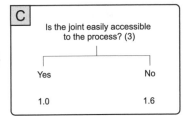

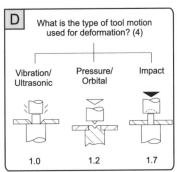

E — Is heat simultaneously applied to the part to be deformed during processing? (5)

| No | Yes |
| 1.0 | 1.3 |

F — Is the deformation process performed on a single component or on two (or more) components simultaneously? (6)

| Single | Two (or more) |
| 1.0 | 1.7 |

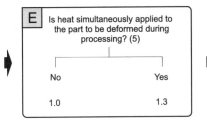

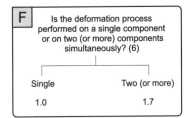

$$a_p = \boxed{A} \times \boxed{B} \times \boxed{C} \quad \boxed{D} \times \boxed{E} \times \boxed{F}$$

Notes

(1)   Automated processes use close control of time for deformation, tool positioning forces to provide consistent joint quality.

(2)   Process capability is dependent on the attention of skilled operators.

(3)   Access for the deformation tool is an important consideration. Awkward positions should be avoided and ideally processing should be performed from above.

(4)   Impact deformation increases the risk of variability.

(5)   The simultaneous application of heat during deformation may create material flow variations.

(6)   The deformation of two (or more) components simultaneously increase the risk of variability in any joint.

C. Adhesive bonding

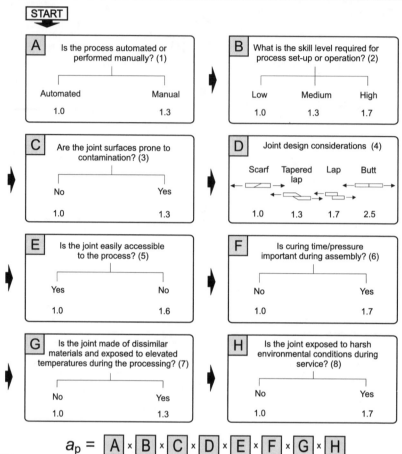

START

A Is the process automated or performed manually? (1)

Automated	Manual
1.0	1.3

B What is the skill level required for process set-up or operation? (2)

Low	Medium	High
1.0	1.3	1.7

C Are the joint surfaces prone to contamination? (3)

No	Yes
1.0	1.3

D Joint design considerations (4)

Scarf	Tapered lap	Lap	Butt
1.0	1.3	1.7	2.5

E Is the joint easily accessible to the process? (5)

Yes	No
1.0	1.6

F Is curing time/pressure important during assembly? (6)

No	Yes
1.0	1.7

G Is the joint made of dissimilar materials and exposed to elevated temperatures during the processing? (7)

No	Yes
1.0	1.3

H Is the joint exposed to harsh environmental conditions during service? (8)

No	Yes
1.0	1.7

$$a_p = \boxed{A} \times \boxed{B} \times \boxed{C} \times \boxed{D} \times \boxed{E} \times \boxed{F} \times \boxed{G} \times \boxed{H}$$

Notes

(1) Automated processes use close control of adhesive preparation, time, temperature and/or pressure to provide optimum adherence properties.

(2) Process capability is dependent on the attention of skilled operators.

(3) Surface preparation of the joint area may be required to remove oxides, grease, moisture, etc. (prior to adhesive application) by mechanical, solvent, etching or degreasing processes to facilitate wetting.

(4) Variations on the basic lap joint are preferred to give large contact areas using uniform thin sections. If at all, the joint should be loaded in shear and not pure tension. Fixtures that keep the joint rigid during processing and subsequent curing are recommended.

(5) Access for adhesive application and/or process heat/pressure is an important consideration. Awkward positions should be avoided and ideally processing should be performed from above.

(6) Sufficient time and/or pressure should be allowed before subsequent operations are performed to allow the joint to acquire strength.

(7) The use of dissimilar materials at high processing temperatures can cause adverse thermal stresses in the joint.

(8) Prolonged exposure to harsh environmental conditions, for example water, ozone, UV and elevated temperatures (>80°C), can cause severe adhesive degradation, reduction in joint strength and distortion. Limiting service temperature is about 200°C and should be observed unless special adhesives are employed.

D. Brazing and soldering

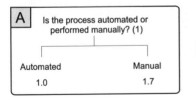

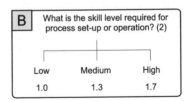

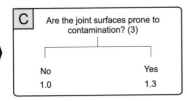

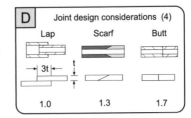

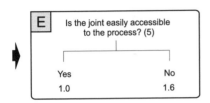

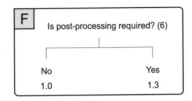

Notes

① Automated processes use close control of timing and/or temperature to provide optimum temperature gradients and uniform heating.

② Process capability is dependent on the attention of skilled operators.

③ Surface preparation of the joint area may be required to remove oxides, grease, etc. (prior to flux applications) by mechanical, pickling or degreasing processes to facilitate wetting.

④ Variations of the basic lap joint are preferred to give a large contact area using thin sections. If at all, the joint should be loaded in shear and not pure tension. Joint clearance for capillary action should be between 0.02 mm and 0.2 mm, otherwise joint strength is reduced. The use of dissimilar materials may cause uneven joint expansion and galvanic corrosion. Provision should also be made for the escape of vapours during processing.

$$a_p = \boxed{A} \times \boxed{B} \times \boxed{C} \times \boxed{D} \times \boxed{E} \times \boxed{F}$$

⑤ Access for the filler material and/or heat is an important consideration. Awkward positions should be avoided and ideally processing should be performed from above. Fixtures that keep the joint rigid during processing and subsequent cooling are preferred to allow filler solidification.

⑥ It is sometimes necessary to remove flux residues after processing due to their corrosive nature. This can be an added source of variability.

E. Resistance welding

START

A — Is the process automated or performed manually? (1)

Automated	Manual
1.0	1.7

B — Are the joint surfaces prone to contamination? (2)

No	Yes
1.0	1.3

C — Joint design considerations (3)

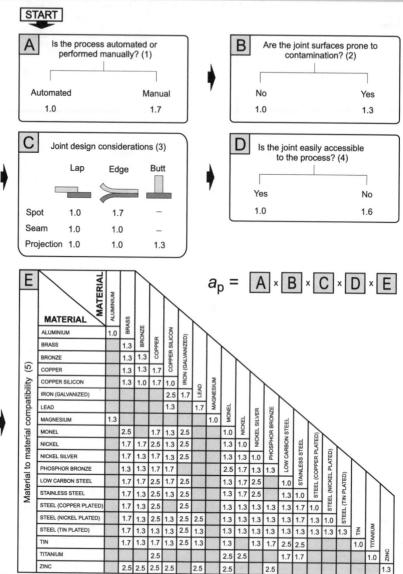

	Lap	Edge	Butt
Spot	1.0	1.7	–
Seam	1.0	1.0	–
Projection	1.0	1.0	1.3

D — Is the joint easily accessible to the process? (4)

Yes	No
1.0	1.6

$$a_p = A \times B \times C \times D \times E$$

E — Material to material compatibility (5)

MATERIAL	ALUMINIUM	BRASS	BRONZE	COPPER	COPPER SILICON	IRON (GALVANIZED)	LEAD	MAGNESIUM	MONEL	NICKEL	NICKEL SILVER	PHOSPHOR BRONZE	LOW CARBON STEEL	STAINLESS STEEL	STEEL (COPPER PLATED)	STEEL (NICKEL PLATED)	STEEL (TIN PLATED)	TIN	TITANIUM	ZINC
ALUMINIUM	1.0																			
BRASS		1.3																		
BRONZE		1.3	1.3																	
COPPER		1.3	1.3	1.7																
COPPER SILICON		1.3	1.0	1.7	1.0															
IRON (GALVANIZED)					2.5	1.7														
LEAD					1.3		1.7													
MAGNESIUM	1.3							1.0												
MONEL		2.5		1.7	1.3	2.5			1.0											
NICKEL		1.7	1.7	2.5	1.3	2.5			1.3	1.0										
NICKEL SILVER		1.7	1.3	1.7	1.3	2.5			1.3	1.3	1.0									
PHOSPHOR BRONZE		1.3	1.3	1.7	1.7				2.5	1.7	1.3	1.3								
LOW CARBON STEEL		1.7	1.7	2.5	1.7	2.5			1.3	1.7	2.5		1.0							
STAINLESS STEEL		1.7	1.3	2.5	1.3	2.5			1.3	1.7	2.5		1.3	1.0						
STEEL (COPPER PLATED)		1.7	1.3	2.5		2.5			1.3	1.3	1.3	1.3	1.3	1.7	1.0					
STEEL (NICKEL PLATED)		1.7	1.3	2.5	1.3	2.5	2.5		1.3	1.3	1.3	1.3	1.3	1.7	1.3	1.0				
STEEL (TIN PLATED)		1.7	1.3	1.3	1.3	2.5	1.3		1.3	1.3	1.3	1.3	1.3	1.3	1.3	1.3	1.0			
TIN		1.7	1.3	1.7	1.3	2.5	1.3		1.3		1.3	1.7	2.5	2.5				1.0		
TITANIUM				2.5					2.5	2.5				1.7	1.7				1.0	
ZINC		2.5	2.5	2.5	2.5		2.5		2.5			2.5								1.3

Notes

(1) Automated processes use close control of pressure, current density, weld time/force, which all reduce variability.

(2) Surface preparation of the joint area may be required to remove thick oxides, grease, paint, etc. prior to welding. Although this operation is beneficial to process performance, it can be an unwanted source of variability.

(3) Variations on the basic lap joint are preferred in spot welding, plus edge joints in seam welding to give large contact areas using uniform thin sections. Avoid very dissimilar thickness of section on the same joint as this leads to heat distortion. Butt joints only apply to projection or stud welding.

(4) Access to the weld is an important consideration. Awkward positions should be avoided and ideally processing should be performed from above.

(5) The compatibility and weldability of materials when using resistance welding has an important influence on weld quality. There is also a risk of galvanic corrosion when using dissimilar materials.

F. Fusion welding

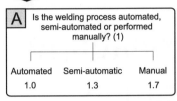

A Is the welding process automated, semi-automated or performed manually? (1)

Automated	Semi-automatic	Manual
1.0	1.3	1.7

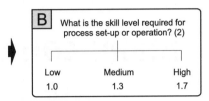

B What is the skill level required for process set-up or operation? (2)

Low	Medium	High
1.0	1.3	1.7

C Values of 'C' relating to material compatibility for selected fusion welding processes (3)

MATERIAL / PROCESS	MMA	MIG	TIG	FCAW	SAW	PAW	OAW	EBW
CARBON STEEL	1.0	1.0	1.7	1.0	1.0		1.0	
STAINLESS STEEL	1.0	1.0	1.7	1.0	1.0	1.0	1.3	1.0
IRON	1.3		1.3				1.0	1.0
NON-FERROUS ALLOYS — ALUMINIUM	1.7	1.0	1.0			1.3	1.7	1.0
NON-FERROUS ALLOYS — COPPER		1.0	1.0			1.3		1.0
NON-FERROUS ALLOYS — MAGNESIUM		1.7	1.0					1.0
NON-FERROUS ALLOYS — NICKEL	1.0	1.0	1.0		1.3	1.0	1.7	1.0
NON-FERROUS ALLOYS — TITANIUM			1.0			1.0		1.0
NON-FERROUS ALLOYS — ZINC							1.3	

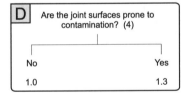

D Are the joint surfaces prone to contamination? (4)

No	Yes
1.0	1.3

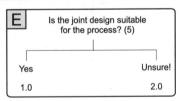

E Is the joint design suitable for the process? (5)

Yes	Unsure!
1.0	2.0

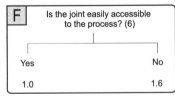

F Is the joint easily accessible to the process? (6)

Yes	No
1.0	1.6

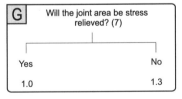

G Will the joint area be stress relieved? (7)

Yes	No
1.0	1.3

$$a_p = A \times B \times C \times D \times E \times F \times G$$

Notes

① Automated processes use close control of welding parameters such as current/power, weld speed, arc length, etc.
② Process capability is dependent on the attention of skilled operators, for example edge preparation, preheating of components and intermittent cleaning of weld area.
③ Certain materials lend themselves more readily to particular materials. Note that material composition should be monitored very closely for impurities.
④ Surface preparation may be required to remove oxides, moisture, grease, etc. Prior to welding they are removed by mechanical, pickling or degreasing processes, the application of which can be an added source of variability.
⑤ Consideration of the joint design is important with regards to processing and service issues relating to:
  - Presence of residual stress due to previous manufacturing processes.
  - Avoidance of very dissimilar section thickness.
  - Avoidance of very thin unsupported sections.
  - Provision for escape of vapours or gases during welding.
  - Uniform heat dissipation through welded parts.
  - Sufficient edge distance allowed.
  - Avoidance of welds meeting at end of runs.
  - The need for fixtures that keep the joint rigid during welding and subsequent cooling to reduce distortion, warping, etc.
  - Type of joint loading in service, either static or dynamic.
  - Post weld inspection.
⑥ Access for the filler material and/or processing heat during welding is an important consideration. Awkward positions should be avoided and ideally processing should be performed from above.
⑦ Stress relieving may be required for the restoration or improvement of the weld's strength properties.

Key to welding processes:
MMA - Manual Metal Arc Welding TIG - Tungsten Inert-gas Welding SAW - Submerged Arc Welding OAW - Oxyacetylene Welding
MIG - Metal Inert-gas Welding FCAW - Flux Cored Arc Welding PAW - Plasma Arc Welding EBW - Electron Beam Welding

Appendix VII

Blank conformability analysis tables

A. Variability risks (q_m and q_a) results table

PRODUCT NAME _____

PRODUCT CODE/ID _____

PRODUCT QUANTITY _____

SHEET No. _____

ANALYSIS REF. _____

ANALYSIS DATE _____

ENGINEER _____

Component/ assembly process reference	Component/ assembly process description	Material/ process	$q_m = t_p \cdot s_p$ (or $q_m = k_p$)									$q_a = h_p \cdot f_p$ (or $q_a = a_p$)												Total fitting process risk, f_p	Additional process risk, a_p	Total risk (q_m or q_a)	Comments
			Characteristic dimension (mm)	Design tolerance (mm)	Surface roughness (μm Ra)	Material process risk, m_p	Geometry process risk, g_p	Adjusted tolerance $= \dfrac{\text{Design tolerance}}{m_p \cdot g_p}$	Tolerance process risk, t_p	Surface roughness process risk, s_p	Surface eng. process risk, k_p	Handling process risk, h_p	Indices for fitting process risk, f_p												
													A	B	C	D	E	F	G	H				

B. Conformability matrix

PRODUCT NAME _____

PRODUCT CODE/ID _____

PRODUCT QUANTITY _____

SHEET No. _____

ANALYSIS REF. _____

ANALYSIS DATE _____

ENGINEER _____

Component/assembly process reference	Component/ assembly process description	Total risk (q_m or q_a)	Failure Mode Description and FMEA Severity Rating (S)											Comments (Including action for suppliers)
Total Failure Mode Isocost %														TOTAL FAILURE COST
Total Failure Mode Cost														

Assembly problems with two tolerances

The analysis of assembly problems with just two tolerances is common and often includes the requirement to find the potential interference or clearance between two components. The following is an example of such a problem, involving the insertion of a deep drawn brass tube and a drilled alloy steel bore, the details of which are given in Figure 1 (Haugen, 1980). It is evident from the dimensions set on the tube and bore that the tolerances assigned are anticipated to be 'worst case'.

An analysis of the brass tube is performed to assess the likely capability using CA. Referring to the process capability map for deep drawing shown in Figure 2, and determining the material, m_p, and geometry, g_p, variability risks gives:

$$\text{Adjusted tolerance} = \frac{\text{Design tolerance } (t)}{m_p \times g_p} = \frac{\pm 0.05}{(1.7 \times 1.2) \times (1.05)} = \pm 0.023 \, \text{mm}$$

A risk value of $A' = 1.3$ is interpolated from the process capability map above and because there are no surface finish constraints, $q_m = 1.3$. An estimate for the shifted standard deviation was given in Chapter 3 as:

$$\sigma'_t = \frac{t \cdot q_m^2}{12} = \frac{0.05 \times 1.3^2}{12} = 0.007 \, \text{mm}$$

Similarly, for the alloy steel bore the process capability map for drilling is shown in Figure 3. Note that the tolerance in the case of a drilled dimension is given as a '+' only:

$$\text{Adjusted tolerance} = \frac{\text{Design tolerance } (T)}{m_p \cdot g_p} = \frac{+0.05}{(1 \times 1.5) \times (1.05)} = +0.032 \, \text{mm}$$

The standard deviation for one half of the tolerance can be estimated by:

$$\sigma'_b \approx \frac{\left(\dfrac{T}{2}\right) \cdot q_m^2}{12} = \frac{0.025 \times 2.5^2}{12} = 0.013 \, \text{mm}$$

Using the algebra of random variables we can solve the probability of interference between the two tolerance distributions, assuming that the variables follow a

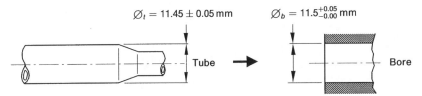

Figure 1 Details of the tube and bore

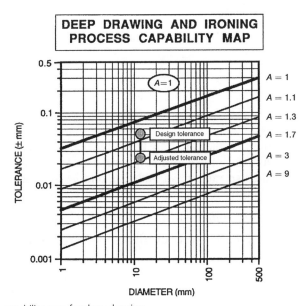

Figure 2 Process capability map for deep drawing

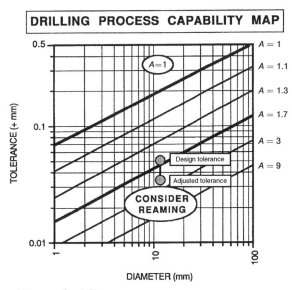

Figure 3 Process capability map for drilling

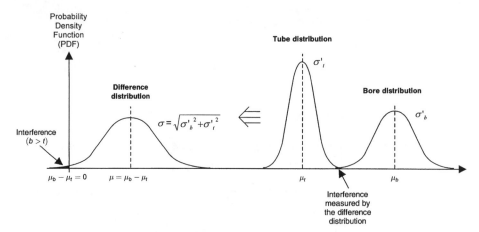

Figure 4 Statistical representation of the interference problem

Normal distribution and that they are statistically independent. The problem is governed, in fact, by the amount of negative clearance. This translates into a difference between the mean of the two random variables, μ_t and μ_b (see Figure 4). The mean of this difference distribution, μ, is given by:

$$\mu = \mu_b - \mu_t = 11.525 - 11.45 = 0.075\,\text{mm}$$

The standard deviation of the difference distribution between the random variables is given by:

$$\sigma = (\sigma_t'^2 + \sigma_b'^2)^{0.5} = (0.007^2 + 0.013^2)^{0.5} = 0.015\,\text{mm}$$

From the Standard Normal Distribution (SND) it is possible to determine the probability of negative clearance, P.

The area under the curve to the left of $\mu_b - \mu_t = 0$ relates to the probability of negative clearance. This area can be found from the SND table (Table 1, Appendix I) by determining the Standard Normal variate, z, where:

$$z = \left(\frac{x - \mu}{\sigma}\right) = \left(\frac{0 - 0.075}{0.015}\right) = -5.00$$

The area $\Phi_{\text{SND}}(z)$ and hence the probability of negative clearance $P \approx 0.0000003$. Therefore, the probability that the parts will interfere on assembly is practically negligible.

Also see Haugen (1980), Kolarik (1995) and Shigley (1986) for examples of this type.

Appendix IX

Properties of continuous distributions

A. Probability Density Functions (PDF)

Normal distribution

$$f(x) = \frac{1}{\sigma\sqrt{2\pi}} \exp\left(-\frac{(x-\mu)^2}{2\sigma^2}\right) \qquad -\infty < x < \infty$$

where μ = mean and σ = standard deviation.

Lognormal distribution

$$f(x) = \frac{1}{\alpha x\sqrt{2\pi}} \exp\left(-\frac{(\ln(x)-\lambda)^2}{2\alpha^2}\right) \qquad 0 < x < \infty$$

where λ = mean and α = dispersion.

2-parameter Weibull distribution

$$f(x) = \left(\frac{\beta}{\theta}\right)\left(\frac{x}{\theta}\right)^{\beta-1} \exp\left(-\left(\frac{x}{\theta}\right)^{\beta}\right) \qquad x \geq 0$$

where θ = characteristic value and β = shape parameter.

($\beta = 1$, equivalent to the Exponential distribution.)

3-parameter Weibull distribution

$$f(x) = \left(\frac{\beta}{\theta - xo}\right)\left(\frac{x - xo}{\theta - xo}\right)^{\beta-1} \exp\left(-\left(\frac{x - xo}{\theta - xo}\right)^{\beta}\right) \qquad x \geq 0$$

where $xo = $ expected minimum value, $\theta = $ characteristic value, and $\beta = $ shape parameter.

($\beta = 1$, and $xo = 0$, equivalent to the Exponential distribution.
$\beta = 1.5$ approximates to the Lognormal distribution.
$\beta = 3.44$ approximates to the Normal distribution.)

Maximum Extreme Value Type I distribution

$$f(x) = \left(\frac{1}{\Theta}\right) \exp\left(-\left(\frac{x - v}{\Theta}\right) - \exp\left(-\left(\frac{x - v}{\Theta}\right)\right)\right) \qquad -\infty < x < \infty$$

where $\Theta = $ scale parameter and $v = $ location parameter.

Minimum Extreme Value Type I distribution

$$f(x) = \left(\frac{1}{\Theta}\right) \exp\left(\left(\frac{x - v}{\Theta}\right) - \exp\left(\frac{x - v}{\Theta}\right)\right) \qquad -\infty < x < \infty$$

where $\Theta = $ scale parameter and $v = $ location parameter.

Exponential distribution

$$f(x) = \left(\frac{1}{\theta}\right) \exp\left(-\left(\frac{x}{\theta}\right)\right) \qquad x \geq 0$$

where $\theta = $ characteristic value.

B. Equivalent mean (μ) and standard deviation (σ)

Lognormal distribution

$$\mu = \exp\left(\lambda + \frac{\alpha^2}{2}\right) \qquad \sigma = (\exp(2\lambda + 2\alpha^2) - \exp(2\lambda + \alpha^2))^{0.5}$$

2-parameter Weibull distribution

$$\mu \approx \theta \cdot \Gamma\left[1 + \frac{1}{\beta}\right] \qquad \sigma \approx \mu \cdot \beta^{-0.926}$$

See Figure 1 for a graph of $\Gamma[1 + 1/\beta]$ for values of $\beta \geq 1$.

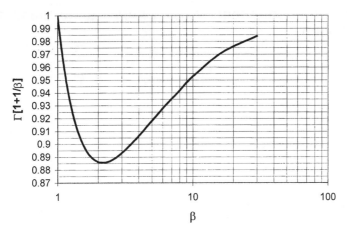

Figure 1 The factor $\Gamma[1 + 1/\beta]$ as a function of β

3-parameter Weibull distribution

$$\mu \approx xo + (\theta - xo) \cdot \Gamma\left[1 + \frac{1}{\beta}\right] \qquad \sigma \approx (\mu - xo) \cdot \beta^{-0.926}$$

Maximum Extreme Value Type I distribution

$$\mu \approx \upsilon + 0.5772157\Theta \qquad \sigma \approx 1.2825498\Theta$$

Minimum Extreme Value Type I distribution

$$\mu \approx \upsilon - 0.5772157\Theta \qquad \sigma \approx 1.2825498\Theta \qquad .$$

Note the parameters for the 3-parameter Weibull distribution, xo and θ, can be estimated given the mean, μ, and standard deviation, σ, for a Normal distribution (assuming $\beta = 3.44$) by:

$$xo \approx \mu - 3.1394473\sigma \qquad \theta \approx \mu + 0.3530184\sigma$$

C. Cumulative Distribution Functions (CDF)

The cumulative distribution function, $F(x)$, can be obtained by integrating the probability density function, $f(x)$, between 0 and the limit of interest, x.

$$F(x) = \int_0^x f(x)\,dx$$

Normal distribution (integrate using numerical techniques or use SND table)

$$F(x) = \frac{1}{\sigma\sqrt{2\pi}} \int_0^x \exp\left(-\frac{(x-\mu)^2}{2\sigma^2}\right)$$

where μ = mean and σ = standard deviation.

Using the SND table in Appendix I:

$$F(x) = \Phi_{\text{SND}}(z)$$

where $z = \left(\dfrac{x-\mu}{\sigma}\right)$.

Lognormal distribution (integrate using numerical techniques or use SND table)

$$F(x) = \frac{1}{\alpha x\sqrt{2\pi}} \int_0^x \exp\left(-\frac{(\ln(x)-\lambda)^2}{2\alpha^2}\right)$$

where λ = mean and α = dispersion.

Using the SND table in Appendix I:

$$F(x) = \Phi_{\text{SND}}(z)$$

where $z = \left(\dfrac{\ln(x)-\lambda}{\alpha}\right)$.

2-parameter Weibull distribution

$$F(x) = 1 - \exp\left(-\left(\frac{x}{\theta}\right)^\beta\right)$$

where θ = characteristic value and β = shape parameter.

3-parameter Weibull distribution

$$F(x) = 1 - \exp\left(-\left(\frac{x-xo}{\theta-xo}\right)^\beta\right) \qquad (x \geq xo)$$

where xo = expected minimum value, θ = characteristic value, and β = shape parameter.

Maximum Extreme Value Type I distribution

$$F(x) = \exp\left(-\exp\left(-\frac{x-\upsilon}{\Theta}\right)\right)$$

where Θ = scale parameter and υ = location parameter.

Minimum Extreme Value Type I distribution

$$F(x) = 1 - \exp\left(-\exp\left(\frac{x-\upsilon}{\Theta}\right)\right)$$

where Θ = scale parameter and υ = location parameter.

Exponential distribution

$$F(x) = 1 - \exp\left(-\left(\frac{x}{\theta}\right)\right)$$

where θ = characteristic value.

Appendix X

Fitting distributions to data using linear regression

A. Cumulative ranking equations

$$F_i = \frac{i}{N}$$

– used for very large samples;

$$F_i = \frac{i - 0.5}{N}$$

– *Hazen formula* (commonly used in engineering giving reliable positions);

$$F_i = \frac{i}{N + 1}$$

– *mean rank* (can be used with small samples, commonly used for estimating the mean and standard deviation);

$$F_i = \frac{i - 0.4}{N + 0.2}$$

– a reasonable approximation for any distribution;

$$F_i = \frac{i - 0.35}{N}$$

– commonly used for the Extreme Value Type I distribution;

$$F_i = \frac{i - 0.3}{N + 0.4}$$

– *median rank* (commonly used for the Weibull distribution);

$$F_i = \frac{i - 0.3175}{N + 0.365}$$

– a reasonable approximation for the Normal distribution;

where F_i = ranked value for ith variable, i = cumulative frequency of variable, and N = population.

B. Linear rectification equations and plotting positions

Distribution	x-axis	y-axis
Normal	x_i	$\Phi_{\text{SND}}^{-1}(F_i)$
Lognormal	$\ln(x_i)$	$\Phi_{\text{SND}}^{-1}(F_i)$
2-parameter Weibull	$\ln(x_i)$	$\ln \ln \left(\dfrac{1}{1 - F_i} \right)$
3-parameter Weibull	$\ln(x_i - xo)$	$\ln \ln \left(\dfrac{1}{1 - F_i} \right)$
Maximum Extreme Value Type I	x_i	$-\ln \ln \left(\dfrac{1}{F_i} \right)$
Minimum Extreme Value Type I	x_i	$\ln \ln \left(\dfrac{1}{1 - F_i} \right)$
Exponential	x_i	$\ln \left(\dfrac{1}{1 - F_i} \right)$

where $x_i = i$th variable, $F_i =$ ranked value for ith variable, $\Phi_{\text{SND}}^{-1} =$ inverse function of the Standard Normal distribution, and $xo =$ expected minimum value (for the 3-parameter Weibull distribution).

C. Distribution parameters from linear regression constants A0 and A1

Normal distribution

$$\mu = -\left(\frac{A0}{A1} \right) \qquad \sigma = \left(\frac{1 - A0}{A1} \right) + \left(\frac{A0}{A1} \right)$$

Lognormal distribution

$$\lambda = -\left(\frac{A0}{A1} \right) \qquad \alpha = \left(\frac{1 - A0}{A1} \right) + \left(\frac{A0}{A1} \right)$$

2-parameter Weibull distribution

$$\theta = \exp \left(-\frac{A0}{A1} \right) \qquad \beta = A1$$

3-parameter Weibull distribution (*xo* determined from best linear fit)

$$\theta = \exp\left(-\frac{A0}{A1}\right) + xo \qquad \beta = A1$$

Maximum Extreme Value Type I distribution

$$\Theta = \frac{1}{A1} \qquad \upsilon = -\left(\frac{A0}{A1}\right)$$

Minimum Extreme Value Type I distribution

$$\Theta = \frac{1}{A1} \qquad \upsilon = -\left(\frac{A0}{A1}\right)$$

Exponential distribution

$$\theta = \frac{1}{A1}$$

Solving the variance equation

A. Partial derivative method

When the function is a combination of more than two statistically independent variables, x_i, then the variance equation below (ignoring second order terms) can be used to determine the variance and hence the standard deviation, σ_ϕ, of the function:

$$\sigma_\phi \approx \left(\sum_{i=1}^{n} \left(\frac{\partial \phi}{\partial x_i} \right)^2 \cdot \sigma_{x_i}^2 \right)^{0.5}$$

In order to solve the equation for σ_ϕ, it is only necessary to find the partial derivative of the function with respect to each variable. This may be simple for some functions, but is more difficult the more complex the function becomes and other techniques may be more suitable.

Suppose we are interested in knowing the distribution of stress associated with the tensile static loading on a rectangular bar (see Figure 1). The governing stress is given by:

$$L = \frac{F}{ab}$$

where L = stress, F = load, and a and b = sectional dimensions of the bar.

The variables F, a and b are all assumed to random in nature following the Normal distribution with parameters:

$$F \sim N(100, 10)\,\text{kN} \qquad a \sim N(0.03, 0.0003)\,\text{m} \qquad b \sim N(0.05, 0.0004)\,\text{m}$$

Using the variance equation, the standard deviation of the stress function becomes:

$$\sigma_L = \left[\left(\frac{\partial L}{\partial F} \right)^2 \cdot \sigma_F^2 + \left(\frac{\partial L}{\partial a} \right)^2 \cdot \sigma_a^2 + \left(\frac{\partial L}{\partial b} \right)^2 \cdot \sigma_b^2 \right]^{0.5}$$

Taking each term separately:

$$\frac{\partial L}{\partial F} = \frac{\partial \left(\dfrac{F}{ab} \right)}{\partial F} = \frac{1}{ab} = \frac{1}{\mu_a \mu_b}$$

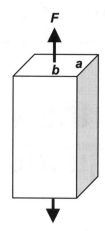

Figure 1 Tensile loading on a rectangular bar

$$\frac{\partial L}{\partial a} = \frac{\partial \left(\dfrac{F}{ab}\right)}{\partial a} = -\frac{F}{a^2 b} = -\frac{\mu_F}{\mu_a^2 \mu_b}$$

$$\frac{\partial L}{\partial b} = \frac{\partial \left(\dfrac{F}{ab}\right)}{\partial b} = -\frac{F}{ab^2} = -\frac{\mu_F}{\mu_a \mu_b^2}$$

Substituting these terms back into the variance equation gives:

$$\sigma_L = \left[\left(\frac{1}{\mu_a \mu_b}\right)^2 \cdot \sigma_F^2 + \left(-\frac{\mu_F}{\mu_a^2 \mu_b}\right)^2 \cdot \sigma_a^2 + \left(-\frac{\mu_F}{\mu_a \mu_b^2}\right)^2 \cdot \sigma_b^2\right]^{0.5}$$

Substituting in values for μ and σ for each variable gives:

$$\sigma_L = \left[\left(\frac{1}{0.03 \times 0.05}\right)^2 \times 10\,000^2 + \left(-\frac{100\,000}{0.03^2 \times 0.05}\right)^2 \times 0.0003^2\right.$$
$$\left. + \left(-\frac{100\,000}{0.03 \times 0.05^2}\right)^2 \times 0.0004^2\right]^{0.5}$$

$$\sigma_L = 6.72\,\text{MPa}$$

The mean stress, μ_L, can be approximated by substituting the mean values for each variable into the original stress function, where:

$$\mu_\phi \approx \phi(\mu_{x_1}, \mu_{x_2}, \ldots, \mu_{x_n})$$

Therefore:

$$\mu_L \approx \frac{F}{ab} = \frac{\mu_F}{\mu_a \cdot \mu_b} = \frac{100\,000}{0.03 \times 0.05}$$

$$\mu_L = 66.66\,\text{MPa}$$

The stress can be approximated by a Normal distribution with parameters:

$$L \sim N(66.66, 6.72)\,\text{MPa}$$

Determining the second order terms in the variance equation will provide more accurate values for the mean and standard deviation of the function under certain circumstances.

B. Finite difference method

Substituting $y = \phi(x_1, x_2, \ldots, x_n)$ into the variance equation for the output of the function, and expanding for n variables gives:

$$\sigma_y \approx \left(\sum_{i=1}^{n} \left(\frac{\partial y}{\partial x_i} \right)^2 \cdot \sigma_{x_i}^2 \right)^{0.5}$$

$$= \left[\left(\frac{\partial y}{\partial x_1} \right)^2 \cdot \sigma_{x_1}^2 + \left(\frac{\partial y}{\partial x_2} \right)^2 \cdot \sigma_{x_2}^2 + \cdots + \left(\frac{\partial y}{\partial x_n} \right)^2 \cdot \sigma_{x_n}^2 \right]^{0.5}$$

The finite difference method can be used to approximate each term in this equation by using the difference equation for the first partial derivative (see Figure 2). The values of the function at two points either side of the point of interest, k, are determined, y_{k+1} and y_{k-1}, which are equally spaced by an increment Δx. The finite difference equation approximates the value of the partial derivative by taking the difference of these values and dividing by the increment range. The terms subscripted by i indicate that only x_i is incremented by Δx_i for calculating y_{k+1} and y_{k-1}, holding the other independent variables constant at their k value points.

$$\left(\frac{\partial y}{\partial x_i} \right)_k \approx \frac{(y_{k+1} - y_{k-1})_i}{2\Delta x_i}$$

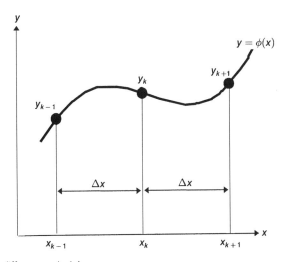

Figure 2 The finite difference principle

The expanded variance equation to determine the standard deviation of the stress in the rectangular bar problem above was:

$$\sigma_L = \left[\left(\frac{\partial L}{\partial F} \right)^2 \cdot \sigma_F^2 + \left(\frac{\partial L}{\partial a} \right)^2 \cdot \sigma_a^2 + \left(\frac{\partial L}{\partial b} \right)^2 \cdot \sigma_b^2 \right]^{0.5}$$

The variables F, a and b were are all assumed to be random in nature following the Normal distribution with parameters:

$$F \sim N(100, 10)\ kN \qquad a \sim N(0.03, 0.0003)\ m \qquad b \sim N(0.05, 0.0004)\ m$$

Consider the point x_k to be the mean value of a variable and x_{k+1} and x_{k-1} the extremes of the variable. The extremes can be determined for each variable by assuming they exist $\pm 4\sigma$ away from the mean which covers 99.99% of situations. For each variable in the problem above, the extremes become:

$$F_{k+1} = 100\,000 + 4(10\,000) = 140\,000\ \text{MPa}$$

$$F_{k-1} = 100\,000 - 4(10\,000) = 60\,000\ \text{MPa}$$

$$\Delta F = 4(10\,000) = 40\,000\ \text{MPa}$$

$$a_{k+1} = 0.03 + 4(0.0003) = 0.0312\ \text{m}$$

$$a_{k-1} = 0.03 - 4(0.0003) = 0.0288\ \text{m}$$

$$\Delta a = 4(0.0003) = 0.0012\ \text{m}$$

$$b_{k+1} = 0.05 + 4(0.0004) = 0.0516\ \text{m}$$

$$b_{k-1} = 0.05 + 4(0.0004) = 0.0484\ \text{m}$$

$$\Delta b = 4(0.0004) = 0.0016\ \text{m}$$

For the first variable, F, in the variance equation:

$$\left(\frac{\partial L}{\partial F} \right) \approx \frac{(y_{k+1} - y_{k-1})}{2\Delta F}$$

Letting the variable F be its maximum value and the variables a and b kept at their mean values gives y_{k+1} when applied to the stress function:

$$y_{k+1} = \frac{140\,000}{0.03 \times 0.05} = 93\,333\,333 \quad \text{and} \quad y_{k-1} = \frac{60\,000}{0.03 \times 0.05} = 40\,000\,000$$

Therefore:

$$\left(\frac{\partial L}{\partial F} \right) = \frac{93\,333\,333 - 40\,000\,000}{2 \times 40\,000} = 666.66$$

For the second variable, a, in the variance equation:

$$\left(\frac{\partial L}{\partial a} \right) \approx \frac{\left(y_{k+1} - y_{k-1} \right)}{2\Delta a}$$

$$y_{k+1} = \frac{100\,000}{0.0312 \times 0.05} = 64\,102\,564$$

$$y_{k-1} = \frac{100\,000}{0.0288 \times 0.05} = 69\,444\,444$$

$$\left(\frac{\partial L}{\partial a}\right) = \frac{64\,102\,564 - 69\,444\,444}{2 \times 0.0012} = -2.226 \times 10^9$$

For the third variable, b, in the variance equation:

$$\left(\frac{\partial L}{\partial b}\right) \approx \frac{(y_{k+1} - y_{k-1})}{2\Delta b}$$

$$y_{k+1} = \frac{100\,000}{0.03 \times 0.0516} = 64\,599\,483$$

$$y_{k-1} = \frac{100\,000}{0.03 \times 0.0484} = 68\,870\,523$$

$$\left(\frac{\partial L}{\partial b}\right) = \frac{64\,599\,483 - 68\,870\,523}{2 \times 0.0016} = -1.335 \times 10^9$$

Substituting into the expanded variance equation gives:

$$\sigma_L = [(666.66^2 \times 10\,000^2) + ((-2.226 \times 10^9)^2 \times 0.0003^2)$$

$$+ ((-1.335 \times 10^9) \times 0.0004^2)]^{0.5}$$

$$\sigma_L = 6.72\,\text{MPa}$$

Again, the mean stress, μ_L can be approximated by substituting the mean values for each variable into the original stress function:

$$\mu_\phi \approx \phi(\mu_{x_1}, \mu_{x_2}, \ldots, \mu_{x_n})$$

Therefore:

$$\mu_L \approx \frac{F}{ab} = \frac{\mu_F}{\mu_a \cdot \mu_b} = \frac{100\,000}{0.03 \times 0.05}$$

$$\mu_L = 66.66\,\text{MPa}$$

The stress can be approximated by a Normal distribution with parameters:

$$L \sim N(66.66, 6.72)\,\text{MPa}$$

If second order partial derivatives are required for a more accurate solution of the variance equation, then these terms can be approximated by:

$$\left(\frac{\partial^2 y}{\partial x_i^2}\right)_k \approx \frac{(y_{k-1} + y_{k+1} - 2y_k)_i}{(\Delta x_i)^2}$$

Solving for the second order terms in the variance equation for both the mean and standard deviation is shown below for the above example:

$$\left(\frac{\partial^2 L}{\partial F^2}\right) = \frac{40\,000\,000 + 93\,333\,333 - 2(66\,666\,666)}{40\,000^2} = 6.25 \times 10^{-10}$$

$$\left(\frac{\partial^2 L}{\partial a^2}\right) = \frac{69\,444\,444 + 64\,102\,564 - 2(66\,666\,666)}{0.0012^2} = 1.4839 \times 10^{11}$$

$$\left(\frac{\partial^2 L}{\partial b^2}\right) = \frac{68\,870\,523 + 64\,599\,483 - 2(66\,666\,666)}{0.0016^2} = 5.3388 \times 10^{10}$$

The variance equation with second order terms is:

$$\sigma_\phi \approx \left(\sum_{i=1}^{n}\left(\frac{\partial\phi}{\partial x_i}\right)^2 \cdot \sigma_{x_i}^2 + \frac{1}{2}\sum_{i=1}^{n}\left(\frac{\partial^2\phi}{\partial x_i^2}\right)^2 \cdot \sigma_{x_i}^4\right)^{0.5}$$

Therefore, substituting in values gives:

$$\sigma_L = \begin{bmatrix} (666.66^2 \times 10\,000^2) + (\frac{1}{2}(6.25 \times 10^{-10})^2 \times 10\,000^4) \\ +((-2.226 \times 10^9)^2 \times 0.0003^2) + (\frac{1}{2}(1.4839 \times 10^{11})^2 \times 0.0003^4) \\ +((-1.335 \times 10^9)^2 \times 0.0004^2) + (\frac{1}{2}(5.3388 \times 10^{10})^2 \times 0.0004^4) \end{bmatrix}^{0.5} = 6.72\,\text{MPa}$$

Using the first and second order terms in the variance equation gives exactly the same answer. For different conditions, say where one variable is not dominating the situation as above for the load, then the use of the variance equation with second order terms will be more effective.

Continuing for the mean value:

$$\mu_\phi \approx \phi(\mu_{x_1}, \mu_{x_2}, \ldots, \mu_{x_n}) + \frac{1}{2}\sum_{i=1}^{n}\frac{\partial^2\phi}{\partial x_i^2} \cdot \sigma_{x_i}^2$$

$$\mu_L = 66.66 \times 10^6 + \frac{1}{2}[(6.25 \times 10^{-10}) \times 10\,000^2 + (1.4839 \times 10^{11})$$

$$\times 0.0003^2 + (5.3388 \times 10^{10}) \times 0.0004^2]$$

$$\mu_L = 66.68\,\text{MPa}$$

Therefore, the stress can be approximated by a Normal distribution with parameters:

$$L \sim N(66.68, 6.72)\,\text{MPa}$$

C. Monte Carlo simulation

Monte Carlo simulation is a numerical experimentation technique to obtain the statistics of the output variables of a function, given the statistics of the input variables. In each experiment or trial, the values of the input random variables are sampled based on their distributions, and the output variables are calculated using the computational model. The generation of a set of random numbers is central to the technique, which can then be used to generate a random variable from a given distribution. The simulation can only be performed using computers due to the large number of trials required.

Rather than solve the variance equation for a number of variables directly, this method allows us to simulate the output of the variance, for example the simulated dispersion of a stress variable given that the random variables in the problem can be characterized.

Typically, the random values, x, from a particular distribution are generated by inverting the closed form CDF for the distribution $F(x)$ representing the random variable, where:

$$x = F^{-1}(u)$$

where $u = $ random number generated from 0 to 1.

Many distributions can be represented in closed form except for the Normal and Lognormal types. The CDF for these distributions can only be determined numerically. For example, the 3-parameter Weibull distribution's CDF is in closed form, where:

$$F(x) = 1 - \exp\left(-\left(\frac{x - xo}{\theta - xo}\right)^{\beta}\right)$$

where $xo = $ expected minimum value, $\theta = $ characteristic value, and $\beta = $ shape parameter.

Therefore, for the 3-parameter Weibull distribution, the inverse CDF with respect to u is:

$$x = xo + (\theta - xo)[-\ln(1 - u)]^{1/\beta}$$

We can use Monte Carlo simulation to determine the mean and standard deviation of a function with knowledge of the mean and standard deviation of the input variables. Returning to the problem of the tensile stress distribution in the rectangular bar, the stress was given by:

$$L = \frac{F}{ab}$$

where $L = $ stress, $F = $ load, and a and $b = $ sectional dimensions of the bar.

The variables L, a and b were all assumed to be random in nature following the Normal distribution with parameters:

$$F \sim N(100, 10)\,\text{kN} \qquad a \sim N(0.03, 0.0003)\,\text{m} \qquad b \sim N(0.05, 0.0004)\,\text{m}$$

Because the Normal distribution is difficult to work with, we can use a 3-parameter Weibull distribution as an approximating model. Given the mean, μ, and standard deviation, σ, for a Normal distribution (assuming $\beta = 3.44$), the parameters xo and θ can be determined from:

$$xo \approx \mu - 3.1394473\sigma \qquad \theta \approx \mu + 0.3530184\sigma$$

However, using these approximations, it is still assumed that the output variable will be a Normal distribution.

For each variable F, a and b in the stress equation, therefore:

$$xo_F = 100\,000 - 3.1394473(10\,000) = 68\,605.5\,\text{N}$$

$$\theta_F = 100\,000 + 0.3530184(10\,000) = 103\,530.2\,\text{N}$$

$$\beta_F = 3.44$$

$$xo_a = 0.03 - 3.1394473(0.0003) = 0.02906\,\text{m}$$

$$\theta_a = 0.03 + 0.3530184(0.0003) = 0.03011\,\text{m}$$

$$\beta_a = 3.44$$

$$xo_b = 0.05 - 3.1394473(0.0004) = 0.04874\,\text{m}$$

$$\theta_b = 0.05 + 0.3530184(0.0004) = 0.05014\,\text{m}$$

$$\beta_b = 3.44$$

The inverse CDF is then used to generate the random numbers following the 3-parameter Weibull distribution, as given earlier. This is shown below in the subroutine for a Monte Carlo simulation using 10 000 trials. The subroutine is written in Visual Basic, but can be easily translated to other computer languages. It requires the declaration of two label objects to display the mean and standard deviation. Running the program gives the mean and standard deviation of the stress as:

$$L \sim N(66.59, 6.74)\,\text{MPa} \quad \text{(for one particular set of trials)}$$

Monte Carlo simulation is useful for complex equations; however, when possible it should be supported by the use of Finite Difference Methods.

Monte Carlo Simulation code (written in Visual Basic)

```
Dim L() As Variant                                              (declaration of variables)
Dim SUM As Variant
Dim VAR As Variant
Dim MEAN As Variant
Dim STD As Variant

Dim F As Variant                                                (declaration of variables
Dim a As Variant                                                 for function)
Dim b As Variant
ReDim L(10000)                                                  (number of trials)

For I% = 1 To 10000
   Let F = 68605.5 + (103530.2 − 68605.5) * ((−Log(1 − Rnd)) ^ 0.2907)  (inverse CDF's for
   Let a = 0.02906 + (0.03011 − 0.02906) * ((−Log(1 − Rnd)) ^ 0.2907)    each variable)
   Let b = 0.04874 + (0.05014 − 0.04874) * ((−Log(1 − Rnd)) ^ 0.2907)
   Let L(I%) = F / (a * b)                                      (function goes here)
Next I%

Let SUM = 0
Let ST = 0
Let VAR = 0

For I% = 1 To 10000                                             (equations to determine
   Let SUM = SUM + L(I%)                                          the mean and
Next I%                                                          standard deviation)

Let MEAN = SUM / 10000

For I% = 1 To 10000
   Let VAR = VAR + (L(I%) − MEAN) ^ 2
Next I%

Let STD = Sqr(VAR / 10000)
Label1.Caption = Format(MEAN, ''#0.000000'')                    (output mean)
Label2.Caption = Format(STD, ''#0.000000'')                     (output standard deviation)
```

D. Sensitivity analysis

Three different methods can be used to solve the variance equation as previously shown. The variance equation is also a valuable tool with which to draw sensitivity inferences, providing the contribution of each variable to the overall variability and so determining the key design variables.

For example, suppose we are interested in knowing the variance contribution of each variable in the stress associated with the tensile loading on a rectangular bar. The governing stress was determined by:

$$L = \frac{F}{ab}$$

where L = stress, F = load, and a and b = sectional dimensions of the bar.

The variables F, a and b have Normal parameters:

$$F \sim N(100, 10)\,\text{kN} \qquad a \sim N(0.03, 0.0003)\,\text{m} \qquad b \sim N(0.04, 0.0004)\,\text{m}$$

Using the variance equation, the standard deviation of the stress function can be found using the method of partial derivatives:

$$\sigma_L = \left[\left(\frac{1}{\mu_a \mu_b} \right)^2 \cdot \sigma_F^2 + \left(-\frac{\mu_F}{\mu_a^2 \mu_b} \right)^2 \cdot \sigma_a^2 + \left(-\frac{\mu_F}{\mu_a \mu_b^2} \right)^2 \cdot \sigma_b^2 \right]^{0.5}$$

Substituting the values in the variance equation gives:

$$\sigma_L = \left[\left(\frac{1}{0.03 \times 0.05} \right)^2 \times 10\,000^2 + \left(-\frac{100\,000}{0.03^2 \times 0.05} \right)^2 \right.$$

$$\left. \times 0.0003^2 + \left(-\frac{100\,000}{0.03 \times 0.05^2} \right)^2 \times 0.0004^2 \right]^{0.5}$$

$$\sigma_L^2 = 4.4444 \times 10^{13} + 4.4444 \times 10^{11} + 2.8444 \times 10^{11}$$

$$\sigma_L^2 = 4.5173 \times 10^{13}$$

This value equals the variance of the stress, or the standard deviation squared. The variance contribution in percent of the load variable, F, to the variance of the stress then becomes:

$$\frac{4.4444 \times 10^{13}}{4.5173 \times 10^{13}} \times 100 = 98.39\%$$

The variance contribution in percent of the dimensional variable, a, to the variance of the stress then becomes:

$$\frac{4.4444 \times 10^{11}}{4.5173 \times 10^{13}} \times 100 = 0.98\%$$

The variance contribution in percent of the dimensional variable, b, to the variance of the stress then becomes:

$$\frac{2.8444 \times 10^{11}}{4.5173 \times 10^{13}} \times 100 = 0.63\%$$

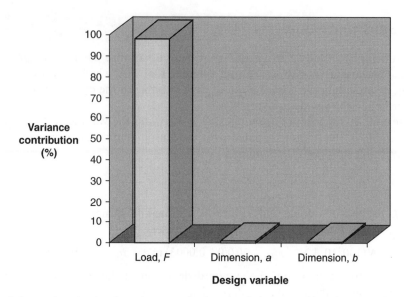

Figure 3 Pareto chart showing the variance contribution of each design variable in the tension bar problem

Plotting this data as a Pareto chart gives Figure 3. It shows that the load is the dominant variable in the problem and so the stress is very sensitive to changes in the load, but the dimensional variables have little impact on the problem. Under conditions where the standard deviation of the dimensional variables increased for whatever reason, their impact on the stress distribution would increase to the detriment of the contribution made by the load if its standard deviation remained the same.

Simpson's Rule for numerical integration

Typically in engineering we are required to find the area under a curve where $y = f(x)$ between limits as shown in Figure 1. Direct integration is sometimes difficult, and the use of numerical integration techniques helps in this respect.

One commonly used technique which gives high accuracy is Simpson's Rule (more correctly called Simpson's $\frac{1}{3}$ Rule). Here the area under the curve is divided into equal segments of width, h. For an even number of segments, m, we can divide the range of interest (MAX − MIN) into ordinates x_0 to x_m, where the number of ordinates is odd. At each ordinate the functional values are then determined, for example $y_0 = f(x_0)$ as shown in Figure 2 using four segments ($m = 4$).

Simpson's method then uses second order polynomials to connect sets of three ordinates representing segment pairs to determine the composite area. The total area, A, under the curve $y = f(x)$ between the limits is then approximated by the following equation:

$$A = \left(\frac{h}{3}\right)\left[f(x_0) + 4\sum_{i=1,3,5,\ldots}^{m-1} f(x_i) + 2\sum_{j=2,4,6,\ldots}^{m-2} f(x_j) + f(x_m)\right]$$

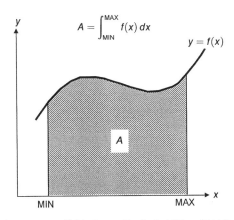

$$A = \int_{MIN}^{MAX} f(x)\, dx$$

Figure 1 Area, A, under the curve $y = f(x)$ between the limits MIN and MAX

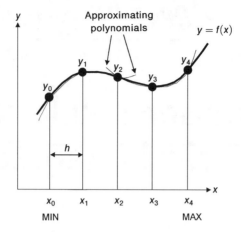

Figure 2 Graphical representation of Simpson's Rule using multiple segments

where:

$$h = \frac{\text{MAX} - \text{MIN}}{m}$$

The higher the value for m, the more accurate the result will be; however, the time for computation will increase. The use of computers makes the task relatively simple though. The subroutine at the end of this appendix is written for Visual Basic, but can be easily translated to other computer languages. It requires the declaration of two textbox objects and one label object to display the area, A.

Example 1 – Numerical integration of the Stress–Strength Interference (SSI) equation

In Section 4.4.1 we described a means of determining the reliability, R, when the loading stress, L, is represented by CDF in closed form (for all the distributions except for the Normal and Lognormal type) and the strength, S, is represented by its PDF.

$$R = \int_0^\infty F(L) \cdot f(S) \, dS$$

If, for example, the loading stress, L, was given by a 3-parameter Weibull distribution, the CDF would be:

$$F(L) = 1 - \exp\left(-\left(\frac{x - xo_L}{\theta_L - xo_L}\right)^{\beta_L}\right)$$

and the strength, S, was represented by a Normal distribution, the PDF would be:

$$f(S) = \frac{1}{\sigma_S\sqrt{2\pi}} \exp\left(-\frac{(x - \mu_S)^2}{2\sigma_S^2}\right)$$

Therefore, substituting these into the reliability equation gives:

$$R = \int_{S=xo_L}^{\infty} \left[1 - \exp\left(-\left(\frac{x - xo_L}{\theta_L - xo_L}\right)^{\beta_L}\right) \right] \cdot \left[\frac{1}{\sigma_S\sqrt{2\pi}} \exp\left(-\frac{(x - \mu_S)^2}{2\sigma_S^2}\right) \right] dx$$

where the limits of integration are from the expected minimum value of the stress, xo_L, to infinity.

Because the loading stress function as an output from the variance analysis is characterized by a Normal distribution, it is advantageous to estimate the parameters for a 3-parameter Weibull distribution, xo and θ, given the mean, μ, and standard deviation, σ, for a Normal distribution (assuming $\beta = 3.44$), where:

$$xo \approx \mu - 3.1394473\sigma \qquad \theta \approx \mu + 0.3530184\sigma$$

Consider the situation where the loading stress on a component is given as $L \sim N(350, 40)$ MPa relating to a Normal distribution with a mean of $\mu_L = 350$ MPa and standard deviation $\sigma_L = 40$ MPa. The strength distribution of the component is $S \sim N(500, 50)$ MPa. It is required to find the reliability under these conditions.

From the above equations:

$$xo_L = 350 - 3.1394473(40) = 224.42 \text{ MPa}$$

$$\theta_L = 350 + 0.3530184(40) = 364.12 \text{ MPa}$$

$$\beta_L = 3.44$$

The reliability equation then becomes:

$$R = \int_{224.42}^{\infty} \left[1 - \exp\left(-\left(\frac{x - 224.42}{364.12 - 224.42}\right)^{3.44}\right) \right] \cdot \left[\frac{1}{50\sqrt{2\pi}} \exp\left(-\frac{(x - 500)^2}{2(50)^2}\right) \right] dx$$

which can be simplified to:

$$R = \int_{224.42}^{\infty} \left[1 - \exp\left(-\left(\frac{x - 224.42}{139.7}\right)^{3.44}\right) \right] \cdot \left[0.00798 \exp\left(-\frac{(x - 500)^2}{5000}\right) \right] dx$$

Using Simpson's Rule outlined above, the maximum limit, ∞, is difficult to work with and an appropriate value reflecting the problem should replace it. For argument's sake, we will give it a value of 700 MPa. Therefore, the reliability can be determined given that:

$$f(x) = \left[1 - \exp\left(-\left(\frac{x - 224.42}{139.7}\right)^{3.44}\right) \right] \cdot \left[0.00798 \exp\left(-\frac{(x - 500)^2}{5000}\right) \right]$$

and the limits are MIN $= 224.42$ and MAX $= 700$.

Now applying Simpson's Rule, let's start with a relatively low number of segments, $m = 10$:

$$h = \frac{\text{MAX} - \text{MIN}}{m} = 47.56$$

and

$$x_0 = 224.42 \qquad\qquad = \text{MIN}$$

$$x_1 = 224.42 + 47.56 = 271.98$$

$$x_2 = 271.98 + 47.56 = 319.54$$

$$x_3 \ldots,$$

$$x_{10} = 700 \qquad\qquad = \text{MAX}$$

From Simpson's Rule:

$$A = \left(\frac{h}{3}\right)\left[f(x_0) + 4\sum_{i=1,3,5,\ldots}^{m-1} f(x_i) + 2\sum_{j=2,4,6,\ldots}^{m-2} f(x_j) + f(x_m)\right]$$

$$= \left(\frac{47.56}{3}\right)\left[\begin{array}{l} 0 + 4(0 + 0.000154 + 0.005985 + 0.004135 + 0.000076) + \\ 2(0.000003 + 0.001756 + 0.007828 + 0.000884) + 0.000003 \end{array}\right]$$

$$= 0.988335$$

which gives the reliability, R, as:

$$R = 0.988335$$

Increasing the number of segments, m, increases the accuracy and using the computer code for $m = 1000$ gives the answer $R = 0.990274$.

Comparing answers with that derived from the *coupling equation*, the Standard Normal variate is:

$$z = -\frac{500 - 350}{\sqrt{50^2 + 40^2}} = -2.34$$

From the SND, Table 1 in Appendix I, the area under the curve, $\Phi(z)$, and hence the probability of failure, P, is found to be 0.009642. The reliability is then given by:

$$R = 1 - P = 1 - 0.009642 = 0.990358$$

The benefit using the approach over other methods is that it can be used to solve the general reliability equation for combination of distribution, including when there are multiple load applications. This is only possible when the loading stress is described in closed form.

Example 2 – Integral transform method

We have a means of determining the interference of two distributions in a similar way as that given above but applying the integral transform method described in Section 4.4.1, where:

$$R = 1 - \int_0^1 H \, dG$$

where H represents the loading stress distribution and G the strength distribution.

Suppose again that both the stress and strength distributions of interest are of the Normal type, where the loading stress is given as $L \sim N(350, 40)$ MPa and the strength distribution is $S \sim N(500, 50)$ MPa. The Normal distribution cannot be used with the integral transform method, but can be approximated by the 3-parameter Weibull distribution where the CDF is in closed form. It was determined above that the loading stress parameters for the 3-parameter Weibull distribution were:

$$xo_L = 224.42 \text{ MPa}$$

$$\theta_L = 364.12 \text{ MPa}$$

$$\beta_L = 3.44$$

Similarly, the strength parameters for the Weibull distribution can be approximated by the same method to:

$$xo_S = 343.03 \text{ MPa}$$

$$\theta_S = 517.65 \text{ MPa}$$

$$\beta_S = 3.44$$

The CDF for the strength distribution is given by:

$$F(x) = 1 - \exp\left(-\left(\frac{x - xo_S}{\theta_S - xo_S}\right)^{\beta_S}\right)$$

which represents the probability of failure, P. The probability of surviving, or the reliability, is therefore $1 - P$, or:

$$G = \exp\left(-\left(\frac{x - xo_S}{\theta_S - xo_S}\right)^{\beta_S}\right)$$

Solving for x to give the inverse of this function gives:

$$x = xo_S + (\theta_S - xo_S)[-\ln G]^{1/\beta_S} \quad \text{or} \quad x = 343.03 + 174.62[-\ln G]^{0.291}$$

The survival equation for the stress distribution is:

$$H = \exp\left(-\left(\frac{x - xo_L}{\theta_L - xo_L}\right)^{\beta_L}\right) \quad \text{or} \quad H = \exp\left(-\left(\frac{x - 224.42}{139.7}\right)^{3.44}\right)$$

Substituting these expressions into equation 4.53 gives the reliability, R, as:

$$R = 1 - \int_0^1 \exp\left(-\left(\frac{\{343.03 + 174.62([-\ln G]^{0.291})\} - 224.42}{139.7}\right)^{3.44}\right) dG$$

where the actual function to be integrated is:

$$f(G) = \exp\left(-\left(\frac{\{343.03 + 174.62([-\ln G]^{0.291})\} - 224.42}{139.7}\right)^{3.44}\right)$$

Let's use 10 segments again, giving:

$$h = \frac{\text{MAX} - \text{MIN}}{m} = \frac{1 - 0}{10} = 0.1$$

and

$$x_0 = 0 \quad = \text{MIN}$$

$$x_1 = 0.1$$

$$x_2 = 0.2$$

$$x_3 \ldots,$$

$$x_{10} = 1 \quad = \text{MAX}$$

Applying Simpson's Rule to the function $f(G)$ gives:

$$A = \left(\frac{h}{3}\right)\left[f(x_0) + 4\sum_{i=1,3,5,\ldots}^{m-1} f(x_i) + 2\sum_{j=2,4,6,\ldots}^{m-2} f(x_j) + f(x_m)\right]$$

$$= \left(\frac{0.1}{3}\right)\left[\begin{array}{l}0 + 4(0 + 0.000001 + 0.000032 + 0.000747 + 0.017962) \\ +2(0 + 0.000005 + 0.000162 + 0.003412) + 0.565802\end{array}\right] = 0.021598$$

$$R = 1 - A = 0.978402$$

Increasing the number of segments, m, again increases the accuracy and using the computer code provided for $m = 1000$ gives the answer $R = 0.991182$. This compares well with the answers given in the previous example, although we are using a 3-parameter Weibull distribution to model the stress and strength, when in fact they are of the Normal type.

Area under a Function Calculated using Simpson's Rule (written in Visual Basic)

```
Dim Ord As Integer                              (declaration of variables)
Dim m As Integer
Dim x( ) As Variant
Dim y( ) As Variant
Dim MAX As Variant
Dim MIN As Variant
Dim b( ) As Integer
Dim c( ) As Variant
Dim Sum As Variant
Dim A As Variant

Let MAX = Text1.Text                            (maximum limit of integration)
Let MIN = Text2.Text                            (minimum limit of integration)

Let Ord = 1001                                  (number of ordinates)
Let m = Ord - 1                                 (number of segments)

ReDim x(m + 1)
For I% = 2 To m
Let x(I%) = MIN + (I% - 1) * ((MAX - MIN) / m)
Next I%

Let x(m + 1) = MAX
```

```
Let x(1) = MIN

ReDim y(m + 1)
For I% = 1 To m + 1
 Let y(I%) = f (x(I%))                          (function of x goes here)
Next I%

ReDim b(m + 1)
For I% = 2 To m Step 2
 Let b(I%) = 4
Next I%

For I% = 3 To (m - 1) Step 2
 Let b(I%) = 2
Next I%

Let b(1) = 1
Let b(m + 1) = 1

ReDim c(m + 1)
For I% = 1 To m + 1
 Let c(I%) = b(I%) * y(I%)
Next I%

Let Sum = 0
For I% = 1 To m + 1
 Let Sum = Sum + c(I%)
Next I%

Let A = (MAX - MIN) * Sum / (3 * m)             (Simpson's Rule)

Let Label1.Caption = Format(A, "#0.00000000")   (output area under function)
```

References

Abbot, H. 1993: The Cost of Getting it Wrong. *Product Liability International*, February.

Abraham, B. and Whitney, J. B. 1993: Management of Variation Reduction Investigations. In: *Advances in Industrial Engineering No. 16 – Quality Through Engineering Design*. Amsterdam: Elsevier Science Publishers.

Albin, S. L. and Crefield, P. J. 1994: Getting Started – Concurrent Engineering for a Medium Sized Manufacturer. *Journal of Manufacturing Systems*, 13(1), 48–58.

Alexander, C. 1964: *Notes on the Synthesis of Form*. Cambridge, MA: Harvard University Press.

Amster, S. J. and Hooper, J. H. 1986: Statistical Methods for Reliability Improvement. *AT&T Journal*, 65(2), 69–76.

Andersson, P. 1993: Design for Quality – As Perceived by Industry. In: *Proceedings ICED '93*, The Hague, 1123–1126.

Andersson, P. 1994: Early Design Phases and their Role in Designing for Quality. *Journal of Engineering Design*, 5(4), 283–298.

Andersson, P. A. 1996: *Process Approach to Robust Design in Early Engineering Design Phases*. PhD Thesis: Department of Machine Design, Lund Institute of Technology, Lund, Sweden.

Andreasen, M. M. 1991: Design Methodology. *Journal of Engineering Design*, 2(4), 321–335.

Andreasen, M. M. and Hein, L. 1987: *Integrated Product Development*. London: IFS Publications and Springer-Verlag.

Andreasen, M. M. and Olesen, J. 1990: The Concept of Dispositions. *Journal of Engineering Design*, 1(1), 17–36.

Arajou, C. S., Benedetto-Neto, H., Campello, A. C., Segre, F. M. and Wright, I. C. 1993: The Utilisation of Product Development Methods: A Survey of UK Industry. *Journal of Engineering Design*, 7(3), 265–278.

Ashby, M. F. and Jones, D. R. H. 1989: *Engineering Materials – Part 1*. Oxford: Pergamon Press.

ASI 1987: *Quality Function Deployment Executive Briefing*. American Supplier Institute.

ASI 1992: *Quality Function Deployment Awareness Senior*. American Supplier Institute.

ASM International 1997a: *Handbook No. 1. Properties and Selection: Irons, Steels and High Performance Alloys*, 10th Edition. OH: ASM International.

ASM International 1997b: *Handbook No. 2: Properties and Selection: Non-ferrous Alloys and Special Purpose Materials*, 10th Edition. OH: ASM International.

Ayyub, B. M. and McCuen, R. H. 1997: *Probability, Statistics and Reliability for Engineers*. Boca Raton: CRC Press.

Baldwin, D. F., Abell, T. E., Lui, M., Defazio, T. L. and Whitney, D. E. 1991: An Integrated Computer Aid for Generating and Evaluating Assembly Sequences for Mechanical Products. *IEEE Transactions on Robotics and Automation*, **7**(1), 78–94.

Barclay, I. and Poolton, J. 1994: Concurrent Engineering – Concept and Practice. *International Journal of Vehicle Design*, **15**(3–5), 529–544.

Barnes, C. J., Dalgleish, G., Jared, G., Swift, K. G. and Tate, S. 1997: Assembly Sequence Structures in Design for Assembly. In: *Proceedings IEEE International Symposium on Assembly and Task Planning*, Marina del Rey, California, 164–169.

Battin, L. 1988: Six Sigma Process by Design – Design and Dimensions. *Group Mechanical Technology*, **1**(1), Government Electronics Group, Motorola Inc., Scottsdale, Arizona.

Ben-Haim, Y. 1994: A Non Probabilistic Concept of Reliability. *Structural Safety*, **14**, 227–245.

Bendell, T., Kelly, J., Merry, T. and Sims, F. 1993: *Quality: Measuring and Monitoring*. London: Century Business.

Benham, P. P. and Warnock, F. V. 1983: *Mechanics of Solids and Structures*. London: Pitman.

Benson, T. 1993: TQM: a child takes a first few faltering steps. *Industry Week*, **242**(7), 16–17.

Bergman, B. 1992: The Development of Reliability Techniques: a retrospective survey. *Reliability and System Safety*, **36**, 3–6.

Bicheno, J. 1994: *The Quality 50: A Guide to Gurus, Tools, Wastes, Techniques and Systems*. Birmingham: Piscie Books.

Bieda, J. and Holbrook, M. 1991: Reliability Prediction, the Right Way. *Reliability Review*, **11**, 8.

Bignell, V. and Fortune, J. 1992: *Understanding Systems Failures*. Manchester: Manchester University Press.

Bjørke, O. 1989: *Computer-aided Tolerancing*, 2nd Edition, NY: ASME Press.

Bloom, J. M. (ed.) 1983: *Probabilistic Fracture Mechanics and Fatigue Methods: Applications for Structural Design and Maintenance*. American Society for Testing and Materials.

Bohnenblust, H. and Slovic, P. 1998: Integrating Technical Analysis and Public Values in Risk-based Decision Making. *Reliability Engineering and System Safety*, **59**, 151–159.

Bolotin, V. V. 1994: Fatigue Life Prediction of Structures. In: Spanos, P. D. and Wu, Y. (eds), *Probabilistic Structural Mechanics: Advances in Structural Reliability Methods*. Berlin: Springer-Verlag.

Bolz, R. W. (ed.) 1981: *Production Processes: The Productivity Handbook*, 5th Edition. NY: Industrial Press Inc.

Bompas-Smith, J. H. 1973: *Mechanical Survival: the use of reliability data*. London: McGraw-Hill.

Booker, J. D. 1994: *Project Specification Quality Questionnaire*. Internal Document – EPSRC GR/J97922, School of Engineering, University of Hull.

Booker, J. D., Dale, B., McQuater, R., Spring, M. and Swift, K. G. 1997: *Effective use of Tools and Techniques in New Product Development*. Manchester: UMIST Press.

Boothroyd, G. and Redford, A. H. 1968: *Mechanized Assembly*. London: McGraw-Hill.

Boothroyd, G., Dewhurst, P. and Knight, W. 1994: *Product Design for Manufacture and Assembly*. NY: Marcel Dekker.

Bowker, A. H. and Lieberman, G. J. 1959: *Engineering Statistics*. Englewood Cliffs, NJ: Prentice-Hall.

Bracha, V. J. 1964: The Methods of Reliability Engineering. *Machine Design*, 30 July, 70–76.

Bralla, J. G. 1986: *Handbook of Product Design for Manufacturing*. NY: McGraw-Hill.

Bralla, J. G. 1996: *Designing for Higher Quality – Design for Excellence*. NY: McGraw-Hill.

Bralla, J. G. (ed.) 1998: *Design for Manufacturability Handbook*, 2nd Edition. NY: McGraw-Hill.

Branan, B. 1991: DFA Cuts Assembly Defects by 80%. *Appliance Manufacture*, November, 12–23.

Braunsperger, M. 1996: Designing for Quality – an integrated approach for simultaneous quality engineering. *Proc. Instn Mech. Engrs*, Part B, **210**(B1), 1–10.

Broadbent, J. E. A. 1993: Evaluating Concept Design for Reliability. In: *Proceedings ICED '93*, The Hague, 1186–1188.

Brown, A. D., Hale, P. R. and Parnaby, J. 1989: An Integrated Approach to Quality Engineering in Support of Design for Manufacture. *Proc. Instn Mech. Engrs*, Part B, **223**, 55–63.

BS 970 1991: *Part 3 – Specification for Wrought Steels for Mechanical and Allied Engineering Purposes*. London: BSI.

BS 1134 1990: *Part 2 – Assessment of Surface Texture – Guidance and General Information*. London: BSI.

BS 4360 1990: *Specification for Weldable Structural Steels*. London: BSI.

BS 4500A 1970: *Specification for ISO limits and fits. Data sheet: selected ISO fits – hole basis*. London: BSI.

BS 5760 1991: *Part 5 – Guide to Failure Modes, Effects and Criticality Analysis (FMEA and FMECA). Reliability of Systems, Equipment and Components*. London: BSI.

BS 6079 1999: *Part 3 – Project Management: Risks*. London: BSI.

BS 6143 1990: *Part 2 – The Guide to the Economics of Quality: Prevention, Appraisal and Failure Model*. London: BSI.

BS 7000 1997: *Part 2 – Guide to Managing the Design of Manufactured Products*. London: BSI.

BS 7850 1992: *Part 1 – Guide to Management Principles*. London: BSI.

BS 7850 1994: *Part 2 – Guidelines for Quality Improvement*. London: BSI.

BS EN 20286 1993: *Part 2 – ISO System of Limits and Fits – Tables of Standard Tolerance Grades and Limit Deviations for Holes and Shafts*. London: BSI.

BS EN ISO 9000 1994: *Quality Management and Quality Assurance Standards*. London: BSI.

BS EN ISO 9001 1994: *Part 1 – Quality Systems – Model for Quality Assurance in Design, Development, Production, Installation and Servicing*. London: BSI.

Burden, R. L. and Faires, J. D. 1997: *Numerical Analysis*, 6th Edition. Pacific Grove, CA: Brooks/Cole.

Burns, R. J. 1994: Reliability: Is it Worth the Effort? – an assessment of the value of reliability tasks and techniques. *Microelectronic Reliability*, **34**(11), 1795–1805.

Bury, K. V. 1974: On Probabilistic Design. *Journal of Engineering for Industry*, November, 1291–1295.

Bury, K. V. 1975: *Statistical Models in Applied Science*. NY: Wiley.

Bury, K. V. 1978: On Product Reliability under Random Field Loads. *IEEE Transactions on Reliability*, **R-27**(4), 258–260.

Bury, K. V. 1999: *Statistical Distributions in Engineering*. Cambridge: Cambridge University Press.

Cable, C. W. and Virene, E. P. 1967: Structural Reliability with Normally Distributed Static and Dynamic Loads and Strength. In: *Proceedings Annual Symposium on Reliability*, 329–336.

Cagan, J. and Kurfess, T. R. 1992: Optimal Tolerance Allocation over Multiple Manufacturing Alternatives. *Advances in Design Automation*, **2**, ASME DE-Vol. 44-2, 165–172.

Calantone, R. J., Vickery, S. K. and Droge, C. 1995: Business Performance and Strategic New Product Development. *Journal of Product Innovation Management*, **12**(3), 214–223.

Carter, A. D. S. 1986: *Mechanical Reliability*, 2nd Edition. London: Macmillan.

Carter, A. D. S. 1997: *Mechanical Reliability and Design*. London: Macmillan.

Cather, H. and Nandasa, P. N. 1995: A Favour Returned. *Manufacturing Engineer*, **74**(4), 45–56.

Chandra, U. 1997: Control of Residual Stresses. In: ASM International, *Handbook No. 20: Materials Selection and Design*, 10th Edition, OH: ASM International.

Chao, L. L. 1974: *Statistics: Methods and Analyses*. NY: McGraw-Hill.

Chase, K. W. and Greenwood, W. H. 1988: Design Issues in Mechanical Tolerance Analysis. *Manufacturing Review*, **1**(1), 50–59.

Chase, K. W. and Parkinson, A. R. 1991: A Survey of Research in the Application of Tolerance Analysis to the Design of Mechanical Assemblies. *Research in Engineering Design*, **3**, 23–37.

Chase, K. W., Gao, J. and Magleby, S. P. 1995: General 2D Tolerance Analysis of Mechanical Assemblies with Small Kinematic Adjustments. *International Journal of Design and Manufacture*, **5**(4), 263–274.

Chase, K. W., Magleby, S. P and Gao, J. 1997: Tolerance Analysis of Two- and Three-Dimensional Mechanical Assemblies with Small Kinematic Adjustments. In: Zhang, H. (ed.), *Advanced Tolerancing Techniques*. NY: Wiley Interscience.

Chrysler Corporation, Ford Motors, General Motors Corporation 1995: *Potential Failure Mode and Effects Analysis (FMEA) – Reference Manual*, 2nd Edition.

Clark, K. B. and Fujimoto, T. 1991: *Product Development Performance Strategy, Organisation and Management in the World Auto Industry*. Boston: Harvard Business School Press.

Clark, K. B. and Fujimoto, T. 1992: *Systematic Automobile Development: Organisation and Management in Europe, Japan and the USA*. Frankfurt/Main and NY: Campus.

Clausing, D. 1994: *Total Quality Development*. NY: ASME Process.

Clausing, D. 1998: Reusability in Product Development. In: *Proceedings EDC '98*. Bury St Edmunds: Professional Engineering Publishing, 57–66.

Collins, J. A. 1993: *Failure of Materials in Mechanical Design*. NY: Wiley-Interscience.

Comer, J. and Kjerengtroen, L. 1996: Probabilistic Methods in Design: introductory topics. *SAE Paper No. 961792*.

Cooper, R. G. and Kleinschmidt, E. J. 1993: Screening New Products for Potential Winners. *Long Range Planning*, **26**(6), 74–81.

Cooper, R. G. and Kleinschmidt, E. J. 1995: Benchmarking the Firm's Critical Success Factors in New Product Development. *Journal of Product Innovation Management*, **12**(5), 374–391.

Craig, M. 1992: Controlling the Variation. *Manufacturing Breakthrough*, **1**(6), 343–348.

Crosby, P. B. 1969: *Cutting the Cost of Quality*. Boston, MA: Industrial Education Institute.

Cruse, T. A. 1997a: Overview of Mechanical System Reliability. In: Cruse, T. A. (ed.), *Reliability-Based Mechanical Design*. NY: Marcel Dekker.

Cruse, T. A. 1997b: Mechanical Reliability Design Variables and Models. In: Cruse, T. A. (ed.), *Reliability-Based Mechanical Design*. NY: Marcel Dekker.

CSC Manufacturing 1995: *Design for Assembly/Manufacturing Analysis Practitioners Manual*, Version 10.5. Solihull: CSC Manufacturing.

Cullen J. M. 1994: Managing the New Product Introduction Process for Quality, Reliability, Cost and Speed. In: *Proceedings IMechE International Conference on Design Competitive Advantage – Making the Most of Design*, Coventry, 23–24 March, Paper C482.

Dale, B. G. (ed.) 1994: *Managing Quality*, 2nd Edition. NY: Prentice-Hall.

Dale, B. and McQuater, R. 1998: *Managing Business Improvement and Quality*. Oxford: Blackwell.

Dale, B. and Oakland, J. S. 1994: Designing for Quality. In: Dale, B. (ed.), *Managing Quality*, 2nd Edition. NY: Prentice-Hall.

Dasgupta, A. and Pecht, M. 1991: Material Failure Mechanisms and Damage Models. *IEEE Transactions on Reliability*, **40**(5), 531–536.

Davies, G. J. 1985: Performance in Service: essential metallurgy for engineers. Bradbury (ed.). London: Van Nostrand-Reinhold.

Dertouzos, M. L., Lester, R. K. and Solow, R. M. 1989: *MIT Commission on Industrial Productivity, Made in America*, Cambridge, MA: MIT Press.

Dhillon, B. S. 1980: Mechanical Reliability: Interference Theory Models. In: *Proceedings Annual Reliability and Maintainability Symposium*, 462–467.

Dieter, G. E. 1986: *Engineering Design: a materials and processing approach*, 1st Metric Edition. NY: McGraw-Hill.

Disney, R. L., Sneth, N. J. and Lipson, C. 1968: The Determination of Probability of Failure by Stress/Strength Interference Theory. In: *Proceedings of Annual Symposium on Reliability*, 417–422.

Ditlevsen, O. 1997: Structural Reliability Codes for Probabilistic Design. *Structural Safety*, **19**(3), 253–270.

Dixon, J. R. 1997: Conceptual and Configuration Design of Parts. In: ASM International, *ASM Handbook No. 20 – Materials Selection and Design,* 10th Edition. OH: ASM International.

Dong, Z. 1989: *Automatic Tolerance Analysis and Synthesis in CAD Environment.* PhD Thesis: State University of NY, Buffalo, NY.

Dong, Z. 1993: Design for Automated Manufacturing. In: Kusiak, A. (ed.), *Concurrent Engineering: Automation, Tools and Techniques.* NY: Wiley.

Dong, Z. 1997: Tolerance Synthesis by Manufacturing Cost Modeling and Design Optimization. In: Zhang, H. (ed.), *Advanced Tolerancing Techniques.* NY: Wiley Interscience.

Dorf, R. C. and Kusiak, A. 1994: *Handbook of Design, Manufacture and Automation.* NY: Wiley Interscience.

Dowling, N. E. 1993: *Mechanical Behavior of Materials.* Englewood Cliffs, NJ: Prentice-Hall.

DTI 1992: *Quality Assurance Programme 1992–1996.* Federal Ministry of Research and Technology, Germany. London: DTI Translation No. 0219-92, DTI.

DTI 1994: *The General Product Safety Regulations.* London: DTI Consumer Safety Unit, HMSO, DTI.

Edwards, K. S. and McKee, R. B. 1991: *Fundamentals of Mechanical Component Design.* NY: McGraw-Hill.

Ellingwood, B. and Galambos, T. V. 1984: General Specifications for Structural Design Loads. In: Shinozuka, M. and Yao, J. (eds), *Probabilistic Methods in Structural Engineering.* NY: ASCE.

Elliot, A. C., Wright, I. C. and Saunders, J. A. 1998: Successful NPD in Engineering Design: meeting customer needs. In: *Proceedings EDC '98.* Bury St Edmunds: Professional Engineering Publishing, 391–398.

Ellis, T. I. A., Dooner, M. and Swift, K. G. 1996: Empirical Studies of Product Introduction in a Project Based Environment. In: *Proceedings CAD/CAM, Robotics and Factory of the Future Conference,* Middlesex University, July, 129–134.

EPSRC 1999: *Risk Assessment and Management – A Retrospective Evaluation of the EPSRC's Recent Research Portfolio.* Swindon: Engineering and Physical Sciences Research Council, UK. http://www.epsrc.ac.uk/

Evans, D. H. 1975: Statistical Tolerancing: The State of the Art, Part III. Shifts and Drifts. *Journal of Quality Technology,* **7**(2), 72–76.

Evbuomwan, N. F. O., Sivaloganathan, S. and Webb, J. 1996: A Survey of Design Philosophies, Models, Methods and Systems. *Proc. Instn Mech. Engrs,* Part B, **210**(B4), 301–320.

Fabrycky, W. J. 1994: Modeling and Indirect Experimentation in System Design and Evaluation. *Journal of NCOSE,* **1**(1), 133–144.

Faires, V. M. 1965: *Design of Machine Elements,* 4th Edition, NY: Macmillan.

Fajdiga, M., Jurejevcic, T. and Kernc, J. 1996: Reliability Prediction in Early Phases of Product Design. *Journal of Engineering Design,* **7**(2), 107–128.

Farag, M. M. 1997: Properties Needed for the Design of Static Structures. In: ASM International, *ASM Handbook No. 20 – Materials Selection and Design,* 10th Edition. OH: ASM International.

Field, S. W. and Swift, K. G. 1996: *Effecting a Quality Change: an engineering approach.* London: Arnold.

Figliola, R. S. and Beasley, D. E. 1995: *Theory and Design for Mechanical Measurements.* New York: Wiley.

Fleischer, M. and Liker, J. K. 1992: The Hidden Professionals: Product Designers and their impact on design quality. *IEEE Transactions on Engineering Management,* **39**(3), 254–264.

Foley, M. and Bernardon, E. 1990: Thermoplastic Composite Manufacturing Cost Analysis for the Design of Cost-Effective Automated Systems. *SAMPE Journal,* **26**(4), 67–74.

Fragola, J. R. 1996: Reliability and Risk Analysis Data Base Development: an historical perspective. *Reliability Engineering and System Safety*, **51**, 125–136.

Fraser, C. J. and Milne, J. S. 1990: *Microcomputer Applications in Measurement Systems*. London: Macmillan.

Freudenthal, A. M., Garrelts, J. M. and Shinozuka, M. 1966: The Analysis of Structural Safety. *Journal of the Structural Division, American Society of Civil Engineers*, **92**, ST1, 267–325.

Furman, T. T. 1981: *Approximate Methods in Engineering Design*. London: Academic Press.

Galt, J. D. A. and Dale, B. G. 1990: The Customer–Supplier Relationship in the Motor Industry: a vehicle manufacturer's perspective. *Proc. Instn Mech. Engrs*, **204**, 179–186.

Garvin, D. A. 1988: *Managing Quality*. NY: Free Press.

Gerth, R. 1997: Tolerance Analysis: A Tutorial of Current Practice. In: Zhang, H. (ed.), *Advanced Tolerancing Techniques*. NY: Wiley Interscience.

Gerth, R. and Hancock, W. M. 1995: Reduction of the Output Variation of Production Systems Involving a Large Number of Processes. *Quality Engineering*, **8**(1), 145–163.

Gill, H. 1990: Adoption of Design Science by Industry – Why so Slow? *Journal of Engineering Design*, **1**, 321–335.

Gilson, J. 1951: *A New Approach to Engineering Tolerances*. London: The Machinery Publishing Company.

Gordon, J. E. 1991: *Structures – or why things don't fall down*. St Ives: Penguin.

Green, R. E. (ed.) 1992: *Machinery's Handbook*, 24th Edition. NY: Industrial Press.

Greenfield, P. 1996: Cogent Argument for Better Design Skills. *Professional Engineering*, January, 13–14.

Grove, D. M. and Davis, T. P. 1992: *Engineering Quality and Experimental Design*. Longman.

Gryna, F. M. 1988: Supplier Relation. In: Juran, J. M. (ed.), *Juran's Quality Control Handbook*. NY: McGraw-Hill.

Hagan, J. T. (ed.) 1986: *Principles of Quality Costs*. American Society of Quality Press.

Hallihan, A. 1997: The Quick Change Artists. *Professional Engineer*, **10**(21), 31–32.

Hammer, W. 1980: *Product Safety Management*. Englewood Cliffs, NJ: Prentice-Hall.

Haque, B. and Pawar, K. S. 1998: Development of a Methodology and Tool for Analysing New Product Design and Development in a Concurrent Engineering Environment. In: *Proceedings EDC '98*, Bury St Edmunds: Professional Engineering Publishing, 669–677

Harry, M. and Stewart, R. 1988: *Six Sigma Mechanical Design Tolerancing*. Schaumberg, Illinois: Motorola University Press.

Haugen, E. B. 1968: *Probabilistic Approaches to Design*. NY: Wiley.

Haugen, E. B. 1980: *Probabilistic Mechanical Design*. NY: Wiley-Interscience.

Haugen, E. B. 1982a: Modern Statistical Materials Selection – Part 1: Some Basic Concepts. *Materials Engineering*, **96**, July, 21–25.

Haugen, E. B. 1982b: Modern Statistical Materials Selection – Part 2: Random Variables and Reliability. *Materials Engineering*, **96**, August, 49–51.

Haugen, E. B. and Wirsching, P. H. 1975: Probabilistic Design: a realistic look at risk and reliability in engineering. *Machine Design*, 15 May.

Henzold, G. 1995: *Handbook of Geometrical Tolerancing, Design, Manufacturing and Inspection*. Chichester: Wiley.

Hopp, T. H. 1993: The Language of Tolerances. In: *Advances in Industrial Engineering No. 16 – Quality Through Engineering Design*. Amsterdam: Elsevier Science Publishers.

Howell, D. 1999: Industry Fails Statistical Analysis. *Professional Engineering*, **12**(1), 20–21.

Huang, G. Q. 1996: Developing DFX Tools. In: Huang, G. Q. (ed.), *Design for X – Concurrent Engineering Imperatives*. London: Chapman & Hall.

Hughes, J. A. 1995: Ethnography, Plans and Software Engineering. In: *Proceedings IEE Colloquium on CSCW and the Software Process*, Savoy Place, London, February.

Hundal, M. S. 1997: Product Costing: a comparison of conventional and activity based costing methods. *Journal of Engineering Design*, **8**(1), 91–103.

Ireson, G. W., Coombs, C. F. Jr and Moss, R. Y. 1996: *Handbook of Reliability Engineering and Management*, 2nd Edition. NY: McGraw-Hill.

Jakobsen, M. M. 1993: A Methodology for Product Development in Small and Medium Sized Companies. In: *Proceedings ICED '93*, The Hague, 17–19 August, 706–709.

Jeang, A. 1995: Economic Tolerance Design for Quality. *Quality and Reliability Engineering International*, **11**, 113–121.

Jenkins, S., Forbes, S., Durrani, T. S. and Banerjee, S. K. 1997a: Managing the Product Development Process – Part I: an assessment. *International Journal of Technology Management*, **13**(4), 359–378.

Jenkins, S., Forbes, S., Durrani, T. S. and Banerjee, S. K. 1997b: Managing the Product Development Process – Part II: case studies. *International Journal of Technology Management*, **13**(4), 379–394.

Jin, Y., Levitt, R. E., Christiansen, T. R. and Kunz, J. C. 1995: Modelling Organisational Behaviour of Concurrent Design Teams. *Artificial Intelligence for Engineering Design Analysis and Manufacturing*, **9**(2), 145–158.

Jones, G. 1978: In: *Proceedings of Conference on Reliability of Aircraft Mechanical Systems and Equipment*. Instn Mech. Engrs, MOD and RaeS, London, 67–70.

Juran, J. M. (ed.) 1988: *Quality Control Handbook*. NY: McGraw-Hill.

Juvinall, R. C. 1967: *Engineering Considerations of Stress, Strain and Strength*. NY: McGraw-Hill.

Kalpakjian, S. 1995: *Manufacturing Engineering and Technology*, 3rd Edition, MA: Addison Wesley.

Kapur, K. 1993: Quality Engineering and Robust Design. In: Kusiak, A. (ed.), *Concurrent Engineering: Automation, Tools and Techniques*. NY: Wiley.

Kapur, K. C. and Lamberson, L. R. 1977: *Reliability in Engineering Design*. NY: Wiley.

Karmiol, E. D. 1965: *Reliability Apportionment*. Preliminary Report EIAM-5, Task II, General Electric, Schenectady, NY, 8 April, 10–22.

Kececioglu, D. 1972: Reliability Analysis of Mechanical Components and Systems. *Nuclear Engineering and Design*, **19**, 259–290.

Kececioglu, D. 1991: *Reliability Engineering Handbook – Volume 1*. NY: Prentice-Hall.

Kehoe, D. F. 1996: *The Fundamentals of Quality Management*. London: Chapman & Hall.

Kirkpatrick, E. G. 1970: *Quality Control for Managers and Engineers*. NY: Wiley.

Kjerengtroen, L. and Comer, J. 1996: Probabilistic Methods in Design: an overview of current technologies. *SAE Paper No. 961793*.

Klit, P., Jensen, F. and Ellevang, P. 1993: Reliable Design Methodology – the use and misuse of reliability data in the design process. In: *Proceedings ICED '93*, The Hague, 1156–1164.

Kluger, P. 1964: Evaluation of the Effects of Manufacturing Processes on Structural Design Reliability. In: *Proceedings 10th National Symposium on Reliability and Quality Control*, IEEE, Washington, 538–544.

Kolarik, W. J. 1995: *Creating Quality*. NY: McGraw-Hill.

Korde, U. 1997: Planning for Variation in Manufacturing Processes – a process tolerance analysis software tool. *SAE Technical Paper No. 970683*.

Kottegoda, N. T. and Rosso, R. 1997: *Statistics, Probability and Reliability for Civil and Environmental Engineers*. NY: McGraw-Hill.

Kotz, S. and Lovelace, C. R. 1998: *Process Capability Indices in Theory and Practice*. London: Arnold.

Kroll, E. 1993: Modeling and Reasoning for Computer-Based Assembly Planning. In: Kusiak, A. (ed.), *Concurrent Engineering: Automation, Tools and Techniques*. NY: Wiley.

Kruger, V. 1996: How Can a Company Achieve Improved Levels of Quality Performance: technology versus employees? *The TQM Magazine*, **8**(3), 11–20.

Kurogane, K. (ed.) 1993: *Cross-Functional Management: Principles and Practical Applications.* Asian Productivity Organization.

Kusiak, A. and Wang, J. R. 1993: Efficient Organising of Design Activities. *International Journal of Production Research*, **31**(4), 753–769.

Kutz, M. 1986: *Mechanical Engineers' Handbook*. NY: Wiley.

Labovitz, G. H. 1988: *Tough Questions Senior Managers Should be Asking about Quality*, The ODI Quality Management Series, Vol. 1.

Larsson, L. E., Jönsson, L. and Malm, S. 1971: Maskinskador – orsaker och economisk betydelse. *Verskstaderna*, April (in Swedish).

Leaney, P. G. 1996a: Design for Dimensional Control. In: Huang, G. Q. (ed.), *Design for X – Concurrent Engineering Imperatives*. London: Chapman & Hall.

Leaney, P. G. 1996b: Case Experience with Hitachi, Lucas and Boothroyd-Dewhurst DFA Methods. In: Huang, G. Q. (ed.), *Design for X – Concurrent Engineering Imperatives.* London: Chapman & Hall.

Lee, W. and Woo, T. C. 1990: Tolerances – Their Analysis and Synthesis. *Journal of Engineering for Industry*, **112**, May, 113-121.

Lee, S., Yi, C. and Suarez, R. 1997: Assemblability Analysis with Adjusting Cost Evaluation based on Tolerance and Adjustability. In: *Proceedings IEEE International Symposium on Assembly and Task Planning*, Marina del Rey, California, 103–108.

Leitch, R. D. 1990: A Statistical Model of Rough Loading. In: *Proceedings 7th International Conference on Reliability and Maintainability*, Brest, France, 8–12.

Leitch, R. D. 1995: *Reliability Analysis for Engineers – an introduction.* Oxford: Oxford University Press.

Lemaire, M. 1997: Reliability and Mechanical Design. *Reliability Engineering and System Safety*, **55**, 163–170.

Lewis, E. E. 1996: *Introduction to Reliability Engineering*, 2nd Edition. NY: Wiley.

Lewis, W. P. and Samuel, A. E. 1991: An Analysis of Designing for Quality in the Automotive Industry. In: *Proceedings ICED '91*, Zurich, 489–500.

Liggett, J. V. 1993: *Dimensional Variation Management Handbook: a guide for quality, Design and Manufacturing Engineers.* Englewood Cliffs, NJ: Prentice-Hall.

Lin, S-S, Wang, H-P and Zhang, C. 1997: Optimal Tolerance Design for Integrated Design, Manufacturing and Inspection with Genetic Algorithms. In: Zhang, H. (ed.), *Advanced Tolerancing Techniques*. NY: Wiley Interscience.

Lincoln, B., Gomes, K. J. and Braden, J. F. 1984: *Mechanical Fastening of Plastics: an Engineering Handbook*. NY: Marcel Dekker.

Lipson, C. and Sheth, N. J. 1973: *Statistical Design and Analysis of Engineering Experiments.* NY: McGraw-Hill.

Lipson, C., Sheth, N. and Disney, R. L. 1967: *Reliability Prediction – Mechanical Stess/ Strength Interference Models (Ferrous)*, RADC-TR-66-710, March (AD/813574).

Lloyd, A. 1994: A Study of Nissan Motor Manufacturing (UK) – Supplier Development Team Activities. *Proc. Instn Mech. Engrs*, **208**, 63–68.

Loll, V. 1987: Load–Strength Modelling of Mechanics and Electronics. *Quality and Reliability Engineering International*, **3**, 149–155.

Lucas PLC 1994: *Identification and Elimination of Critical Characteristics.* Engineering and Design Procedures, No. QP219, Lucas Automotive Electronics, Solihull.

Maes, M. A. and Breitung, K. 1994: Reliability Based Tail Estimation. In: Spanos, P. D. and Wu, Y. (eds), *Probabilistic Structural Mechanics: Advances in Structural Reliability Methods*. Berlin: Springer-Verlag.

Mager, T. R. and Marschall, C. W. 1984: *Development of Crack Arrest Data Bank Irradiated Reactor Vessel Steels*. 2, EPRI NP-3616, Palo Alto, CA, July.

Mahadevan, S. 1997: Physics-Based Reliability Models. In: Cruse, T. A. (ed.), *Reliability-Based Mechanical Design*. NY: Marcel Dekker.

Mansoor, M. 1963: The Application of Probability to Tolerances used in Engineering Designs. *Proc. Instn Mech. Engrs*, **178**(1), 29–51.

Marbacher, B. 1999: Metric Fasteners. In: Bickford, J. H. and Nassar, S. (eds), *Handbook of Bolts and Bolted Joints*. NY: Marcel Dekker.

Margetson, J. 1995: *Discussion Meeting on 'Engineering, Statistics and Probability'*, 2 May, Inst. Mech. Engrs. HQ, London.

Maylor, H. 1996: *Project Management*. London: Pitman Publishing.

McCord, K. R. and Eppinger, S. D. 1993: *Managing the Integration Problem in Concurrent Engineering. Sloan School of Management*, MIT, Report No. Sloan WP3495-93-MSA, August, Cambridge, MA.

McLachlan, V. N. 1996: In Praise of ISO 9000. *The TQM Magazine*, **8**(3), 21–23.

McQuater, R. and Dale, B. 1995: *Using Quality Tools and Techniques Successfully*. UMIST and TQM International Ltd.

Meeker, W. Q. and Hamada, M. 1995: Statistical Tools for the Rapid Development and Evaluation of High Reliability Products. *IEEE Transactions on Reliability*, **44**(2), 187–198.

Meerkamm, H. 1994: Design for X – A Core Area of Design Methodology. *Journal of Engineering Design*, **5**(2), 145–163.

Metcalfe, A. V. 1997: *Statistics in Civil Engineering*. London: Arnold.

Michaels, J. V. and Woods, W. P. 1989: *Design to Cost*. NY: Wiley.

Miles, B. L. and Swift, K. G. 1992: Design for Manufacture and Assembly. In: *Proc. 24th FISITA Congress*, Inst. Mech. Engrs. HQ, London.

Miles, B. and Swift, K. G. 1998: Design for Manufacture and Assembly. *Manufacturing Engineer*, October, 221–224.

Mischke, C. R. 1970: A Method of Relating Factor of Safety and Reliability. *Journal of Engineering for Industry*, August, 537–542.

Mischke, C. R. 1980: *Mathematical Model Building; an introduction for engineers*. Ames, IO: Iowa State Press.

Mischke, C. R. 1989: Stochastic Methods in Mechanical Design: Part 1: property data and Weibull parameters. In: *Proceedings 8th Biannual Conference on Failure Prevention and Reliability*, Design Engineering of ASME, Montreal, Canada, 1–10 Sept.

Mischke, C. R. 1992: Fitting Weibull Strength Data and Applying it to Stochastic Mechanical Design. *Journal of Mechanical Design*, **114**, 35–41.

Modarres, M., 1993: *What Every Engineer Should Know about Reliability and Risk Analysis*. NY: Marcel Dekker.

Mood, A. M. 1950: *Introduction to the Theory of Statistics*. NY: McGraw-Hill.

Morrison, S. J. 1997: Statistical Engineering: the key to quality. *Engineering Management Journal*, **7**(4), 193–198.

Morrison, S. J. 1998: Variance Synthesis Revisited. *Quality Engineering*, **11**(1), 149–155.

Mørup, M. 1993: *Design for Quality*. PhD Thesis, Institute for Engineering Design, Technical University of Denmark, Lyngby.

Murty, A. S. R. and Naikan, V. N. A. 1997: Machinery Selection – process capability and product reliability dependence. *International Journal of Quality and Reliability Management*, **14**(4), 381–390.

Nagode, M. and Fajdiga, M. 1998: A General Multi-Modal Probability Density Function Suitable for the Rainflow Ranges of Stationary Random Processes. *Int. Journal of Fatigue*, **20**(3), 211–223.

NASA Engineering Management Council (EMC) 1995: *Preliminary NASA Standard, General Requirements for Structural Design and Test Factors of Safety*. 1 August (http://amsd-www.larc.nasa.gov/amsd/refs/fac_saf.html).

Nelson, W. 1982: *Applied Life Data Analysis*. NY: Wiley.

Nelson, D. L. 1996: Supplier Reliability and Quality Assurance. In: Ireson, W. G., Coombs, C. F. and Moss, R. Y. (eds), *Handbook of Reliability Engineering and Management*, 2nd Edition. NY: McGraw-Hill.

Newland, D. E. 1975: *An Introduction to Random Vibrations and Spectral Analysis*. London: Longman.

Nichols, K. 1992: Better, Cheaper, Faster Products – by Design. *Journal of Engineering Design*, **3**(3), 217–228.

Nichols, K., Pye, A. and Mynott, C. 1993: *UK Product Development Survey*. London: The Design Council.

Nicholson, C. E., Heyes, P. F. and Wilson, C. 1993: Common Lessons to be Learned from the Investigations of Failures in a Broad Range of Industries. In: Rossmanith, H. P. (ed.), *Structural Failure, Product Liability and Technical Insurance*. Elsevier.

Niehaus, F. 1987: Prospects for Use of Probabilistic Safety Criteria. In: Wittman, F. H. (ed.), *Structural Mechanics in Reactor Technology*. Rotterdam: Balkema.

Nixon, F. 1958: Quality in Engineering Manufacture. In: *Proceedings of the Conference on Technology of Engineering Manufacture*, IMechE, London, 14 March, 561–569.

Noori, H. and Radford, R. 1995: *Production and Operations Management: Total Quality and Responsiveness*. NY: McGraw-Hill.

Norell, M. 1992: *Advisory Tools and Co-operation in Product Development*. PhD Thesis, TRITA-MAE-1992:7, Department of Machine Elements, The Royal Institute of Technology, Stockholm, Sweden.

Norell, M. 1993: The Use of DFA, FMEA, QFD as Tools for Concurrent Engineering in Product Development Processes. In: *Proceedings ICED '93*, The Hague.

Norell, M. and Andersson, S. 1996: Design for Competition: The Swedish DFX Experience. In: Huang, G. Q. (ed.), *Design for X – Concurrent Engineering Imperatives*. London: Chapman & Hall.

Norton, R. L. 1996: *Machine Design: An Integrated Approach*. Englewood Cliffs, NJ: Prentice-Hall.

Oakley, M. 1993: Managing Design: Practical Issues. In: Roy, R. and Wield, D. (eds), *Product Design and Technological Innovation*, Open University Press.

O'Connor, P. D. T. 1995: *Practical Reliability Engineering*, 3rd Edition Revised. Chichester: Wiley.

Olson, W. 1993: The Most Dangerous Vehicle on the Road. *Wall Street Journal*, 9th February, NY.

Osgood, C. C. 1982: *Fatigue Design*, 2nd Edition. Oxford: Pergamon Press.

Ostrowski, S. L. 1992: Extra Profits from Managed Quicker Time to Market. In: *Proceedings Design Matters*, Inst. Mech. Engrs HQ, Birdcage Walk, London, September.

Pahl, G. and Beitz, W. 1989: *Engineering Design*. London: Design Council.

Parnaby, J. 1995: Design of the New Product Introduction Process to Achieve World Class Benchmarks. *Proc. IEE Sci. Meas. Tech.*, **142**(5), 338–344.

Parker, A. 1997: Engineering is not Enough. *Manufacturing Engineer*, December, 267–271.

Parry-Jones, R. 1999: Engineering for Corporate Success in the New Millennium. *The Royal Academy of Engineering – The 1999 Engineering Manufacture Lecture*.

Personal Communication 1998: Materials Testing Committee. London: BSI.

Phadke, M. D. 1989: *Quality Engineering Using Robust Design*. Englewood Cliffs, NJ: Prentice-Hall.

Pheasant, S. 1987: *Ergonomics – standards and guidelines for designers* (PP7317). London: BSI.

Pilkey, W. D. 1997: *Stress Concentration Factors*, 2nd Edition. NY: Wiley.

Pitts, G. and Lewis, S. M. 1993: Design Modelling the Taguchi Way. *Professional Engineering*, April, 32–33.

Plunkett, J. J. and Dale, B. G. 1988: Quality Costs: a critique of some economic cost of quality models. *International Journal Production Research*, **26**(11), 1713–1726.

Plunkett, J. J. and Dale, B. G. 1991: *Quality Costing*. London: Chapman & Hall.

Poolton, J. and Barclay, I. 1996: Concurrent Engineering Assessment: a proposed framework. *Proc. Instn Mech. Engrs*, Part B, **210**(B4), 321–328.

Prasad, B., Morenc, R. S. and Rangan, R. M. 1993: Information Management for Concurrent Engineering: Research Issues. *Concurrent Engineering: Research and Applications*, **1**, 3–20.

Priest, J. W. 1988: *Engineering Design for Producibility and Reliability*. NY: Marcel Dekker.

Pugh, S. 1991: *Total Design Integrated Methods for Successful Product Engineering*. Wokingham: Addison-Wesley.

QS 9000 1998: *Quality System Requirements QS 9000*. London: BSI.

Radovilsky, Z. D., Gotcher, J. W. and Slattsveen, S. 1996: Implementing Total Quality Management: statistical analysis of survey results. *International Journal of Quality and Reliability Management*, **13**(1), 10–23.

Rao, S. S. 1992: *Reliability Based Design*. NY: McGraw-Hill.

Rice, R. C. 1997: Statistical Aspects of Design. In: ASM International, *ASM Handbook No. 20 – Materials Selection and Design*, 10th Edition. OH: ASM International.

Rosenau, Jr, M. D. and Moran, J. J. 1993: *Managing the Development of New Products*. Van Nostrand Reinhold.

Rosyid, D. M. 1992: Elemental Reliability Index-Based System Design for Skeletal Structures. *Structural Optimization*, **4**(1), 1–16.

Ruiz, C. and Koenigsberger, F. 1970: *Design for Strength and Production*. London: Macmillan.

Russell, R. S. and Taylor, B. W. 1995: *Production and Operations Management: Focusing on Quality and Competitiveness*. Englewood Cliffs, NJ: Prentice-Hall.

Sadlon, R. J. 1993: *Mechanical Applications in Reliability Engineering*. NY: RAC, Griffiss AFB.

Sanchez, J. M. 1993: Quality by Design. In: Kusiak, A. (ed.), *Concurrent Engineering: automation, tools and techniques*. NY: Wiley.

Sanderson, A. C. 1997: Assemblability Based on Maximum Likelihood Configuration of Tolerances. In: *Proceedings IEEE International Symposium on Assembly and Task Planning*, Marina del Rey, CA, Wiley, 96–102.

Schatz, R., Shooman, M. and Shaw, L. 1974: Application of Time Dependent Stress–Strength Models of Non-Electrical and Electrical Systems. In: *Proceedings Reliability and Maintainability Symposium*, 540–547.

Schonberger, R. 1992: Total Quality Management Cuts a Broad Swathe – through manufacturing and beyond. *Organisational Dynamics*, Spring, 16–27.

Shah, J. J. 1998: Design Research and Industry Relevance. In: *Proceedings EDC '98*. Bury St. Edmunds: Professional Engineering Publishing, 31–41.

Shigley, J. E. 1986: *Mechanical Engineering Design*, 1st Metric Edition. NY: McGraw-Hill.

Shigley, J. E. and Mischke, C. R. (eds) 1989: *Mechanical Engineering Design*, 5th Edition. NY: McGraw-Hill.

Shigley, J. E. and Mischke, C. R. (eds) 1996: *Standard Handbook of Machine Design*, 2nd Edition. NY: McGraw-Hill.

Shimada, J., Mikakawa, S. and Ohashi, T. 1992: Design for Manufacture, Tools and Methods: the Assemblability Evaluation Method (AEM). In: *Proceedings FISITA '92 Congress*, London, No. C389/460.

Shingo, S. 1986: *Zero Quality Control: Source Inspection and the Poka Yoke System*. Cambridge, MA: Productivity Press.

Shinozuka, M. and Tan, R. 1984: Probabilistic Load Combinations and Crossing Rates. In: Shinozuka, M. and Yao, J. (eds), *Probabilistic Methods in Structural Engineering*. NY: ASCE.

Siddal, J. N. 1983: *Probabilistic Engineering Design: principles and applications*. NY: Marcel Dekker.

Sigurjonsson, J. B. 1992: *A Contribution to a Theory for Selecting Production Methods*. PhD Thesis, Technical University of Norway, Trondheim, Norway.

Smith, C. O. 1976: *Introduction to Reliability in Design*. NY: McGraw-Hill.

Smith, C. O. 1997: *Reliability in Design*. In: ASM International, *ASM Handbook No. 20 – Materials Selection and Design*, 10th Edition. OH: ASM International.

Smith, D. J. 1993: *Reliability, Maintainability and Risk: Practical Methods for Engineers*, 4th Edition. Oxford: Butterworth-Heinemann.

Smith, H. Jr 1995: Applications in Aircraft Structures. In: Sundararajan, C. (ed.), *Probabilistic Structural Mechanics Handbook: Theory and Industrial Applications*. NY: Chapman & Hall.

Soderberg, R. 1993: Tolerance Allocation Considering Customer and Manufacturer Objectives. *ASME Advances in Design Automation*, **Vol. 2**, ASME, DE-Vol. 65-2, 149–157.

Soderberg, R. 1995: *On Functional Tolerances in Machine Design*. PhD Thesis, Chalmers University of Technology, Gothenberg, Sweden.

Speckhart, F. H. 1972: Calculation of Tolerance Based on Minimum Cost Approach. *ASME Journal of Engineering for Industry*, **94**(2), 447–453.

Spotts, M. F. 1973: Allocation of Tolerances to Minimise Cost of Assembly. *ASME Journal of Engineering for Industry*, **95**, 762–764.

Stephenson, J. A. and Wallace, K. M. 1996: Design for Reliability of Mechanisms. In: Huang, G. Q. (ed.), *Design for X – Concurrent Engineering Imperatives*. London: Chapman & Hall.

Straker, D. 1995: *A Toolbook for Quality Improvement and Problem Solving*. London: Prentice-Hall.

Sturgis, R. W. 1992: *Tort Cost Trends: An International Perspective*. Seattle: Tillinghast Corporation.

Suh, N. P. 1990: *The Principles of Design*. NY: Oxford University Press.

Sullivan, L. P. 1987: Quality Function Deployment. *Quality Progress*, May, 39–50.

Sum, R. N., Jr 1992: Activity Management: A Survey and Recommendations for the DARPA Initiative in Concurrent Engineering. *DARPA Technical Paper*, 17 January.

Sundararajan, C. and Witt, F.J. 1995: Stress-Strength Interference Method. In: Sundararajan, C. (ed.), *Probabilistic Structural Mechanics Handbook: theory and industrial applications*. NY: Chapman & Hall.

Swift, K. G. and Allen, A. 1994: Product Variability Risks and Robust Design. *Proc. Instn Mech. Engrs*, Part B, **208**, 9–19.

Swift, K. G. and Booker, J. D. 1997: *Process Selection: from design to manufacture*. London: Arnold.

Swift, K. G., Raines, M. and Booker, J. D. 1998: Techniques in Capable and Reliable Design. In: *Proceedings EDIProD '98*, Subutka, Poland, IX1-29.

Swift, K. G., Raines, M. and Booker, J. D. 1997: Design Capability and the Costs of Failure. *Proc. Instn Mech. Engrs*, Part B, **211**, 409–423.

Taguchi, T., Elsayed, E. A. and Thomas, H. 1989: *Quality Engineering in Production Systems*. NY: McGraw-Hill International Edition.

Timoshenko, S. P. 1966: *Strength of Materials Part II – Advanced Theory and Practice*, 3rd Edition. NY: D. van Nostrand.

Timoshenko, S. P. 1983: *History of Strength of Materials*. NY: Dover.

Towner, S. J. 1994: Four Ways to Accelerate New Product Development. *Long Range Planning*, **27**(2), 57–65.

Ullman, D. G. 1992: *The Mechanical Design Process*. NY: McGraw-Hill.

Unwin, W. C. 1906: *The Elements of Machine Design – Part 1*. London: Longman.

Urban, G. L. and Hauser, J. R. 1993: *Design and Marketing of New Products*. Englewood Cliffs, NJ: Prentice-Hall.

Urry, S. A. and Turner, P. J. 1986: *Solving Problems in Solid Mechanics Volume 2*. Harlow, Essex: Longman Scientific and Technical.

Vasseur, H., Kurfess, T. R. and Cagan, J. 1992: A Decision-Analytic Method for Competitive Design for Quality. *Advances in Design Automation*, **1**, DE-Vol. 44-1, ASME, 329–336.

Verma, A. K. and Murty, A. S. R. 1989: A Reliability Design Procedure for Arbitrary Stress–Strength Distributions. *Reliability Engineering and System Safety*, **26**, 363–367.

Versteeg, M. F. 1987: External Safety Policy in the Netherlands: an approach to risk management. In: *Proceedings of the Technical Committee Meeting, Status, Experience, and Future Prospects for the Development of Probabilistic Safety Criteria*, Vienna, 27–31 January.

Villemeur, A. 1992: *Reliability, Availability, Maintainability and Safety Assessment – Volume 2*. Chichester: Wiley.

Vinogradov, O. 1991: *Introduction to Mechanical Reliability: a designer's approach*. NY: Hemisphere.

Vrijling, J. K., van Hengel, W. and Houben, R. J. 1998: Acceptable Risk as a Basis for Design. *Reliability Engineering and System Safety*, **59**, 141–150.

Walsh, W. J. 1992: Developing the Meta-Quality Product. *Research Technology Management*, **35**(5), 44–49.

Waterman, N. A. and Ashby, M. F. (eds) 1991: *Elsevier Materials Selector*. Essex: Elsevier Science Publishers.

Watson, B., Radcliffe, D. and Dale, P. 1996: A Meta-Methodology for the Application of DFX Design Guidelines. In: Huang, G. Q. (ed.), *Design for X – Concurrent Engineering Imperatives*. London: Chapman & Hall.

Weber, M. A. and Penny, R. K. 1991: Probabilistic Stress Analysis Methods. *Stress Analysis and Failure Prevention*, DE-Vol. 30, ASME, 21–27.

Weibull, W. 1951: A Statistical Distribution Function of Wide Applicability. *Journal of Applied Mechanics*, **73**, 293–297.

Welch, R. V. and Dixon, J. R. 1991: Conceptual Design of Mechanical Systems. In: *Proceedings ASME Design Theory and Methodology Conference*, **31**, 61–68.

Welch, R. V. and Dixon J. R. 1992: Representing Function Behaviour and Structure During Conceptual Design. In: *Proceedings ASME Design Theory and Methodology Conference*, **42**, 11–18.

Welling, M. and Lynch, J. 1985: Probabilistic Design to Techniques Applied to Mechanical Elements. *Naval Engineers Journal*, May, 116–123.

Wheelwright, S. C. and Clark, K. B. 1992: *Revolutionising Product Development*. NY: The Free Press.

Woodson, W. E., Tillman, B. and Tillman, P. 1992: *Human Factors Design Handbook*. NY: McGraw-Hill.

Wright, C. J. 1989: *Product Liability – The Law and Its Implications for Risk Management*. London: Blackstone Press.

Wu, Z., Elmaraghy, W. H. and Elmaraghy, H. A. 1988: Evaluation of Cost-Tolerance Algorithms for Design Tolerance Analysis and Synthesis. *Manufacturing Review*, **1**(3), October, 168–179.

Wyatt, C. M., Evans, S. and Foxley, K. 1998: The Integration of Vehicle Manufacturers and their Suppliers Prior to Product Development. In: *Proceedings EDC '98*. Bury St Edmunds: Professional Engineering Publishing, 613–620.

Yokobori, T. P. 1965: *Strength, Fracture and Fatigue of Materials*. Groningen: Holland: Noordhoff.

Young, W. C. 1989: *Roarke's Formulas for Stress and Strain*, 6th Edition. NY: McGraw-Hill.

Young, T. M. and Guess, F. M. 1994: Reliability Processes and Corporate Structures. *Microelectronics and Reliability*, **34**(6), 1107–1119.

Zhu, T. L. 1993: A Reliability Based Safety Factor for Aircraft Composite Structures. *Computers and Structures*, **48**(4), 745–748.

Bibliography

Bolz, R. W. (ed.) 1981: *Production Processes: The Productivity Handbook*, 5th Edition. NY: Industrial Press.

Bralla, J. G. (ed.) 1998: *Design for Manufacturability Handbook*, 2nd Edition. NY: McGraw-Hill.

Dallas, D. B. (ed.) 1976: *Society of Manufacturing Engineers – Tool and Manufacturing Engineers Handbook*, 3rd Edition. NY: McGraw-Hill.

Degarmo, E. P., Black, J. T. and Kohser, R. A. 1988: *Materials and Processes in Manufacturing*, 7th Edition. NY: Macmillan.

Dieter, G. E. 1986: *Engineering Design: a materials and processing approach*, 1st Metric Edition. NY: McGraw-Hill.

Green, R. E. (ed.) 1992: *Machinery's Handbook*, 24th Edition. NY: Industrial Press.

Kalpakjian, S. 1995: *Manufacturing Engineering and Technology*, 3rd Edition. MA: Addison Wesley.

Linberg, R. A. 1983: *Processes and Materials of Manufacture*. MA: Allyn and Bacon.

Schey, J. 1987: *Introduction to Manufacturing Processes*, 2nd Edition. NY: McGraw-Hill.

Todd, R. H., Allen, D. K. and Alting, L. 1994: *Manufacturing Processes Reference Guide*. NY: Industrial Press.

Walker, J. M. (ed.) 1996: *Handbook of Manufacturing Engineering*. NY: Marcel Dekker.

Waterman, N. A. and Ashby, M. F. (eds) 1991: *Elsevier Materials Selector*. Essex: Elsevier Science Publishers.

BS 1134 1990: *Assessment of Surface Texture – Part 2: Guidance and General Information*. London: BSI.

BS 4114 1984: *Specification for Dimensional and Quantity Tolerances for Steel Drop and Press Forgings and for Upset Forgings Made on Horizontal Forging Machines*. London: BSI.

BS 6615 1985: *Dimensional Tolerances for Metal and Metal Alloy Castings*. London: BSI.

BS 7010 1988: *A System of Tolerances for the Dimensions of Plastic Mouldings*. London: BSI.

PD 6470 1981: *The Management of Design for Economic Manufacture*. London: BSI.

Index